Genes, Genesis and God

Values and Their Origins in Natural and Human History

Can the phenomena of religion and ethics be reduced to the phenomena of biology? Holmes Rolston says no, and in this sweeping account of the subject written with considerable verve and clarity he challenges the sociobiological orthodoxy that would naturalize science, ethics, and religion. The book argues that genetic processes are not blind, selfish, and contingent, and that nature is not value-free.

The author examines the emergence of complex biodiversity through evolutionary history. Especially remarkable in this narrative is the genesis of human beings with their capacities for science, ethics, and religion. A major conceptual task of the book is to relate cultural genesis to natural genesis. There is also a general account of how values are created and transmitted in both natural and human cultural history.

The book is thoroughly up to date on current biological thought and is written by one of the most well-respected figures in the philosophy of biology and religion. It is likely to provoke considerable controversy amongst a wide range of readers in such fields as philosophy, religious studies, and biology, as well as being suitable for courses on science and religion.

Holmes Rolston III is University Distinguished Professor and Professor of Philosophy at Colorado State University. He is the author of many books. *Genes, Genesis and God* constituted the Gifford Lectures, University of Edinburgh, November 1997.

Genes, Genesis and God

Values and Their Origins in Natural and Human History

The Gifford Lectures
University of Edinburgh, 1997–1998

HOLMES ROLSTON III

PUBLISHED BY THE PRESS SYNDICATE OF THE UNIVERSITY OF CAMBRIDGE
The Pitt Building, Trumpington Street, Cambridge, United Kingdom

CAMBRIDGE UNIVERSITY PRESS
The Edinburgh Building, Cambridge CB2 2RU, UK   http://www.cup.cam.ac.uk
40 West 20th Street, New York, NY 10011-4211, USA   http://www.cup.org
10 Stamford Road, Oakleigh, Melbourne 3166, Australia

First published 1999

Printed in the United States of America

Typeface Palatino 10.25/13 pt.   *System* DeskTopPro$_{/UX}$®   [BV]

A catalog record for this book is available from the British Library

Library of Congress Cataloging-in-Publication Data
Rolston, Holmes, 1932–
Genes, genesis and God : values and their origins in natural and human history /
Holmes Rolston III.
p. cm.
Includes bibliographical references and index.
ISBN 0–521–64108–X (hardback)
1. Ethics, Evolutionary.   2. Religion.   3. Values.   I. Title.
BJ1311.R65   1998
171'.7 – dc21   98–20715
CIP

ISBN 0 521 64108 X   hardback
ISBN 0 521 64674 X   paperback

Contents

Contents

Contents

Contents

Preface

The Earth is remarkable, and valuable, for the genesis that occurs on it. Genesis is astronomical first; Earth must be set in its cosmological precedents and environments. There it is remarkable that something appears out of nothing, that this something appears, as cosmologists are now saying, "fine-tuned" for constructing a complex world. But the capacity of this something to generate, its "nature," is especially revealed in the complexity and diversity of the events that take place on Earth.

Ours is an age of many doubts, but no one doubts that there has been remarkable genesis on our planet, no one including those who doubt "creation," since this hints of a Creator. Nor do those who, in seeming sophistication, doubt whether "nature" exists, the latter term being (they may complain) some sort of socially constructed category, or filter, with which to view the phenomena – in this respect not unlike the "Creator God," a way of framing up a worldview, only now a modern, Western, secular frame. For the puzzled, there are, in broadest outline, two complementary or competing explanations of this genesis: a scientific account, for which we take the title word "genes," and a religious account for which the symbolic word is "God." The term "genesis" mediates between the dual accounts, keeping the naturalistic accounts in dialogue with other philosophical and metaphysical possibilities for the explanation of this Earthen fertility.

In this genesis, the Earth is striking for both the nature and the culture that occur on it. Almost to overwork the alliteration, we can place the term "genius" in the sequence between genes and God. There appears, nascently in the higher animals and flourishing in *Homo sapiens*, a "genius" or "spirit" of extraordinary ingenuity and intelligence. In German, the term needed is *Geist*. Genes, genesis,

Geist, and God. Though the term "genius" runs some risk of arrogance, it does register (recalling etymology again) this existential spiritedness, these inventive ("ingenious") human capacities of mind and spirit that become so notable for the building of cumulative transmissible cultures. Such phenomena too demand explanation; human cultural life is an outcome of the genes, a part of their genesis. Especially brilliant parts of this cultural genius, which figure large in this study because they are such critical cases for explanation, are the human capacities for religion, ethics, and science.

Long before culture arrives, the nature on Earth is already spectacular, not so much in the geological as in the biological phenomena. Although there is generative creativity in the physical sciences (illustrated when life emerges from physics and chemistry), in biology there appears a means for creating, storing, testing, and transmitting the novel emergent discoveries that make developing life possible – the genetic information vital to all living things. The principal achievement of biological science in this century has been its unfolding of how the secret of life is coded into the genes, an achievement matching the discovery by Darwin in the last century of the evolutionary history of this life – deep history in deep time.

This molecular biology, on microscales, underlies the macrobiology of natural history, though it is equally true that the macrobiology shapes what information is generated, stored, and transmitted in the genes. From the evolution of life onward, genesis is linked with genes. Cosmologists may want to know about the big bang or black holes or quarks, and how to get something out of nothing. The history of life on Earth takes a narrative form, a developmental history made possible by and accumulating in the genetic know-how. Biologists want to know how the life loops get started; what is the interplay among inevitability, probability, and contingency in the evolution of life; and how to explain more arising out of less.

If visitors from space were to file a report about Earth, volume 1 might cover the geological and biological phenomena, but volume 2 would require assessing the anthropological and sociological events, an account of culture. Such a two-volume division might seem historically biased, since evolutionary history has been going on for billions of years, whereas human cultural history is only a hundred thousand or so years old. But if this were a report that sought to describe what of note has taken place on Earth, and to evaluate it, the human phenomenon, with its myriad cultures, is phenomenal enough to warrant the second volume.

Certainly nature and culture are currently in tandem, with, from here onward, culture increasingly determining what natural history shall continue. Volume 2 would be even more historical than volume 1. The human genesis is linked with cumulative transmissible cultures, with continuities to nature, and yet also radically different from anything previously realized in natural history.

A major conceptual task, here undertaken, is to relate cultural genesis to natural genesis. Darwinian evolutionary biology is a brilliant achievement, the more so when coupled with that of genetic and molecular biology. Unfortunately, biology has been less successful in relating itself to culture. Despite vigorous efforts and some promising developments, I will be forced to conclude that this effort today is sometimes wrong-headed. There is a genuine novelty that emerges with culture, now superimposed on the wild nature out of which humans once emerged. It is important to see (so far as one can) how biological phenomena gave rise to culture, but it is just as critical to realize how culture exceeds biology, just as it is vital to see how biology exceeds physics and chemistry.

The general account presented here will revolve around values created and transmitted in both natural history and human cultural history. Few persons will deny that, one way or another, much that has occurred on Earth is valuable ("able to be valued" – if not "valuable in itself"), and this despite the fact that nature is often taken, alike by natural scientists and humanist philosophers, to be "value-free." As soon, however, as one begins to give a more systematic account of such appearance of the valuable, this will be hotly contested. I will interpret the Earth story, or, more pluralistically, the developing stories, on this remarkable planet, as the genesis of value, for which, in biological evolution, the genes are critical in a setup remarkably propitious for life, and, later and equally remarkably, as the culturing of value, for which the genes, however necessary, are insufficient.

The physical sciences are true all over the universe; the biological sciences are true all over the Earth. But the distinctively human sciences, such as anthropology, sociology, political science, economics, and (for the most part) psychology, study only one species, *Homo sapiens*. It may seem strange to devote several sciences to just one species. One reason is our special human needs; perhaps another is our arrogance; but one is, as our specific name indicates, our "wisdom," also evidence of the radical difference being human makes. Unlike coyotes or bats, humans are not just what they are by nature; they come into the world by nature quite unfinished and become what they become by culture. Though the genes remain indispensable, they no longer carry all the genesis. There is generative creativity in culture, a second level of genesis.

The products of culture are myriad – languages, rituals, tools, clothing, houses, plowed fields, villages, and on through churches and scriptures, computers and rockets – all coupled with ideas; and the home of ideas is human minds. Other animals, the higher ones, may also have minds, although it has proved difficult to discover what sort of minds they have, or even, for many vertebrates, whether they have minds at all. Meanwhile, the human mind is the only mind that permits the building of complex transmissible cultures, first oral and later literate. Humans are the only species who think about their ideas, who teach their ideas to the next generation, and who make creative ideological achievements that can be transmitted from generation to generation.

Three of the most notable products of human minds are science, ethics, and religion – emergent phenomena in culture. None is found in natural history; indeed, there will be controversy as soon as one asks whether there are even precursors there. Science, in its current form at least, is late in literate cultures; there have been ethics and religion in every classical culture, oral and literate. When human minds turn reflectively to give account of the human place in nature and in culture, they now do so predominantly by employing what has been learned in science, ethics, and religion, at the same time that these three great achievements in human life must themselves be given account of. For humans can reason, and they can value, and they can reason about values, including both those in natural and those in cultural history. Among their cultural achievements, science, ethics, and religion are principal carriers of value.

We know no minds that do not inhabit selves. These are somatic, organismic selves, objective bodies with skin boundaries, and also, in humans, psychological selves, existential selves, subjective egos, the "I" located within, from which perspective one encounters the world. The relation of such human selves to the world has been a perennial assignment of philosophy, but this task has become especially formidable in our own century, with its revelations about evolutionary history and the genetic basis of life. The inquiry that follows is an effort to locate the human self, with its genius, in its genesis in nature and culture, in a value-laden world. I do this by working a way first through natural history, and then through cultural history, particularly through science, ethics, and religion.

I shall use these three great domains for the generating, conserving, and distributing of values as test cases, demanding their incorporation into the larger picture of what is taking place on our planet. The more comprehensive model is the generating and testing of value, which take place through the generating of information, first in nature and later in culture. Here evolutionary history is interpreted as the genesis of natural value, which is conserved, enriched, and distributed over time. Such values in nature can and ought to be conserved, enriched, and appreciated by humans using their capacities for science, ethics, and religion.

The questions here become ultimate ones, though they are born in the phenomena of natural history and of human culture. The religions, including those of the monotheistic West (with which this argument is principally concerned), have steadily thought to detect a Beyond in the midst of the here and now. They have found neither nature nor culture to be in and of itself either final or fully self-explanatory. They have claimed a Presence immanent and transcendent, stirring in the Earth history. The evidence for such presence is the striking emergence, or genesis, of information and of value. There are genes, there is genesis, but explanations are not over until one has reckoned with the question of God. That claim, in what follows, takes the form of whether the phenomena of religion and ethics, in their powers of self-transformation, can be reduced to phenomena of biology, that is, whether such culture can be reduced to nature, and nature in turn found to be its own explanation. If not, perhaps explanations must rise to something beyond.

I have chosen a theme that invites sometimes passionate reactions, and where ideology frames the problems. Our "selves" and what we value are at stake, as well as our world picture. I also realize that I am often speaking to religion's cultured despisers, including its scientifically cultured despisers, but I ask only that such readers hear me out as I build my case. I hope to narrate, argue for, and evaluate these storied achievements in which so much of value is created and conserved, generated and shared. The question, from this perspective, is whether biology forbids, or discourages, or permits, or even invites religious inquiry.

The claim here is that any study in self-identity proves to be a study in one's location in the world and in the location of value. One thing I cannot doubt, as Descartes insists, is that I exist, as a thinking self, *cogito, ergo sum*. As a thinking self, neither can I doubt that I value, *valeo, ergo sum*. Such dynamic valuing, though an indubitable given, is a considerable challenge to interpret philosophically, to say nothing of scientifically. Such inquiry is the driving concern of this analysis, growing out of my conviction that the place of valuing in natural and cultural history has not yet been adequately interpreted. Indeed, and alas, it has too often been misinterpreted. In terms of human intellectual history, at the close of Darwin's century, ending also the century of molecular biology, facing a new century, indeed a new millennium, we urgently need an account of human selves and their values in this value-laden world of natural and cultural history.

Further, failing such insight, one may mistakenly transfer cultural phenomena back into biological phenomena and misinterpret what is going on, calling the genetic defense of organismic life "selfish," for example. That misunderstanding can, in turn, be brought back into culture, finding all human behavior pervaded with genetic self-interest and taking this to be the dominant determinant in all human affairs. Finally, all this can lead to a misvaluing of what is legitimately to be appreciated in both nature and culture. We fail to understand what is of value in each domain, and how these values are transmitted and shared.

The root idea in our English word "nature" is "giving birth," found also in such words as "native," "natal," "nation" (those born and bred in a country), from a Latin root going back to a Greek one, *natans*, being born. Though no longer evident in English, this is from the same root as "gene" (*gi[g]nomai*, to generate, give birth), with various *gna* forms, surviving in such words as "genesis," "pregnant," "progeny," "Gentile." The essential idea in "nature" is a kind of generative creativity, so remarkably exemplified in the events studied by the biological sciences. The root idea in culture is of deliberately tending, selectively cultivating, resourcefully modifying spontaneously wild events. For humans it is not enough to be born and to develop physiologically and behaviorally in one's ecology;

one must be cultivated within a society with its ideological heritage. Nurture is added to nature.

We often forget how everyday experience can demand certain things of the sciences. Science must save the phenomena, and if physics presents a theory of quarks that implies that I cannot wave to my friend, so much the worse for that theory. Astrophysics and microphysics must permit and deliver the world as something we can recognize at native range, reinterpret this range though they do at other scales. Likewise with biology, and if microbiology presents a theory of neurological synapses that implies that I cannot wave to my friend, so much the worse for that theory too.

Both the levels of biology that are in focus here, evolutionary biology and genetics, can present theories that need to be reconciled with our native range experiences, especially our cultural experiences. They too have to save the appearances. Alternately put, we have to figure out what kind of "appearances" there are. That might mean the theory revises our account of the appearances, as with sunsets, after astronomy. But our experience of the appearances can as well revise the theory, as – according to the argument here – we must do with some widely prevailing ideas in biology after ethics and religion. Sometimes an appearance means more out of less, as with life where none was before. No one can deny that science, ethics, and religion have appeared in history, and what does their appearance mean for our theories of genes, genesis, and God?

Integrated accounts ("grand narratives," if that is what I present here) are out of style. But fashions come and go; there is no particular reason to prefer plural, dis-integrated accounts, or relatively non-grand stories, piecemeal explanations. In fact, Darwinian theory, much in style, is itself a rather grand narrative; the popular selfish gene theory, now claimed as orthodoxy in biology, entails an orienting metaphysics of nature. If so, one might want to test against these an alternative view, not one that casts Darwinism down but one that casts the individualistic aspects of an ultra-Darwinism in a different light. Science can be colored by the preoccupations of the age, and its views need larger analysis. Only thus can we find out whether some accounts are better than others. That is the task undertaken here.

I express my great appreciation to the University of Edinburgh and the Gifford Lecture Committee for the privilege and opportunity of delivering these lectures there, November 1997. I trust that this inquiry does stand centrally in the scope intended by Lord Gifford, when he endowed this seminal series, now well over a century ago.

Chapter 1

Genetic Values: Diversity and Complexity in Natural History

Any account of genesis on Earth must place genes on the scene of global natural history. Nothing is more central to the contemporary neo-Darwinian view than an emergence over time of diversity and complexity, and genes are critical in this historic composition. Yet these developing phenomena, evident and indisputable though they are as fact of the matter, are subject to vigorous dispute about what is going on, a scientific issue laden with deeper philosophical significance.

1. NATURAL HISTORY: DIVERSITY AND COMPLEXITY

Something is learned across evolutionary history: how to make more diverse and more complex kinds. These events on Earth stand in marked contrast with events on other planets, such as the gases that swirl around Jupiter or the winds that blow on Venus. Even on Earth there is no such learning with the passing of cold and warm fronts; they just come and go. With the rock cycle, orogenic uplift, erosion, and uplift again, there is no natural selection. Nothing is competing, nothing is surviving, nothing has adapted fit. Climatological and geomorphological agitations continue in the Pleistocene period more or less as they did in the Precambrian. But the life story is different, because in biology, unlike physics, chemistry, geomorphology, or astronomy, something can be learned.

In result, where once there were no species on Earth, there are today five to ten million. On average and environmental conditions permitting, the numbers of life forms start low and end high. Seeking a philosophy of biology, Ernst Mayr realizes that many life-forms do not progress and that "higher" is a troublesome word. Still, he is forced to ask:

1

And yet, who can deny that overall there is an advance from the prokaryotes that dominated the living world more than three billion years ago to the eukaryotes, with their well-organized nucleus and chromosomes as well as cytoplasmic organelles; from the single-celled eukaryotes to plants and animals with a strict division of labor among their highly specialized organ systems; and, within the animals, from ectotherms that are at the mercy of climate to the warm-blooded endotherms, from types with a small brain and low social organization to those with a very large central nervous system, highly developed parental care, and the capacity to transmit information from generation to generation? (1991, p. 62; 1988, pp. 251–252)

Edward O. Wilson concludes his study of the diversity of life:

Biological diversity embraces a vast number of conditions that range from the simple to the complex, with the simple appearing first in evolution and the more complex later. Many reversals have occurred along the way, but the overall average across the history of life has moved from the simple and few to the more complex and numerous. During the past billion years, animals as a whole evolved upward in body size, feeding and defensive techniques, brain and behavioral complexity, social organization, and precision of environmental control – in each case farther from the nonliving state than their simpler antecedents did.
   More precisely, the overall averages of these traits and their upper extremes went up. Progress, then, is a property of the evolution of life as a whole by almost any conceivable intuitive standard, including the acquisition of goals and intentions in the behavior of animals. It makes little sense to judge it irrelevant. ... In spite of major and minor temporary declines along the way, in spite of the nearly complete turnover of species, genera, and families on repeated occasions, the trend in biodiversity has been consistently upward. (1992, pp. 187 and 194)

John Bonner, in a detailed study of the evolution of complexity, summarizes his findings:

There has ... been an extension of the upper limit of complexity during the course of evolution. ... There has also been an increase in the complexity of animal and plant communities, that is, there has been an increase in the number of species over geological time, and this has meant an increase in species diversity

2

in any one community. . . . One can conclude that evolution usually progresses by increases in complexity. . . . As evolution proceeded on the surface of the earth, there has been a progressive increase in size and complexity. (1988, pp. 220, 228, and 245)

E. C. Pielou concludes a long study of diversity, "Thus worldwide faunal diversification has increased since life first appeared in a somewhat stepwise fashion, through the development and exploitation of adaptations permitting a succession of new modes of life" (1975, p. 149). Life appears in the seas; moves onto the land, then into the skies. Terrestrial communities developed from the Silurian onward. In the Tertiary there was a marked increase in diversity due to the rise of warm-blooded vertebrates (mammals and birds), more than making up for the decrease in reptiles and amphibians. When vertebrates took to the air, there was introduced an entirely new mode of life.

There were setbacks, notably in the Permian-Triassic and again in the wave of mammal extinctions in the middle (pre-Pleistocene) Quaternary. But there was recovery. Many factors figure in, including climates and continental drift. Sometimes, the change due to organic evolution is overwhelmed by the change due to climatic cooling or drying out. The change due to organic evolution may be accelerated or decelerated by continental drift; continents fused together may provide a bigger area that supports more species, or they may provide more competition that eliminates species that previously evolved on separated continents. If the tectonic plates drift together and form a supercontinent, the supercontinent may saturate with species and suppress further speciation (some think), and if afterward the continents drift apart, this may add to the provinciality of the world and facilitate by isolation the evolution of diversity. On the whole, organic evolution has "the result that the present diversity of the world's plants and animals is (or was just before our species appeared) probably greater than it has ever been before" (Pielou 1975, p. 150).

George Gaylord Simpson, after surveying the fossil record extensively and noting that there are exceptions, concludes: "The evidence warrants considering general in the course of evolution . . . a tendency for life to expand, to fill in all available spaces in the liveable environments, including those created by the process of that expansion itself. . . . The total number and variety of organisms existing in the world has shown a tendency to increase markedly during the history of life" (1967, pp. 242, and 342). R. H. Whittaker finds, despite "island" and other local saturations and equilibria, that on continental scales and for most groups "increase of species diversity . . . is a self-augmenting evolutionary process without any evident limit." There is a natural tendency toward increased "species packing" (1972, p. 214). This is also called "bootstrapping in ecosystems," feed-forward loops that generate new niches that reinforce each other and open up new opportunities for species specialization (Perry et al. 1989). M. J. Benton, in a quantitative analysis of the fossil record, concludes "that the diversity of both marine and continental life increased exponentially since the end of the Precambrian" (1995, p. 52).

Complexity and diversity can sometimes be independent variables: beetles or grasses can become more diverse without becoming more complex. But the two are not always unrelated. Cumulatively over the millennia, as a result of the genetic capacity to acquire, store, and transmit new information, complexity can increase. There are advantages in specialized cells or organs, the efficiencies of the division of labor, which couples more complex and more diverse forms of life. In some situations, diversity increase has the result of stimulating complexity. A diverse environment is heterogeneous, and species are favored that are multiadaptable, and not just well adapted to one homogeneous environment. Such adaptability requires complexity, capacities to search out better environments and migrate to them, and, once there, abilities to invade successfully, to prey on or resist predation by – or to find and share resources with – the different kinds of organisms that can live in both wet and dry, cold and hot, grassland and forested environments. Complexity sometimes helps in dealing with the challenges and opportunities offered by diversity. Complexity helps in tracking changing environments.

Reptiles can survive in a broader spectrum of humidity conditions than can amphibians, mammals in a broader spectrum of temperature than can reptiles. Once there was no smelling, swimming, hiding, defending a territory, gambling, making mistakes, or outsmarting a competitor. Once there were no eggs hatching, no mothers nursing young. Once there was no instinct, no conditioned learning. Once there was no pleasure, no pain. But all these capacities got discovered by the genes. Once there was no capacity to make intentional reference, but this capacity arose, as when vervet monkeys learned to give different alarm calls to indicate the approach of dif-

4

ferent predators: leopards, eagles, snakes. Once there was no meta-meric segmentation, as in worms; once there was no pentameric seg-mentation, as in starfish. But all these phenomena appear, gradually, but eventually, without precedent if one looks further along their developmental lines.

J. W. Valentine, after a long survey of evolutionary history, con-cludes for marine environments that both complexity and diversity increase through time. First, with regard to diversity:

> A major Phanerozoic trend among the invertebrate biota of the world's shelf and epicontinental seas has been towards more and more numerous units at all levels of the ecological hierarchy. This has been achieved partly by the progressive partitioning of ecospace into smaller functional regions, and partly by the inva-sion of previously unoccupied biospace. At the same time, the expansion and contraction of available environments has con-trolled strong but secondary trends of diversity. . . . The bio-sphere has become a splitter's paradise. (Valentine 1969, p. 706)

When the landmasses fragment, speciation is favored; when they coalesce, previously endemic offshore marine faunas and floras merge and decimate each other in competition. So there are ups and downs in numbers of families and species, due to the contingencies of drift; nevertheless, biologically, the trend is up.

Complexity also increases:

> A sort of moving picture of the biological world with its selective processes that favor increasing fitness and that lead to "biologi-cal improvement" is projected upon an environmental back-ground that itself fluctuates. . . . The resulting ecological images expand and contract, but, when measured at some standardized configuration, have a gradually rising average complexity and exhibit a gradually expanding ecospace. (Valentine 1973, p. 471)

This double tendency in the biological system is disrupted but not overwhelmed by continental drift.

These summary conclusions can be illustrated in graph form, in a series of graphs that provocatively illustrate at once the vicissitudes and the progress of evolutionary history. Valentine graphs diversity in kinds of marine invertebrates (Fig. 1.1). There are steep climbs and drops, with a rise overall from 0 to 450 families.[1]

---

[1] Numbers of genera and species cannot be reliably estimated from the fossil rec-ord.

Figure 1.1. Diversity level of families of well-skeletonized shelf invertebrates. Reprinted with permission from *Evolutionary Paleoecology of the Marine Biosphere*, by James W. Valentine, Prentice-Hall, 1973, p. 387. This graph also appeared in James W. Valentine, "Patterns of Taxonomic and Ecological Structure of the Shelf Benthos during Phanerozoic Time," *Palaeontology*, vol. 12, part 4, 1969, p. 692.

Raup and Sepkoski (1982) add the marine vertebrates and graph a rise, again with climbs and drops, especially at times of catastrophic extinctions, from 0 to perhaps 750 families (Fig. 1.2). A common interpretation of the somewhat flat midportion of the graph (Ordovician to Permian periods) is that Earth's tectonic plates were configured to fuse the landmasses, resulting in a saturation of kinds of species that had at that point evolved on the continental shelves. Since marine life is primarily on the continental shelves, it may be

6

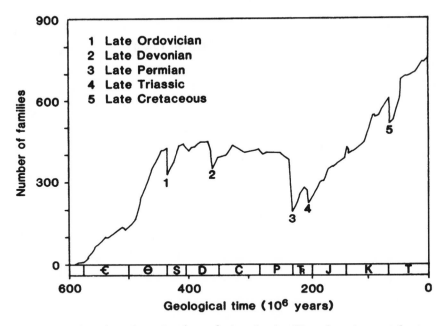

Figure 1.2. Standing diversity through time for families of marine vertebrates and invertebrates, with catastrophic extinctions. Reprinted with permission from David M. Raup and J. John Sepkoski, Jr., "Mass Extinctions in the Marine Fossil Record," *Science* 215 (19 March 1982): 1501–2503, p. 1502. Copyright 1982 American Association of the Advancement of Science.

especially susceptible to contingencies in continental drift. Also, during this relatively flat part of the marine curve, life moves onto the land and greatly diversifies there, from the Silurian period onward (not shown in these graphs). That requires also considerable evolution of complexity, since the terrestrial environment is more demanding.

Plants develop steadily on the landmasses, with species turnover resulting in increased diversity. Andrew H. Knoll graphs (Fig. 1.3) this record for local ecosystems over evolutionary time. In the Paleozoic there is a general rise, and after that a plateau. "The history of diversity within floras from subtropical to tropical mesic floodplains is marked by several periods of rapid increase separated by extended periods of more or less unchanging taxonomic richness." After the mid-Mesozoic, with the rise of the angiosperms (flowering plants), there is a steady climb in regional floras. Knoll concludes "that spe-

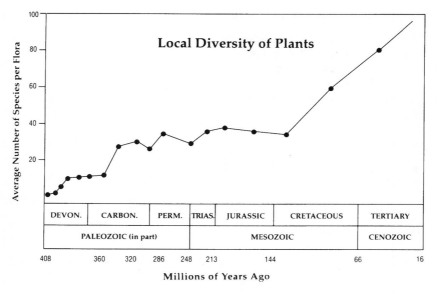

Figure 1.3. Average number of plant species found in local floras. From *The Diversity of Life* by Edward O. Wilson. Copyright © 1992 by Edward O. Wilson. Reprinted by permission of Harvard University Press. This graph is adapted from Andrew H. Knoll, "Patterns of Change in Plant Communities Through Geological Time," from *Community Ecology*, Jared Diamond and Ted J. Case, eds. Copyright © 1986, by Harper Collins Publishers. Reprinted by permission of Addison–Wesley Educational Publishers.

cies numbers within subtropical to tropical communities have been rising continually since the Cretaceous and that a plateau has yet to be established" (1986, pp. 140 and 132).

For animals, the story of terrestrial life may be less contingent than that in marine shelves, since terrestrial animal species are quite mobile and have often crossed land bridges between the continents, resulting in a different pace of competition and selection for different traits. It is in the vertebrates, most of all, that advance is difficult to deny. The sea, though required for the inception of life and though long an environment for diverse forms of life, is not an environment that has ever produced big brains; dramatic cerebral evolution has always been terrestrial, because the more challenging land environment seems to demand more neural power. Even today the "minds" in the sea (whales, dolphins) were once formed on land and returned to the sea. In the global picture, complementing the marine one, if

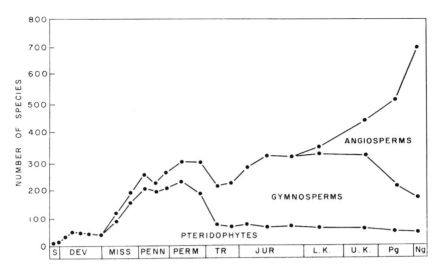

Figure 1.4. Species diversity changes in vascular plants. Reprinted with permission from K. J. Niklas, "Large Scale Changes in Animal and Plant Terrestrial Communities," in D. M. Raup and D. Jablonski, eds., *Patterns and Processes in the History of Life*, Springer–Verlag, 1986, p. 385.

one examines the top trophic rungs for complexity and if one adds the increases of diversity cumulating in both marine and terrestrial environments, the increase of complexity and diversity will be still more evident.

In the composition of the floras and faunas, certain forms can later be less numerous than before, but, climatic conditions permitting, overall biodiversity gradually and sometimes rather spectacularly rises (Fig 1.4 and Fig. 1.5). Here too the later-coming forms are often more complex than the earlier ones they replace. Mammals with their warm blood and higher energy requirements develop metabolisms and behavioral skills not present in cold-blooded reptiles and amphibians. Angiosperms advance over, and may displace, gymnosperms. Fortunately for overall biodiversity, these earlier groups, in reduced numbers (and with species turnover), continue to enrich present faunas and floras.

Norman D. Newell has graphed the numbers of all families, terrestrial and marine, vertebrate and invertebrate, increasing through evolutionary time (Fig. 1.6).

It is difficult to produce a graph of increasing complexity, since

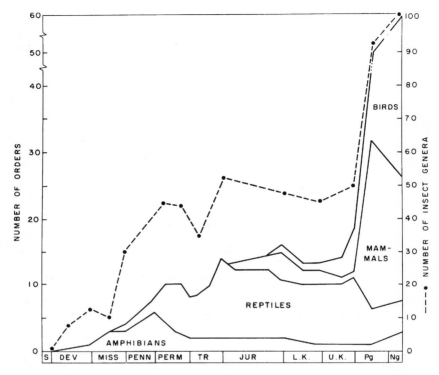

Figure 1.5. Changes in the composition of vertebrate orders and numbers of insect genera. Reprinted with permission from K. J. Niklas, "Large Scale Changes in Animal and Plant Terrestrial Communities," in D. M. Raup and D. Jablonski, eds., *Patterns and Processes in the History of Life*, Springer–Verlag, 1986, p. 390.

complexity (unlike a numerical count of families) may not be any single thing to graph. There is unlikely to be any single parameter measuring it that always increases (progresses?) over the course of natural history. Nor does complexity always coincide with advancement, because sometimes complexity is a disadvantage. The over-specialized frequently become extinct. Nevertheless increases in capacities for centralized control (neural networks with control centers, brains surpassing mere genetic and enzymatic control), increases in capacities for sentience (ears, eyes, noses, antennae), increases in capacities for locomotion (muscles, fins, legs, wings), increases in capacities for manipulation (arms, hands, opposable thumbs), increases in capacities for acquired learning (feedback loops, synapses, mem-

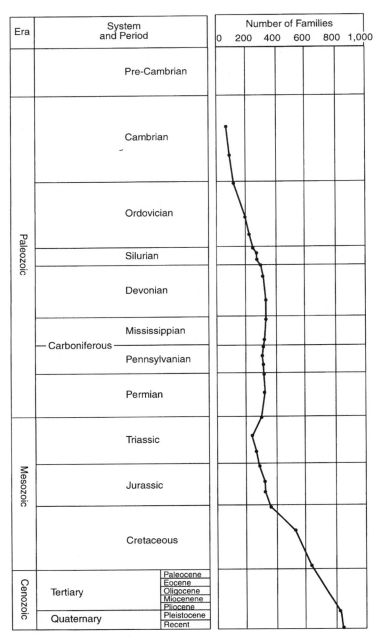

Figure 1.6. Number of major families of fossil animals increasing through time. Reprinted with permission from Norman D. Newell, "Crises in the History of Life," *Scientific American*, 208 (no. 2, February 1963): 76–92, p. 80.

ory banks), increases in capacities for communication and language acquisition – all these take increased complexity. Nothing seems more evident over the long range than that complexity has increased; in the Precambrian there were microbes; in the Cambrian period trilobites were an advanced life-form; the Pleistocene period produced persons.

Trends, which are a sine qua non of historical interpretation, are never directly observable and may be difficult to detect in a limited span of time or range of observation. They show up statistically, but even statistics deals poorly with developing cybernetic trends, where there is information buildup with trial and error learning making critical discoveries (such as photosynthesis, neurons, endoskeletons).

The lower forms remain, too; there must be trophic pyramids, food chains. Even the microbes have a remarkable diversity (Service 1997). There cannot be higher forms, all by themselves. They must be superposed on lower forms, embedded in communities. So there can seem only changing diversity, not increased complexity, if one looks at the monocots and dicots, the crustaceans and flatworms. Perhaps one should not expect much progress in the invertebrates, not much past that in the arthropods. In all the understories, which must remain occupied if there is to be a biotic community, there is mainly turnover, perhaps with some increased fitness for survival. But if we are to have the whole story of what is going on, one must look at the uppermost forms. These do seem to get built up over time.

A word recently coined by Humberto R. Maturana and Francisco J. Varela to describe such phenomena in nature is "autopoiesis" (*autos*, self, and *poiein*, to produce) (Maturana and Varela 1980). The idea, however, is an old one: "The earth produces of itself [Greek: automatically]" (Mark 4.28). The "auto" here should not be taken to posit a "self," but rather an innate principle of the spontaneous origination of order, that is, of genesis. That, we recall, is the root meaning of the word "nature," to generate or give birth. Organisms, which do have somatic selves, are self-organizing, but so are species lines, in which such organismic selves are contained. Ecosystems are spontaneously organizing: the species get arranged into interdependencies; novel niches appear and species arise to fill them. Even the planet, globally, is a prolific system.

Stuart Kauffman concludes a long study of the origins of order in evolutionary history: "Since Darwin, biologists have seen natural selection as virtually the sole source of that order. But Darwin could

not have suspected the existence of self-organization, a recently discovered, innate property of some complex systems. . . . Selection has
molded, but was not compelled to invent, the native coherence of
ontogeny, or biological development. . . . We may have begun to understand evolution as the marriage of selection and self-
organization" (1991, summarizing Kauffman 1993; 1995; Rosen 1991;
Salthe 1993). Analyzing computer, mathematical, and biological
models, Kauffman finds that natural selection can drive ordered systems to the edge of chaos because that is where the greatest possibility for self-organization, and survival in changing environments, occurs. "Evolution has tuned adaptive gene regulatory systems to the
ordered region and perhaps to near the boundary between order and
chaos." "Networks on the boundary between order and chaos may
have the flexibility to adapt rapidly and successfully" (Kauffman
1991). If so, we will not be surprised to find that in these "poised
systems" creativity is often entwined with chance and chaos. The
construction of order is most probable at the edge of disorder.

> Evolution is a complex combinatorial optimization process in
> each of the coevolving species in a linked ecosystem, where the
> landscape of each actor deforms as the other actors move. Within
> each organism, conflicting constraints yield a rugged fitness
> landscape graced with many peaks, ridges, and valleys. . . . Such
> order has beauty and elegance, casting an image of permanence
> and underlying law over biology. Evolution is not just "chance
> caught on the wing." It is not just a tinkering of the ad hoc, of
> bricolage, of contraption. It is emergent order honored and
> honed by selection. (1993, p. 644)

Francisco J. Ayala concludes, "Progress has occurred in nontrivial
senses in the living world because of the creative character of the
process of natural selection" (1974, p. 353). Theodosius Dobzhansky
says:

> Nobody has been able to propose a satisfactory definition of
> what counts as evolutionary progress. Nevertheless, viewing ev-
> olution of the living world as a whole, from the hypothetical
> primeval self-reproducing substance to higher plants, animals
> and man, one cannot avoid the recognition that progress, or ad-
> vancement, or rise, or ennoblement, has occurred. . . . Seen in ret-
> rospect, evolution as a whole doubtless had a general direction,
> from simple to complex, from dependence on to relative inde-

pendence of the environment, to greater and greater autonomy of individuals, greater and greater development of sense organs and nervous systems conveying and processing information about the state of the organism's surroundings, and finally greater and greater consciousness. (1974, pp. 310–311)

If in natural history we define progress as "increase in the ability to gather and process information about the environment" (Ayala 1988, p. 92; 1974, p. 344), then again and again, evolution produces phenomena that rise above the former levels with breakthroughs in achievement and power. Evolutionary progress is the systemic generation of increased richness in value, such as diverse species, each of value in itself; or more complex skills, such as the capacity for acquired learning; or better adaptation for survival, such as thicker fur in a cooling climate. Such breakthroughs might not be something that one can catch in graph form; they might not be something that rises in steady ascent on a landscape where there are hills up and down, and circuitous ways up, over, down, and around them, even tunnels and bridges, where organisms manage to survive in a convoluted and adventurous world. Narratives are not statistical affairs; narrative plots do not plot well on line graphs. But narratives may be required to tell the progressing story of increasing diversity and complexity in natural history.

In nature there are, if we consult physics and chemistry, two kinds of things, matter and energy, but if we consult biology there is a third kind of thing: information. All three are required for the genesis so evident in Earthen natural history. At this level, neither matter nor energy can be created or destroyed, though, at the more fundamental levels of atomic and astronomical physics, the one can be transformed into the other. Matter throughout natural history has been energetically structurally transformed. This happens in physics and chemistry with impressive results, as with the construction of the higher elements in the stars or the composition of crystals, rocks, mountains, rivers, canyons on Earth. But the really spectacular constructions that are manifest in biological diversity and complexity do not appear without the simultaneous genesis of information how to compose and maintain such structures and processes. It is this information that is recorded in the genes, and such information, unlike matter and energy, can be created and destroyed. Such genetic information is the key to all progress in biological nature. Making sense of that genesis, with its results alike in nature and in culture, from both scientific and philosophical perspectives is the task that lies ahead of us. Does the epic have any plot?

2. CONTINGENT NATURAL HISTORY?

We noted that this developing diversity and complexity, evident though both are, are subject to diverse interpretations. Despite the conclusions of the various scientists already cited, and on the basis of the textbook theory of natural selection, hardnosed scientists are reluctant to see any progress[2] in the evolutionary epic, because this theory, as usually interpreted, does not entitle them to see any. John Maynard Smith says, "There is nothing in neo-Darwinism which enables us to predict a long-term increase in complexity." But he goes on to suspect that this is not because there is no such long-term increase, but that Darwinism is inadequate to explain it. We need "to put an arrow on evolutionary time" but get no help from evolutionary theory.

> It is in some sense true that evolution has led from the simple to the complex: prokaryotes precede eukaryotes, single-celled precede many-celled organisms, taxes and kineses precede complex instinctive or learnt acts. I do not think that biology has at present anything very profound to say about this. . . . We can say little about the evolution of increasing complexity. (1972, pp. 89 and 98–99)[3]

---

[2] The idea of progress, say the postmodernists and others, is an ideological illusion, historically generated in Enlightenment Europe, and, in such discussions as we are engaged in here, superimposed by some scientists onto natural history, though denied by others. Such imposing is myth making to bolster up the good feelings of dominant Europeans about their connections with natural history, which is otherwise rather fearful, red in tooth and claw. See Ruse (1996). Readers will have to judge, if they can find within themselves the capacity to critique such allusions of illusion, whether (for example) the graphs used here, all made by scientists, are illusion-making projections. If not, do they reveal anything discovered about objective, nonhuman natural history, and what has been created ("projected") by evolutionary forces, even though perhaps not adequately explained by natural selection theory?

[3] Darwin was pulled both ways. He concluded *On the Origin of Species*, "As natural selection works solely by and for the good of each being, all corporeal and mental endowments will tend to progress towards perfection" (1859, 1964, p. 489). "The inhabitants of each successive period in the world's history have beaten their predecessors in the race for life, and are, in so far, higher in the scale of nature; and this may account for that vague yet ill-defined sentiment, felt by many pa-

The received theory says only that the better adapted survive, and (despite the use of the word "better") this adaptation leaves entirely open the question whether the survivors are better in any sense involving progressive worth. It does not even say that the survivors must be more complex, perceptive, sentient, specialized, or that the ecosystems in which later-coming forms are components will be more diverse or stable than the earlier ones. Later-coming grasses are not any better than earlier, now extinct ones; they are just different. Some life-forms (cockroaches, marine shellfish) survive over long periods little changed from their ancestors. In other cases, surviving life-forms have lost organs – eyes, legs, wings – and become parasites. (It does seem, though, that such parasites typically depend on whatever skills they lose remaining in their hosts.) In climates growing colder or drier fewer species may live there later. There are fewer dinosaurs now than in Cretaceous times, fewer birds than in Pleistocene times.

By this account, whether in fertile or harsh environments, species are simply buffeted about by their changing environment. If the environment just drifts through tectonic changes, climatic changes, continental drift, and so on, then neither can the life-forms that inhabit such an environment have direction. At the molecular, genotypic level, those species that survive do so on the basis of random variations, ventured from below and unrelated to the needs of the organism. At the molar, phenotypic level, species must be "better" adapted, but if the environment that they track better is drifting, then they do not progress toward complexity or diversity, or anything else: they just track drift – the species are as aimless as the geomorphic processes. The only form of progress that natural selection can promote is progress in capacity to survive, and that is an indepen-

laeontologists, that organisation on the whole has progressed" (1859, 1964, p. 345; cf. 1872a, 1962, p. 355; 1874, 1895, pp. 145 and 619).

On the other hand, "After long reflection, I cannot avoid the conviction that no innate tendency to progressive development exists" (Darwin 1872b), and Darwin penned himself a memo: "Never use the words *higher* and *lower*" (Darwin 1858, p. 114).

Darwin lived in an age that thought highly of progress and his interpretation of natural history may be colored by his cultural era. His belief in progress may have waned in later life. One can, of course, defend progress in natural history whether or not belief in cultural progress is in or out of style.

dent variable with regard to increasing complexity or increasing diversity.

Evolution takes place wherever there is any change in gene frequency. If so, it may be said that such change has nothing to do with the selection of the advanced. There is just wandering up or down the ranges of life's complexity, across less or more of the ranges of life's diversity. There are local trends (cushion plants in alpine environments; thick leaves in deserts; repeated evolution of horns), but natural selection theory is unable to predict any long-term or big-scale outcomes. From the point of view of the theory, this resulting increasing diversity and complexity are contingent.

Despite Maynard Smith's modest conclusion that biology has little to say about these longer-range trends in evolutionary history, others draw the stronger conclusion that evolutionary history is a random walk. With Maynard Smith, biology has no explanation; with Jacques Monod the explanation is pure chance:

> Chance *alone* is at the source of every innovation, of all creation in the biosphere. Pure chance, absolutely free but blind, at the very root of this stupendous edifice of evolution: this central concept of modern biology is no longer one among other possible or even conceivable hypotheses. It today is the *sole* conceivable hypothesis, the only one that squares with observed and tested fact. And nothing warrants the supposition – or the hope – that on this score our position is likely ever to be revised. . . .
>
> When one ponders on the tremendous journey of evolution over the past three billion years or so, the prodigious wealth of structures it has engendered, and the extraordinarily effective teleonomic performances of living beings, from bacteria to man, one may well find oneself beginning to doubt again whether all this could conceivably be the product of an enormous lottery presided over by natural selection, blindly picking the rare winners from among numbers drawn at utter random. [Nevertheless,] a detailed review of the accumulated modern evidence [shows] that this conception alone is compatible with the facts. . . . Man knows at last that he is alone in the universe's unfeeling immensity, out of which he emerged only by chance. (Monod 1972, pp. 112–113, 138, and 180)

Stephen Jay Gould agrees, emphasizing the philosophical implications of what biological science has found: "We are the accidental result of an unplanned process . . . the fragile result of an enormous concatenation of improbabilities, not the predictable product of any definite process" (1983, pp. 101–102). "Natural selection is a theory of *local* adaptation to changing environments. It proposes no perfecting principles, no guarantee of general improvement" (1977a, p. 45). Natural selection provides no reason to believe in "innate progress in nature"; none of the local adaptations is "progressive in any cosmic sense" (1977a, p. 45).

"There are no intrinsic trends towards increasing (or decreasing) diversity. Ecological roles are, in a sense, 'preset' by the nature of environments and the topological limits to species packing; they are filled soon after the Cambrian explosion. Thereafter, inhabitants change continually, but the roles remain" (1977b, p. 19). As in a rotating kaleidoscope, there is change without development, steady turnover, but not really different from the astronomical panorama of the cycling planets and revolving galaxies. The system is without value heading. Any values are produced by luck. "Almost every interesting event of life's history falls into the realm of contingency" (1989, p. 290). That is "a claim about the nature of reality" "denying that progress characterizes the history of life as a whole, or even represents an orienting force in evolution at all" (1996, p. 3).

Michael Ruse surveys the conclusions of evolutionary biologists at great length. "A major conclusion of this study is that some of the most significant of today's evolutionists are Progressionists, and . . . we find (absolute) progressivism alive and well in their work" (1996, p. 536). Nevertheless, they are all wrong, because, biased, they are reading progress into the evolutionary record. They have slipped into "pseudo-science." "For nigh two centuries, evolution functioned as an ideology, as a secular religion, that of Progress" (p. 526). In fact, he argues, today more "mature" scientists, unbiased, have "expelled progress" from evolutionary history (p. 534). "Evolution is going nowhere – and rather slowly at that" (Ruse 1986, p. 203).

Evolutionary history wanders in the first place because of atomic and molecular chance, unrelated to the needs of the organism. There is selection operating over this chance, of course, but that selection does not introduce any ordered direction into the chance variation, because it is not selection for advancement, only selection for survival. The biggest events (the coming of mammals and humans) not less than the smallest events (the microscopic mutations) are accidental or random with respect to anything the theory can predict or retrospectively explain. It might first seem that in one part of the theory, the supply side, internal to the organism, one finds randomness, but that in another part of the theory, the retention side, external to the organism, one might find progress, because the "better" are selected. In the genes, there is record keeping. From among the myriad trials that come momentarily into existence, the fittest are selected to stay. The new events occur at random with respect to their direction but are preserved for the direction they take.

But when we look more closely at even the retention side – so this claim runs – randomness is equally present there. There is no direction in the microevolution (random variants), and no direction in the macroevolution either (selection headed nowhere), a twice-compounded randomness. Selection is for survival, yes, but there is only changing genetics that records changing morphology and behavior that reflects drifting environments. This does give local trends (hair growing whiter as environments grow colder). But there is no covering law, or trend, enabling one to say that microbes, or mammals, or men could statistically be expected. They just occur as historical events, and the theory is surprised by them, although in retrospect they are consistent with the theory. Among the equally fit, some are more complex, some less so, and although survival might have been possible without advancing complexity, there is nevertheless advancing complexity in some few forms, consistent with, but not required by, the principle of natural selection.

We can say that if life starts out simply, there is nowhere to go but up. So some development of diversity and complexity is not surprising. But life does not steadily and irreversibly have to go up. "Nowhere to go but up" is true at the launching, but not thereafter. There are down, stable, and out, and many forms take these routes. The evolutionary process might have achieved a few simple, reliable forms, needing little modification, and stagnated thereafter, as has sometimes happened in little-changing habitats. Nor is there any account of why the life process, if it happens to ascend, will not happen to descend, earlier more complex, later simpler, devolution after evolution, since up or down is immaterial to survival. Life might have gone extinct; many life-forms did. Nor does it help here to appeal to time to guarantee complexity. Time does nothing to cure randomness, not unless there is some further principle (that natural selection does not supply) that locks in the upstrokes.

We nowhere here wish to deny that there is contingency in natural history; to the contrary we will enlist this in the service of the genesis of value. But is this the whole story? True, much in evolutionary history can seem contingent, if one considers only the fortunes of this or that lineage, which is typically the focus of analysis. But the history begins to look different when one considers the evolution of skills, irrespective of what lineage they happen to be in. Assuming more or less the same Earth-bound environments, if evolutionary history were to occur all over again, things would be different. Still, there would likely again be organisms reproducing, genotypes and phenotypes, natural selection over variants, multicelluar organisms with specialized cells, membranes, organs; there would likely be plants and animals: photosynthesis or some similar means of solar energy capture in primary producers such as plants, and secondary consumers with sight, and other sentience such as smell and hearing; mobility with fins, limbs, and wings, such as in animals. There would be predators and prey, parasites and hosts, autotrophs and heterotrophs, ecosystemic communities; there would be convergence and parallelism. Coactions and cooperations would emerge. Life would probably evolve in the sea, spread to the land and the air.

Play the tape of history again; the first time we replayed it the differences would strike us. Leigh Van Valen continues:

> Play the tape a few more times, though. We see similar melodic elements appearing in each, and the overall structure may be quite similar. . . . When we take a broader view, the role of contingency diminishes. Look at the tape as a whole. It resembles in some ways a symphony, although its orchestration is internal and caused largely by the interactions of many melodic strands. (Van Valen 1991)

One clue, already supplied by Kauffman, may lie in realizing that genetic creativity is stimulated at "the boundary between order and chaos." At such a boundary, interpreters who prefer to emphasize the chaos can face in that direction and see only the contingency. But in a more complete account, one needs also to see the order maintained at the edge of chaos. Indeed, this is an order generated in such an environment, and, more profoundly still, made possible by it. Life at such a boundary needs, above all, information, for it is such information by which it can form, or inform, matter and energy into the living molecules by which life is generated, regenerated, and maintained. In this prolific natural history, normal extinction and turn-over are essential to the creative process.

Even atypical extinction rates are incorporated into the creative process. Niles Eldredge reviews the geological epochs of accelerated extinction to find that they are also times of accelerated speciation, as though to endorse Kauffman's claim that creativity is increased at the boundary between chaos and order:

> The particularly compelling aspect of this account is that the fac-tors underlying species extinction – namely, habitat disruption, fragmentation and loss – are the very same as those convention-ally cited as causes of speciation. Thus the causes of extinction may also serve as the very wellspring of the evolution of new species. (Eldredge 1995a, p. 79)

This effect can be coupled with the finding by geneticists that at times of stress, genetic mutation rates are increased.[4]

There have also been a few (perhaps five, as numbered in Fig. 1.2) rare but devastating catastrophic extinctions. These decimated diver-sity, presumably also with adverse results on complexity. The late Permian and late Cretaceous extinctions are the most startling. Each catastrophic extinction is succeeded by a recovery. Although natural events, these extinctions so deviate from the trends that many pale-ontologists look for causes external to the evolutionary ecosystem. If caused by supernovae, collisions with asteroids, oscillations of the solar system above and below the plane of the galaxy, or other extra-terrestrial upsets, such events are accidental to the evolutionary eco-system. If the causes were more terrestrial – cyclic changes in cli-mates or continental drift – the biological processes that characterize Earth nevertheless prove to have powers of recovery. Uninterrupted by accident, or even interrupted so, the biological forces steadily in-crease the numbers of species.

David M. Raup, the paleontologist who has best documented these catastrophic extinctions, has also reflected philosophically on them. He finds it striking that, though seemingly catastrophic, these periodic cutbacks prepare the way for more complex diversity later on. Evolution can tend to stagnate, unless there are crises and upsets (an insight also reached independently by Kauffman). The cata-

---

[4] See Section 4.

strophic extinctions first seem quite a bad thing, an unlucky disaster. But in fact they were good luck. Indeed, were it not for them we humans would not be here, nor would any of the mammalian complexity. Life on Earth is quite resilient in its capacity to track shifting environments, and, though there is lots of turnover, normal geological processes lack the power to cause significant extinctions in major groups that are widespread. But just such a resetting of natural history is productive – rarely but periodically. We should think twice before judging these catastrophic extinctions to be a bad thing.

Raup explains:

> Without species extinction, biodiversity would increase until some saturation level was reached, after which speciation would be forced to stop. At saturation, natural selection would continue to operate and improved adaptations would continue to develop. But many of the innovations in evolution, such as new body plans or modes of life, would probably not appear. The result would be a slowing down of evolution and an approach to some sort of steady state condition. According to this view, the principal role of extinction in evolution is to eliminate species and thereby reduce biodiversity so that space – ecological and geographic – is available for innovation. (1991, p. 187)

Raup argues for what he calls "extinction-driven evolution." Natural selection fosters diversity, but natural selection acting steadily and without interruption would saturate and stagnate natural history. Extinctions subtract from the biodiversity but at the same time provide the space for more innovative biocomplexity. This is not only true of the normal extinction turnover, but is especially true during catastrophic extinctions. There is a big shakeup; this is, if you like, at random; it is, we must say, catastrophic, but we must also say that the system is creatively stimulated by the catastrophe. The result is innovation beyond stagnation. The randomness is integrated into the creative system. Catastrophic extinction, though quite rare, "has been the essential ingredient in producing the history of life that we see in the fossil record" (1991, p. 189).

The storied character of natural history is increased, not diminished, by the catastrophic extinctions and by the element of chance operating during normal extinction turnover. Once "we thought that stable planetary environments would be best for evolution of advanced life," but now we think instead that "planets with enough environmental disturbance to cause extinction and thereby promote speciation" are required for such evolution (Raup 1991, p. 188). So large-scale fluctuations are vital to the dynamics of large systems.

Although it was once thought that much biodiversity was lost in the catastrophic extinctions, more recent studies suggest a different picture. Sean Nee and Robert M. May find, on the basis of a mathematical analysis of fossil extinctions: "A large amount of evolutionary history can survive an extinction episode. . . . A substantial proportion of the tree of life could survive even such a large extinction as occurred in the Late Permian." Some paleontologists have figured that up to 95 percent of marine (though not land) species perished in this extinction, though this is now thought to be an overestimate. But even if this had been so, "approximately 80 percent of the tree of life can survive even when approximately 95 percent of species are lost." Mass extinction more often cuts off the twigs of the tree of life (the species), so to speak, than the main branches (the families, orders, classes), which persist in species that do survive. "Much of the tree of life may survive even vigorous pruning" (1997, pp. 692–694; Myers 1997).

In fact, as Raup claims, the pruning results in vigorous regrowth and the production of entire new branches of life; the main achievements of evolutionary history persist through the pruning. Put in graph form (Fig. 1.7), the major extinctions in species make only transient dips in the proliferation of the number of families on Earth over evolutionary history, combining the results of graphs such as that by Newell (Fig. 1.6) with that by Raup and Sepkoski (Fig. 1.2). The conclusion that life proliferates and elaborates over time seems inescapable. The secret lies in the genes, to which we next turn.

3. SEARCHING GENES

Living organisms must track drifting environments, sometimes chaotic environments, but the life process is drifting through an information search and locking onto discoveries. It is cybernetic or hereditary, as geomorphic processes are not; there is no cumulation of information in the hydrologic, climatological, orogenic cycles, but there is in the birth, life, death, genetic cycles. That is why biology is historical in ways impossible in physics or geophysics. Genesis becomes genetic, and, later, neural.

P. T. Saunders and M. W. Ho (1976) argue that there is an increase in the amount of genetic information stored in the organism. Kimura

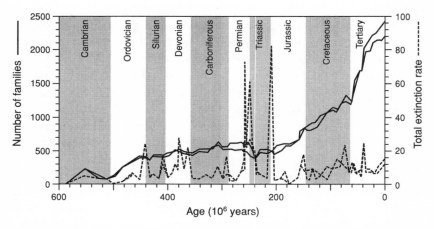

Figure 1.7. Proliferation of number of families on Earth, continuing through major extinctions. The double lines in both the number of families and the extinction rate represent maximum and minimum estimates (Benton 1995). Reprinted with permission from Norman Myers, "Mass Extinction and Evolution," *Science* 278 (24 October 1997): 597–598, p. 598. Copyright 1997 American Association for the Advancement of Science.

(1961) estimates that the higher organisms have accumulated genetic information from the Cambrian to the present at an average rate of 0.29 bit per generation. Generation times differ widely; genetic information involves quality as well as quantity; some of it is redundant. There is at present no way of reliably measuring the amount of significant genetic information in any one organism or species. Although both complexity and information resist quantification, the two are related and there does seem to be increased ability to gather and to process information about the environment, not as a general characteristic of life but as an achievement at the top trophic levels[5] (Nitecki 1988; McShea 1991).

One must understand genes as a phenomenon of searching, using variations generated in the encounter of the organism with changing environments to do this. The perpetuated gene, coding for a form of life, is only part of the story. "The role recognized by modern biology for an individual organism . . . is to transmit its genes to future generations to the maximum possible extent" (Williams 1988, pp. 385–

---

[5] McShea finds a clear consensus among evolutionary biologists that there has been increasing complexity over evolutionary time, though he suspects this arises from cultural bias interpreting the fossil evidence. Ruse agrees (1996).

386; 1993). This is its inbuilt program, but there is also the playing out of this program in the ecosystem with its interactions, checks and balances, creativity over time, and fuller history. Genes promote the survival of their kind, and ecosystems supply a satisfactory place for that to occur, so that there are repeated turnovers of individuals – birth, death, birth, death – and persistence of the kind over millennia. But what also and even more often happens is the arrival of new kinds. On the big scale it is not individual organisms that survive; it is not even species. There is speciation, the generation of new kinds, not simply reproduction to the maximum possible extent of existing kinds. There is historical development, superposed on survival. Organisms have powers that transcend their own self-defensive powers; they actualize themselves, but they actualize much more. There is descent with modification, and this sometimes results in ascent with modification.

The first microbes sought only to make more microbes to the maximum possible extent, but some transformed into trilobites, who sought to make trilobites to the maximum possible extent, but some transformed into lizards, who sought to make more lizards to the maximum possible extent, but some transformed into warblers. In the fuller narrative, genes evolve, develop, and code organisms that sometimes remain in their places, sometimes go extinct, but sometimes pass over into something else. The perpetuated gene is where information, the memory of the past with which the organism faces the future, is stored. The story is of selves preserving their self-identity, and, further, the story is of transformation as much as of evolutionary stability. The genes are the locus of this innovative evolutionary change.

Organisms in the world run on self-interest, but, as the world story runs on, genes are caught up in genesis, in the production of new ranges, orders, and kinds of selves – metazoans beyond protozoans, conditioned behavior beyond instinct, neurons where before was only stimulus–response, brains where before were only neurons, experience where before was only biochemical vitality, personality where before was only animal experience. The genes conserve order but also introduce novelty. Genes code a coping, and the coping is a defense of these values gained and dynamically transformed over time. What is conserved is what has proved valuable, tested, and transmitted intergenerationally. In result, with exploratory variations, what is selected is promising and seminal. Natural history is a story of how significant values endure through a context of suffering, stress, perpetual perishing and regeneration.

An organism arrives in the world with a genotype, shaped by a long evolutionary history. The genotype is the past delivered to the present, en route to the future. Nor is it just the genetic past; the genotype is what it is because it records the macroscopic evolutionary past. This macroscopic past is not one that the organism has lived on its own; it is the interactant past that ancestors have lived with others of that species and of other species. The history of ancestral skin-out interactions is driving the skin-in biochemistries, morphological development, and instinctive behaviors, as the genotype unfolds into a phenotype. That past history is continued with a present chapter; the genotype becomes a phenotype as it projects this coded past onto the contemporary environment. The story is reenacted as natural selection acts again in the new organism, not on the genotype, but on the historically maturing phenotype, into which the local history is now incorporated. The organism copes, and natural selection evaluates and records that coping, coded now for the generation to come.

Karl Popper concludes:

> Animals, and even plants, are problem-solvers. And they solve their problems by the method of competitive tentative solutions and the elimination of error. . . . Just like theories, organs and their functions are tentative adaptations to the world we live in. . . . A new tentative solution – a theory, an organ, a new kind of behaviour – may discover a new virtual ecological niche and thus may turn a virtual niche into an actual one. New behaviour or organs may also lead to the emergence of new problems. And in this way, they [plants and animals] may influence the further course of evolution, including the emergence of new biological values. (1972, p. 145)

Survival of the fittest is a subroutine in spinning a bigger story. There is survival of the searchers.

4. SMART GENES

The claim that natural history is random often couples with a claim that the genes are "blind" (Dawkins 1986). This claim, though metaphor, has initial plausibility because there is no intentionality in the genes. They do not "see" where they are going. By the standard scientific accounts, random variations bubble up from the genetic level, and these genotypic variations are expressed in variant phenotypes. Organisms compete for a place in drifting environments, struggling to hold a place against other lives, and those few variations that are accidentally useful are selected; the most, harmful, are discarded. Further variations are neutral; to them even natural selection is blind, since they produce no differential survival rates. These can sometimes remain, though unselected. There can also be random genetic drift, variation in gene frequency from one generation to another due to chance fluctuations.

Such lack of intentionality, or "blindness," is frequently extrapolated to draw more philosophical conclusions. George Williams asserts: "The evolutionary process is immensely powerful and oppressive, . . . it is abysmally stupid" (1988, p. 400; 1993). Charles Darwin exclaimed once that the process is "clumsy, wasteful, blundering, low, and horribly cruel"[6] (Darwin, quoted in de Beer 1962, p. 43). On the other hand, Donald J. Cram, accepting the Nobel prize for his work deciphering how complex and unique biological molecules recognize each other and interlock, concludes: "Few scientists acquainted with the chemistry of biological systems at the molecular level can avoid being inspired. Evolution has produced chemical compounds that are exquisitely organized to accomplish the most complicated and delicate of tasks." Organic chemists can hardly "dream of designing and synthesizing" such "marvels" (1988, p. 760).

We want to reconsider this alleged "blindness," both the science and the philosophical interpretations, in the light of how genes have in fact generated this prolific natural history. Nonintentional though the genetic processes may be, the genes do successfully both maintain their own kind and also steadily generate novelty. In this defense of life and in this search for innovations, might not genetic creativity in fact be a rather sophisticated problem-solving process? Talk of a genetic "strategy" has become commonplace among biologists, not thereby implying consciousness, but strongly suggesting a problem-solving skill.

Removing (or, if you like, shifting) the metaphor, more precisely

---

[6] In other moods, Darwin can find the process impressive, beautiful, and quite creative. Also, see note 3.

the question is whether an organism is "informed," whether information is present as needed for the organism's competence in its ecological niche. Every organism has considerable information about how to make a way through the world. It has a "program," and in that sense a "blind" plant has know-how about the life it is set to defend. All biology is cybernetic; the information storage in deoxyribonucleic acid (DNA), the know-how for life, is the principal difference between biology and chemistry or physics. Past achievements are recapitulated in the present, with variations, and these results get tested today and then folded into the future (Campbell 1982; Wicken 1987).

Well, it may be replied, the stored information is not so blind, but the method of discovering any new genetic information is blind. Because genes do not "see" where they are going, the variations are accidental and groping, and for this "blind" is a convenient metaphor. As organisms move from earlier genetic achievements to the discovery of later ones, there are certainly elements of random exploration. But is that all there is to be said? Consider how these elements of trial and error are incorporated in a larger generative process.

In reproduction the genetically originated novelties are formed in a shuffle that, although from one perspective may be said to be blind to the organismic needs, is far from chaotic and is only more or less random. Any and all variations are not equally probable. Genetic and enzymatic controls on the variation process limit the range of trials. There are different mutation rates at different genetic locations. Mutators and antimutators increase or trim the mutation rates as a function of population stresses (Tamarin 1996, pp. 472–474; Cairns, Overbaugh and Miller 1988; Gardner and Snustad 1981, pp. 330–331). Specific mutations are nondirected, but the rate and place at which they occur are partially regulated. In that probabilistic sense, adaptive mutation takes place (Drake 1991). For example, an enzyme produced under stress conditions "not only enhances the absolute rate of genetic change, it also alters the spectrum of the resulting mutations." "Components exist for feedback between the generators of genetic diversity and the environment that selects among variants." "Natural selection acts beyond particular alleles. It also favors genetic metabolism that generates alleles with a high probability of passing the tests of environmental selection." The result, according to David S. Thaler, is "the evolution of genetic intelligence" (1994).

There is a tendency for genes to sort in pretested blocks, a sort of

modular construction and reconstruction, facilitated by the organization of the DNA into exons, units of genetic material expressed in
structures, and introns, intervening sequences, that allow exon shuf-
fling, also with various enzymes to facilitate this (Tamarin 1996,
pp. 262–263). Repair mechanisms proofread and snip out certain genetic errors, and thus eliminate some variation, altering the
variation-versus-fidelity ratio by up to five orders of magnitude
(Mary K. Campbell 1991, pp. 580–583; Friedberg 1985; Friedberg,
Walker, and Siede 1995). The genetic program has the capacity to
reject some of the random recombinants on the basis of information
already present in the genetic coding. Individual genetic sets are
adept at pumping out their own disorder. But they do not pump out
all novelty; that would cease evolutionary development and lead to
extinction. There is a shakeup of the genes under environmental
stress, so that the fastest evolution toward variant forms, often more
highly organized forms, takes place almost explosively after major
geologic crises.[7]

Mutation is usually kept slight and conservative. Chaotic mutations that code for nothing do not even begin to produce the biomolecular units or subunits of proteins, enzymes, or lipids that were
previously coded for in the unmutated gene. Other radical mutations
that do produce structures, but nonfunctional ones, immediately
abort. The only mutations that really get tested are those just incremental enough to fit cooperatively within the whole organismic organization, or, rarely and surprisingly, those with bigger, quantum
leaps that still fit cooperatively enough to allow their trial in the life
of a phenotype. So there is constraint as to what random variations
are much or at all tested. On the one hand, changing environments,
especially if they are unpredictably changing, will select the more
evolvable species; the less evolvable will go extinct. On the other
hand, what worked well in the past, or a variant rather much like it,
will probably work well in the future.

Adaptation is imperfect; there are ways in which any organism
could be better adapted, and the variants produced are not all
equally and perfectly fitted. But if no improvement were possible, or
the variants all equal, evolution would cease, natural selection would
stop selecting for the better adapted ones, nor could life track changing environments. There must be room to make "mistakes," which

---

[7] See the observations by David Raup and Niles Eldredge, discussed earlier.

are more like "retakes," taking another tack, zigzagging the direction a bit, testing whether some of the trials are better informed methods of survival. Natural selection selects to leave the successful information in place, so far as this is possible under local genetic and ecological constraints. In this sense, it is the maladaptation that permits adaptation, the imperfection that drives the world toward perfection.

The challenge is to get as much versatility coupled with as much stability as is possible, but this is a matter of optimizing twin maxima. On the one hand there is selective advantage in using as much of past knowledge as is possible, even in keeping that which might be useful; on the other there is advantage in quickly breaking through to something new where this is required (with the risk of mis-takes this involves). Many variations are not eliminated but made recessive, transmitted by infrequently expressed genetic potential that is kept as versatility, subsequently favored if environments alter. The dominant/recessive phenomenon in genetics is a way of storing variability that is not usually expressed in a stable environment, but that is nevertheless there when the environment shifts. There may be a large number of such recessive alleles waiting in the population. Should these shifts come, the species is ready to deal with a broader range of environmental conditions than the usual environment requires (Ayala 1978).

When rabbits vary, whether through expression of variant alleles or through mutation, some are not immobile, and some travel at twice their previous velocity. Rather, some run a little faster, some a little slower, and the extra or reduced speed must integrate with benefits and costs to all the other vital rabbit metabolisms and behaviors. The variation that really emerges for natural selection to operate on is subject to prior constraint by the accomplished successes of the organism. The process is blind in that the particular variations are not generated by the needs of the organism, yet only those variations are tested that are more or less functional. Probing is restricted to the cutting edges. The organism typically only searches the nearby space for possible directions of development.

Contemporary geneticists are insisting that thinking of this process as being "blind" misperceives it. Species often move through likely possibility space much faster than would be expected on the "blind" assumption (and sometimes they do not). Genes in these species have substantial solution-generating capacities. Though not deliberated in the conscious sense, the process is cognitive, somewhat like that of computers, which, likewise without felt experience, can run problem-solving programs. There is a vast array of sophisticated enzymes to cut, splice, digest, rearrange, mutate, reiterate, edit, correct, translocate, invert, and truncate particular gene sequences. There is much redundancy (multiple and variant copies of a gene in multigene families) that shields the species from accidental loss of a beneficial gene, provides flexibility – both overlapping backup and unique detail – on which these enzymes can work.

John H. Campbell, a molecular geneticist, writes, "Cells are richly provided with special enzymes to tamper with DNA structure," enzymes that biologists are extracting and using for genetic engineering. But this "engineering" is already going on in spontaneous nature:

> Gene-processing enzymes also engineer comparable changes in genes in vivo. Cells deliberately manipulate the structures of their gene molecules for phenotypic and possibly evolutionary goals. . . . We have discovered enzymes and enzyme pathways for almost every conceivable change in the structure of genes. The scope for self-engineering of multigene families seems to be limited only by the ingenuity of control systems for regulating these pathways. (1983, pp. 408–409)

These pathways may have "governors" that are "extraordinarily sophisticated." "Self-governed genes are 'smart' machines in the current vernacular sense. Smart genes suggests smart cells and smart evolution. It is the promise of radically new genetic and evolutionary principles that is motivating today's study" (1983, pp. 410 and 414).

Despite Campbell's use of "deliberately," biologists do not think that such self-engineering is deliberate in the conscious sense, but rather in the programmed sense of a computer on problem-solving search (Latin: *deliberatio*, well weighed), that is, systematically ventured and tested. "Smart" in the vernacular sense means "clever" or "ingenious," and that is beginning to seem a better metaphor for genes than that they are "blind." "Smart," like "selfish," could be a misleading metaphor, but something creative is going on. "Selection" has been usefully extended from human to natural selection; "information," from human to genetic affairs. "Cognitive" or "learning" capacities may not be restricted to organismic individuals.

31

In a study of whether species as historical lines, using various genetic strategies to solve problems, can be considered "intelligent," Jonathan Schull concludes:

> Plant and animal species are information-processing entities of such complexity, integration, and adaptive competence that it may be scientifically fruitful to consider them intelligent.... Plant and animal species process information via multiple nested levels of variation and selection in a manner that is surprisingly similar to what must go on in intelligent animals. As biological entities, and as processors of information, plant and animal species are no less complicated than, say, monkeys. Their adaptive achievements (the brilliant design and exquisite production of biological organisms) are no less impressive, and certainly rival those of the animal and electronic systems to which the term "intelligence" is routinely (and perhaps validly) applied today. (1990, p. 63)

Analogies with the artificial intelligence in computers are particularly striking. Such cognitive processing is not conscious, but that does not mean it is not intelligent, where there are clever means of problem solving in a phyletic lineage. Schull continues:

> Gene pools in evolving populations acquire, store, transmit, transform, and use vast amounts of fitness-relative information. ... The information-processing capacities of these massively parallel distributed processing systems surpasses that of even the most sophisticated man-made systems. ... It seems likely that an evolving species is a better simulation of "real" intelligence than even the best computer program likely to be produced by cognitive scientists for many years. (1990, pp. 64 and 74)

So it seems that if we recognize that there are smart computers, we must also recognize that there are even smarter genes.

Phenotypes can be more or less labile, or plastic. What features they develop from a possible repertoire for which they are genetically endowed depends on what environments they find themselves in. *Nemoria* caterpillars that eat oak catkins grow body shapes that mimic the catkins; those that eat leaves develop body shapes that mimic twigs (Greene 1989). Insects sometimes survive in environments for which they are less than optimally adapted (feeding on a less nutritious plant). This genetically based plasticity enables phenotypes to handle changing environments, or to explore novel nearby environments, although not yet genetically changing. The plasticity itself is an adaptation for provisional trials, without yet solidifying these trial directions in genetic coding, not unless there is found some selection pressure that makes worthwhile fixing that behavior in genetic coding. The species tests various possibility spaces, and thereby hangs on until such time as better genes that equip it for more adequate performance in the shifting or novel environment (better capacities to digest the once less desirable plant, which may be increasing its numbers in the warming climate) do appear. Such plasticity allows better searching.

Augmenting this may be "maternal effects," whereby, at times, acquired characteristics are inherited for several generations (Landman 1991). Beetles eating seeds of the less desirable species, which have a harder seed coat, may, once some of them do break through the coat, lay bigger eggs that give their young a head start, enabling these young to eat their way into more of the seeds. Beetle larvae, otherwise genetically identical, whose mothers did not eat such plants and who were hatched from smaller eggs, can seldom survive on them, although they do quite well if feeding on the more desirable species (Fox, Thakar, and Mousseau 1997). That too enables species to occupy possibility spaces, until such time as genes do appear for a better adapted fit.

The protein-coding portions of the genes account for only about 3 percent of the DNA in the human genome; the other 97 percent encodes no proteins. Geneticists once thought this might be "junk DNA," but "geneticists are beginning to formulate a new view of the genome. Rather than being considered a catalogue of useful genes interspersed with useless junk, each chromosome is beginning to be viewed as a complex 'information organelle,' replete with sophisticated maintenance and control systems." That summarizes the work of Eric Lander, from his research on the Human Genome Project. This DNA seems to be able to regulate and control somatic processes without even making protein enzymes, and molecular biologists misunderstood it until they realized this. Much of it is not waste at all, but it "is turning out to play vital roles in normal genome function" (Nowak 1994).

Not only does such problem solving take place, but the genes, over the millennia, get better at it. Christopher Wills concludes, "There is an accumulated wisdom of the genes that actually makes them better at evolving (and sometimes makes them better at not evolving) than were the genes of our distant ancestors. . . . This wisdom consists both of the ways that genes have become organized in the course of evolution and the ways in which the factors that change the genes have actually become better at their task" (1989, pp. 6–8). Blind genes accumulating wisdom? Perhaps the contingent variation is part of the wisdom in the process.

Is all this problem solving accidental to evolution? It would be a rather anomalous result of nothing but blind genes, driven to reproduce nothing but their own kind, if there had appeared novel kinds steadily over many millennia but only by drifting into them. The genes seek only survival, but the story is of arrivals. There seems to be something present in the environmental system in which these genes are embedded that not only irritates them, producing an agitated effort at competitive survival, but also induces them, sometimes, to pass over into something higher. Species increase their kind, but ecosystems increase kinds. Evolution tries out mutations, variations, and that means diversity and, sometimes, complexity. The graphs we have reviewed do not particularly look like graphs of genes (much less "selfish" genes) keeping themselves intact; they rather suggest a creative upflow of life transmitted across a long continuing turnover of kinds, across a long history that includes struggling toward more diverse and more complex forms of life.

5. GENETIC ALGORITHMS

In certain kinds of problem-solving searches, so far from disparaging the seemingly accidental groping of genes under natural selection pressures, computer scientists may deliberately (now in the conscious sense) seek to imitate a similar process on their unconscious computers. Some sophisticated computer programs use what are called "genetic algorithms" (Holland 1992; 1980; 1975; Davis 1987; Goldberg, 1989; Mühlenbein, Gorges-Schleuter, and Krämer 1988; Whitley, Starkweather, and Bogart 1990; Koza 1992; Forrest 1993; Mitchell 1996).[8] An "algorithm" is a set of instructions or rules that is repeated to solve a problem. In simpler computing programs these algorithms can be precisely and logically specified. But in more complex programs, they cannot, because they are not known. Nor can

[8] See especially the journal *Evolutionary Computation* (MIT Press), also *Artificial Life* and *Adaptive Behavior* (MIT Press).

there be random searches because all possible solutions to a problem are so numerous that it would take a computer millions of years to check them all.

"Genetic" algorithms involve combining and recombining partial solutions to a problem in order to generate improved solutions. They are "search algorithms based on the mechanics of natural selection and natural genetics" (Goldberg 1989, p. 1). The model for such programs is biological: sexual mating and strings of genes on chromosomes that can be shuffled and selected. The underlying metaphor is natural selection, and the field is sometimes called evolutionary computation. The "genotype" is the bits coding the program, written microscopically on tape and disk; the "phenotype" is what the program does in ordinary life. Scientists may want to program a computer to search for the optimal set of values to solve certain multivalued problems where the values interact with each other, such as solving certain sets of mathematical equations, or detecting patterns against a background of noise, or predicting the weather, or scheduling the most effective work and meeting times for many dozens of employees in a manufacturing plant, each of whom has different time slots available, a different pay scale, and each of whom contributes different skills to the production process, many of which have to operate together or sequentially.

The computer will generate at random some "bit strings," or "genotypes," analogous to information coded on chromosomes, which are possible values in solution. These sequences are its initial "population." It will then test members of the population for effectiveness at a solution, rank them for what the scientists call their "fitness," and select the fittest. The computer will then generate new possible solutions, stimulating variations, "mutations," on the highest-ranking ones; inhibiting the lower-ranking ones; evaluate the new possibilities for their "fitness", and put them in competition with the previous, partially effective solutions. The computer also "mates" the various solutions, that is, cuts up and splices portions of bit strings that seem to code the most effective values, and then tests these "offspring" for their fitness.

The computer works with coadjusted clusters that probably (but not inevitably) move together during crossover. It may vary the "population size" of the set of solution values that it mates. It will discard solutions with low fitness. If two or more sets of solutions that have little in common (widely separated local optima) begin to appear, the computer will preserve these multiple solution tracks but try an occasional cross-mixing of segments from the different local optima, some of which will result in offspring that have enough fitness to remain in the working population. Such outbreeding prevents getting trapped in local optima that are suboptimal solutions globally.

The computer will continue with lesser probability (which may be varied during the program) occasionally to explore unlikely solutions. Even in large and complicated search spaces, genetic algorithms tend to converge on solutions that are globally optimal or nearly so. Simple bit strings can encode complicated structures, and reiterated transformations of partial solutions have a striking power to improve them. Computer searches for optimal solutions that would take a computer an estimated billion years, if done completely at random, can be accomplished by genetic algorithms in a few hours. Genetic algorithms have been used to find the most effective management of complex cross-country gas pipelines, to engineer better jet engine turbines (analyzing 100 variables, each with a range of values, a search space with over 10^{387} points), and they are being used in the design of the information superhighway. They are more expert than so-called expert systems, just because of the exploratory mutation and mating recombinations.

In these genetic algorithm programs, mimicking the chromosomes, a set of values, coded on the "genes," is being optimized through the concerted interactions of numerous information storage sites. What is being tested is the intensification or inhibition of one part of a solution (one value) coupled with the contributions of many others, all integrated (not just aggregated) in solution. Genetic algorithms only explore software possibilities, but researchers are now also developing software programs that mutate, recombine, regenerate, and test the computer chips themselves, evolving new hardware over many generations, promising the most powerful computing yet (Taubes 1997). And this is more nearly like what genes do, since they revise both morphology and behavior, evolving, so to speak, both new hardware and new software.[9]

---

[9] A note of caution is in order, worrying about how well computers can simulate natural systems: "Verification and validation of numerical models of natural systems is impossible. This is because natural systems are never closed and because

Genetic problem solving, then, does not seem so tinkering,[10] jury-rigged,[11] and blind. To the contrary, it is remarkably like what some of the smartest scientists are doing. "Nature creates highly complex problem-solving entities via evolution," concludes John Koza, surveying the possibilities for the programming of computers by means of natural selection (1992, p. 6). Stephanie Forrest finds that the "use of genetic algorithms suggests a computational view of evolution in which the mechanisms of natural selection, inheritance, and variation serve primarily to transmit and process information" (1993, p. 872). Herbert A. Simon, a cybernetics theorist, compares scientific problem solving with natural selection, to find that, on the cutting edges of science,

> the process ordinarily involves much trial and error. Various paths are tried; some are abandoned, others are pushed further. Before a solution is found, many paths of the maze may be explored. The more difficult and novel the problem, the greater is likely to be the amount of trial and error required to find a solution. At the same time, the trial and error is not completely random or blind; it is, in fact, rather highly selective.... Human problem solving, from the most blundering to the most insightful, involves nothing more than varying mixtures of trial and error and selectivity. (1969, pp. 95 and 97)

John Holland, after studying such algorithms for forty years, concludes, "Living organisms are consummate problem solvers.... Pragmatic researchers see evolution's remarkable power as something to be emulated.... By harnessing the mechanisms of evolution, researchers may be able to 'breed' programs that solve problems even when no person can fully understand their structure" (Holland 1992, p. 66). We will return to this similarity of the genetic search with scientific searching in Chapter 4.

---

model results are always non-unique.... Confirmation is inherently partial" (Oreskes, Shrader-Frechette, and Belitz 1994, p. 641). "The computational setting is highly simplified compared with the natural world.... Necessarily, they have abstracted out much of the richness of biology" (Forrest 1993, pp. 872 and 877). "Explicit fitness evaluation is the most biologically unrealistic aspect of GAs [genetic algorithms]" (Mitchell and Forrest 1994, p. 282). See later, on computer simulations of the emergence of altruism (Chapter 5, Section 2[1]).

[10] Recalling Jacob (1977).
[11] Recalling Gould (1980, pp. 20–21).

6. INTRINSIC AND INCLUSIVE GENETIC VALUES

Using a positive axiological paradigm, genes can be interpreted as loci of intrinsic value, expressed and defended in individuals and also inclusively present and distributed in family, population, and species lines. If one is working from a humanist or psychological view of what "value" can mean, this perspective can seem unfamiliar. On a sentientist and experiential account, when considering all the flora and most of the fauna (microbes, protozoans, insects, nematodes, mollusks, and crustaceans with little central nervous system) – all these organisms, with their genes, are not able to value because they are not able to feel anything. These organisms include over 98 percent of the species of life, or, counted by individuals or cumulative biomass, all but a tiny fraction of living things (Ruppert and Barnes 1994). Nothing "matters" to them because there is "nobody there," no experiential self. There is no valuer evaluating anything, nobody taking an interest in what they are doing. Such organisms, driven by their genes, do not have any options among which they are choosing. A minimally sentient awareness is required for value.

Consider, however, a more biologically based concept of value. Biologists regularly speak of the "selective value" or "adaptive value" of genetic variations (Ayala 1982, p. 88; Tamarin 1996, p. 558). Plant activities, such as dispersing seeds or producing thorns, have "survival value." Bees sting and do their waggle dance. Natural selection picks out whatever traits an organism has that are valuable to it, relative to its survival. Biologically, rather than psychologically, it is difficult to dissociate the idea of value from natural selection. When natural selection has been at work gathering these traits into an organism, coding them into genes, that organism is able to value on the basis of those traits. It is a valuing organism, even if the organism is not a sentient valuer, much less a conscious evaluator. Those traits, though picked out by natural selection, are innate in the organism, that is, stored in its genes. In our terms, these are intrinsic values.

Those who continue to insist on a sentientist or humanist theory of value must argue away all such defense of life under natural selection as not dealing with "real" value at all, but mere function. Those arguments are, in the end, more likely to be stipulations than reasoned arguments. If one stipulates that valuing must be felt valuing, that there must be somebody there, some subject of a life, then all flora and most fauna are not able to value, and that is so by one's definition. But does that definition, faced with the facts of biology, remain plausible? Perhaps the sentientist definition covers correctly but narrowly certain kinds of higher animal valuing, namely, that done by sentient animals, and omits all the rest.

Plants, for example, are quite alive. Like all other organisms, they are self-actualizing. Plants are unified entities of the botanical though not of the zoological kind; that is, they are not unitary organisms highly integrated with centered neural control, but they are modular organisms, with a meristem that can repeatedly and indefinitely produce new vegetative modules, additional stem nodes and leaves when there is available space and resources, as well as new reproductive modules, fruits and seeds. Plants repair injuries and move water, nutrients, and photosynthate from cell to cell; they store sugars; they make tannin and other toxins and regulate their levels in defense against grazers; they make nectars and emit pheromones to influence the behavior of pollinating insects and the responses of other plants; they emit allelopathic agents to suppress invaders; they make thorns, trap insects, and so on. They can reject genetically incompatible grafts.

This description of plant activities does not suppose any intentional pursuit of desires. There may be some metaphorical elements in expressions such as "defending life" or "repairing injuries," but we also take this to be a rather literal account, effectively descriptive of what is going on. To say that the genome is a set of "conservation molecules," or that the plant has a "good of-its-own" is not to be dismissed as mere metaphor. That rather seems the plain fact of the matter.

A plant, like any other organism, sentient or not, is a spontaneous, self-maintaining system, sustaining and reproducing itself, executing its program, making a way through the world, checking against performance by means of responsive capacities with which to measure success. Something more than merely physical causes, even when less than sentience, is operating within every organism. In its genetic set, there is *information* superintending the causes; without it the organism would collapse into a sand heap. The information is used to preserve the plant identity. Perhaps in physics and chemistry matter and energy cannot be lost, only transformed, but in biology this information can be and often is lost, and the plant activities promote its vital conservation. Though things do not matter *to* plants, a great deal matters *for* them. We ask, of a failing plant, What's the matter *with* that plant? If it is lacking sunshine and soil nutrients, and we arrange for these, we say, The plant is benefiting from them, and "benefit" is – everywhere else we encounter it – a value word. Biologists speak regularly of the beneficial genetic mutations and their phenotypic expressions in morphology and behavior, with their adaptive value. Harmful mutations indicate as well that values are at stake.

We are developing here this more biological, more genetic sense of value. We can approach this thinking of an organism's good-of-its own, its good-of-its-kind. As an heir to its portion of the diversity and complexity generated in evolutionary natural history, any particular organism, with its genes, defends that organism's good inhering in itself, in its "self" – a somatic though not a psychological self – which in reproduction is passed (in part) to an offspring "self," which is also such good defended in kin. Every organism inherits its portion of the past genetic line, by which it is self-constituted, and it must also be self-projecting, pushing itself forward. That is the beauty of life, the means of genesis, not something suspect. Self-development, self-defense, is the essence of biology, the law of the wilderness, though there is more to be said when such individuals are located in families, in populations, in species lines, and in ecosystems.

Why is the organism not valuing what it is making resources of? – even when not consciously so, for we do not want to presume that there is only conscious value or valuing. That should be debated, not assumed. Life is organized vitality, which may or may not have an experiential psychology. A valuer is an entity able to feel value? Yes, psychologically, and only the higher organisms can do so. A valuer is an entity able to defend value? Yes, biologically, and all organisms defend their lives.

Approach this idea with another set of metaphors. Think of a genetic set, with its cybernetic identity and information, as essentially a set of *linguistic* molecules, a *logical* set (Searls 1992). The genetic set is a *propositional* set – to choose a provocative term – recalling how the Latin *propositum* is an assertion, a set task, a theme, a plan, a proposal, a project, as well as a cognitive statement. From this the genetic set is also a motivational set, since these life motifs are set to drive the movement from genotypic potential to phenotypic expression. Given a chance, these molecules seek organic self-expression. They thus proclaim a life way, and with this an organism, unlike an inert rock, claims the environment as source and sink, from which to abstract energy and materials and into which to excrete them. Life thus arises out of Earthen sources (as do rocks), but life turns back on its sources to make resources out of them (unlike rocks). Rocks do not of themselves give rise to other rocks; rivers do not reproduce themselves and make offspring. But oaks make other oaks. An acorn becomes an oak; the oak rises from the ground and stands on its own.

So far this can seem only a description of the logic of life. Value more evidently appears when one recognizes that the genetic set is a *normative set*; it distinguishes between what *is* and what *ought to be*. The genome is a set of *conservation* molecules (if also, in another sense, a set of developmental molecules). The organism is an axiological, evaluative system. So the oak grows, reproduces, repairs its wounds, and resists death. The physical state that the organism seeks, idealized in its programmatic form, is a valued state. *Value* is present in this achievement. One is not dealing simply with an individual defending its solitary life but with an individual in a species lineage and having situated fitness in an ecosystem. Still, one needs to affirm that the living individual, the "self," taken as a point experience in the web of interconnected life, is per se an intrinsic value.

A life is defended for what it is in itself, without necessary further contributory reference, although, given the structure of all ecosystems and given the necessity for reproduction, such lives necessarily do have further reference. Organisms have their own standards, fit into their niche though they must. They promote their own realization, at the same time that they track an environment. They have a technique, a know-how. Every organism has a *good-of-its-kind*; it defends its own kind as a *good kind*. As soon as one knows what a giant sequoia tree is, one knows the biological identity that is sought and conserved.

The tree is valuable in the sense that it is able to value itself. If we cannot say this, then we will have to ask, as an open question, Well, the tree has a good of its own, but is there anything of value to it? This tree was injured when the elk rubbed its velvet off its antlers, and the tannin secreted there is killing the invading bacteria. But is this valuable to the tree? Botanists say that the tree is irritable in the biological sense; it responds with the repair of injury. The bee is making use of the nectar in the flower, but is the honey valuable to the bee? Few of us doubt that bees are irritable when they sting. Such capacities can be "vital," now a better word than "biological," and a description with values built into it. These are observations of value in nature with just as much certainty as they are biological facts; that is what they are: facts about value relationships in nature. We are really quite certain that organisms use their resources, and one is overinstructed in philosophy who denies that such resources are of value to organisms instrumentally. But then, why is the tree not defending its own life just as much a fact of the matter as its use of nitrogen and photosynthesis, or honey, to do so?

Bacteria, insects, crustaceans – including also the sentient creatures, the mice and chimpanzees – are projects of their own, each a life-form to be defended for what it is intrinsically. An intrinsic value, from the perspective of biology, is found where there is a constructed, negentropic, cybernetic identity that is defended in such a somatic organismic self with an integrity of its own. Using its genes, the organism is acting "for its own sake," or, more philosophically put, "to protect its intrinsic value." These are "axiological genes."

But the life that the organismic individual has is something passing through the individual as much as something it intrinsically possesses. All such selves have their identity in kinship with others, not on their own. This individual and familial identity is placed in a species line that must be historically maintained in the death and regeneration process, with both information stored at the genotypic level and morphology and behavior expressed at the phenotypic level. A species is another level of biological identity reasserted genetically over time: sequoia–sequoia–sequoia, bee–bee–bee. Identity need not attach solely to the centered or modular organism; it can persist as a discrete pattern over time. The individual is subordinate to the species, not the other way around. The genetic set, in which is coded the *telos*, is as evidently the property of the species as of the individual through which it passes. A consideration of species strains any value theory fixed on individual organisms, much less on sentience or persons. But the result can be biologically more sound.

Reproduction is typically assumed to be a need of individuals, but since any particular individual can flourish somatically without reproducing at all, indeed may be put through duress and risk or spend much energy reproducing, by another logic we can interpret reproduction as the species keeping up its own kind by reenacting itself again and again, individual after individual. It stays in place by its replacements. In this sense a female grizzly bear does not bear cubs to be healthy herself, any more than a woman needs children to be healthy. Rather, her cubs are *Ursus arctos*, threatened by nonbeing, recreating itself by continuous performance. A species in reproduction defends its own kind. A female animal does not have mammary glands nor a male animal testicles because the function of these is to preserve its own life; these organs are defending the line of life bigger than the somatic individual. The lineage in which an individual exists dynamically is something dynamically passing through it, as much as something it has. The locus of the intrinsic value – the value that is really defended over generations – seems as much in the form of life, the species, as in the individuals, since the individuals are genetically impelled to sacrifice themselves in the interests of reproducing their kind. Value is something dynamic to the specific form of life.

The species line is the *vital* living system, the whole, of which individual organisms are the essential parts. The species too has its integrity, its individuality. Processes of value that we earlier found in an organic individual reappear at the specific level: defending a particular form of life, pursuing a pathway through the world, resisting death (extinction); regeneration maintaining a normative identity over time, creative resilience discovering survival skills. It is as logical to say that the individual is the species' way of propagating itself as to say that the embryo or egg is the individual's way of propagating itself. The value resides in the dynamic form; the individual inherits this, exemplifies it, and passes it on. If, at the specific level, these processes are just as evident, or even more so, what prevents value from existing at that level? The appropriate survival unit is the appropriate location of valuing.

All such value is deeply embedded in the historical evolutionary ecosystem. The species lineage is woven into a supporting, stimulating, biotic community. The system is a kind of field with characteristics as vital for life as any property contained within particular organisms. The ecosystem is the depth source of individual and species alike; it has systemic value. The molecular configurations of DNA are what they are because they record at the microscopic level the story of a particular form of life in the macroscopic, historical ecosystem. What is generated arises from molecular mutations, but what survives is selected for adaptive fit in an ecosystem. One cannot make sense of biomolecular life without understanding ecosystemic life, the one level as vital as the other.

Values are intrinsic, instrumental, and systemic, and all three are interwoven, no one with final priority over the others in significance, although systemic value is foundational. Each locus of intrinsic value defends its self and kind as a good of its own, and yet each such organism gets folded into instrumental value within the system. There are no intrinsic values, nor instrumental ones either, without the encompassing systemic creativity. Properly understood, the story at the microscopic genetic level reflects the story at the species–eco-systemic level, with the individual a macroscopic midlevel between. The genome is a kind of map coding the species; the individual is an instance incarnating it. The ecosystem is the generative matrix out of which all life comes.

From such an axiological perspective, we can incorporate what theoretical biologists have come to call "inclusive fitness" (Hamilton 1964). ("Inclusive" is an interesting term for such "fitness," partly because of the parallels in social circles, recommending "inclusive" language or "inclusive" politics.) The prevailing account of the behavior of individuals toward family members goes up to the family level at the same time that it goes down to the genetic level. If one takes the gene's-eye view, as one must when the interests at stake are transmitting information and reproducing in families, one has to think of a gene as being present not only in a single cell but in all cells where there are copies of it. Since genes are a kind of information, this is somewhat like asking, Where is the book *War and Peace*? It is wherever there is a copy. So a particular gene is copresent in myriad cells within any one individual. That particular gene may be likewise copresent in relatives, copies within kin in a different skin.

Facing out, the organism finds that it is sometimes facing in, finding a similar self in others. External relations here turn out to be internal relations. Expanding the concept of the self, the survival and reproduction of a relative are partly equivalent in evolutionary effect to one's own survival and reproduction. The individual fitness is held partially in common with kin on all sides, all those "blood relations" in whom there are partial copies of "my genes," of whose genes "my genes" are partial copies. From the "point of view of a gene" (so to speak), or from the point of view of "my self," it does

44

not matter whether the descendants (gene copies) are mine immediately, as a result of my individual fitness, or in my family (inclusive fitness, within two brothers, or eight cousins, and so forth). If I fail entirely to reproduce copies of any of my own individual genes, it is just as well to have copies transmitted over there in my cousins. Narcissism and nepotism are all the same. What I value is found to be present both here and there.

Fitness is now spread across the whole family; some is within me, some in brothers, cousins, parents, children. Some insist on interpreting this from the perspective that the individual acts "selfishly" in his or her own interests (as we will later see in considerable detail), but "selfish" is now being stretched to cover benefits to father, mother, niece, nephew, cousin, children, aunts, uncles, and so on, however far one chooses to look along the indefinitely extended lines of relationship, lines that fan out eventually to all my conspecifics. In this more complete picture of what is going on, the "my" that once was located from the skin in has been so reallocated that it is now an "our." Individuals and their kin need to be distinguished from but related to the populations and species of which they are members. We will return to such a perspective under the theme of genetic identity and organisms in their communities of kin and kind (Chapter 2, Sections 1 and 3).

One must be careful here to set aside any issues of culture or morality and consider this simply biologically, as such genetic dispositions might operate in the nonhuman world. If one gene can locate its interests in a peer gene elsewhere among kin, buried though these genes are in the networked genomes they inhabit, this expands the concept of intrinsic value from something located in any particular self. It distributes value more inclusively. An individual somatic self is helping other selves, but, in turn and in return, their in-common genes also mean that they are helping this first individual self.

Each organism is in pursuit of, that is, values, its own *proper* life (Latin: *proprius*, one's own), which is all that the (nonhuman) individual organism either can or ought to pursue. It turns out, however, that any such "own proper life" is not exclusively individually owned, but is scattered about in the family, and that the individual competently defends its so-called self wherever and to the extent that this is manifested in the whole gene pool. This means that values can be held intrinsically only as they are inclusively distributed, and that places us in a position to reconsider this process by which diversity and complexity are generated.

7. DISTRIBUTED AND SHARED GENETIC VALUES

The more neutral word here is "distributed," but now that the individual self has become implicated into an "inclusive" fitness we can introduce, rather provocatively, the word "shared" with which to interpret this genetic "allocating" and "proliferating." "Share" has the Old English and Germanic root *sker*, to cut into parts, surviving in "shears," "plowshare," and "shares" of stock. As used here, to "share" is to distribute in parts the self's genetic information, thereby conserving it. Genes do generate; they reproduce or communicate what survival value they possess; they share (= distribute in portions) their information, literally, although preconsciously and premorally. The central feature of genes is that they can be copied and expressed, again and again. They replicate. Their power to send information through to the next generation is what counts. The genetic information gets allocated and reallocated, portioned out, and located in various places. Whatever the process, rather obviously genetic information has been widely distributed, communicated, networked, recycled, and shared throughout natural history.

Take two examples, the first at basic metabolic levels. Some genes code for making cytochrome *c* molecules, and these are found in organisms ranging from yeast to persons: that is, they are extremely widely shared. They are vital in the energy metabolism of all higher plants and animals and go back some 1.5 billion years, to the early history of life. Cytochrome *c* molecules do evolve through various nucleotide substitutions but are comparatively stable molecules. The primary structure is identical in humans and chimpanzees, which diverged about 10 million years ago; there is only one replacement between humans and monkeys, whose most recent common ancestor lived 40 to 50 million years ago. Even between humans and yeast the code is more than half the same, and, where it is different, the differences are often inconsequential in function (Dickerson 1971; Fitch and Margoliash 1967). Similar observations could be made regarding genes that make adenosine triphosphate (ATP), biotin, riboflavin, hematin, thiamine, pyridoxine, vitamins K and B_{12}; or those involved in fatty acid oxidation, glycolysis, and the citric acid cycle; or those that make actin and myosin. These metabolic skills are quite extensively shared by living organisms – or, if you like, quite "inclusively" distributed because their know-how has been transmitted over the millennia on the genes.

As a second example, restricted to primates, consider what *Homo sapiens* holds in common with chimpanzees. Mary-Claire King and Allan Wilson find that the difference in the protein coding sequences of DNA for structural genes in chimpanzees and humans is quite small. "The average human protein is more than 99 percent identical in amino acid sequence to its chimpanzee homolog" (King and Wilson 1975). Jared Diamond gives the figure as 98.4 percent (1992). Differences between the two species lie largely in regulatory genes.[12] E. O. Wilson recognizes this:

> We are literally kin to other organisms. . . . About 99 percent of our genes are identical to the corresponding set in chimpanzees, so that the remaining 1 percent accounts for all the differences between us. . . . Furthermore, the greater distances by which we stand apart from the gorilla, the orangutan, and the remaining species of living apes and monkeys (and beyond them other kinds of animals) are only a matter of degree, measured in small steps as a gradually enlarging magnitude of base-pair differences in DNA. (1984, p. 130; Sibley and Ahlquist 1984)

That means that the vital structural information for making the advanced primate body has been widely shared for millions of years. Similar points could likewise be made with the basic vertebrate body plan, or hearts, livers, kidneys, and so on.

We use the word "share" both as a descriptive term and as a deliberate corrective to the more fashionable word "selfish," frequently applied to such genes. "The selfish gene" is vivid imagery (Dawkins 1989). But imagery needs philosophical analysis, especially

---

[12] The similarity can be overemphasized. The regulatory genes, which govern behavior, among other things, are not included here. Many regulatory genes will also be similar in chimpanzees and humans, but many will not. Further, there is much more room for differences than the 99 percent identity in amino acid sequences recognizes. Only about 3 percent of the human genome codes for proteins; 97 percent does not, and this other DNA varies so widely between species, even between organisms of the same species, even between cells of the same organism, that it is difficult to interpret. This has been misinterpreted as "junk DNA," but geneticists increasingly see it as vital to life (Nowak 1994).

imagery that colors worldviews, even more if this seems to have scientific sanction. When scientists speak of ant wars, or queen bees and their slaves, or immunoglobulins as carrying on a "battle within" us against invading microbes, they borrow words from one domain of experience and transfer them to another. A careful analyst needs to be cautious about overtones also transferred. A great deal depends on the metaphors one chooses, since these so dramatically color the way we see the natural world. One must be careful not to let negative moral words, borrowed from culture, discolor nature. Something like this happened before in Darwinism when "survival of the fittest" was the paradigm, interpreted as "nature red in tooth and claw," but biologists now prefer to restate this as "adapted fit," a better description, since fitness takes place in various ways, only one of which is combative or aggressive. "Adapted fit" colors events differently than "survival of the fittest."

In the chapter to follow, under the themes of genetic identity, we pursue this issue in detail; here we can begin to project this more comprehensive scientific and philosophical picture. The genesis of biodiversity and complexity, so striking in natural history, is possible only as information found out by these searching genes is widely distributed, carried on from one generation to the next in such a way that it cumulates; is tested in experience, discarded where it is found to be less fit; selected and conserved where it is found to be more fit. That has happened with cytochrome c molecules and with primate protein structures. Genes must find a method of distributing and elaborating, of proliferating what values they contain and conserve. That process makes possible the genesis of life, the accumulation of all those values inherent in biodiversity and complexity.

Along with the word "distribute," en route to the word "share," consider another, relatively neutral word: Genes "divide." They "divide" in order to "multiply." Life must be enclosed in cells; yet cell division is required for cell multiplication, for ongoing life. The cell division requires genetic division. "Dividers" are required to partition out their goods, and this multiplies such goods. Such division and distributing, replicating, recycling, together with adapted fitness, place each gene where it belongs, in a commons in which it participates. The gene is engaged in dispersing vital information, in transmitting its intrinsic values. Communicated information, transmitted when a gene reproduces, has in fact been *re-produced*, produced again. Genes, in their most fundamental character, are bits of valuable information, coding bytes, in a world where vital information, the secret of life, has to be dispersed if life is to continue. Genes are a flow phenomenon.

There is nothing pejorative about either biological conservation or, what is the same thing, biological division. The dividing, reproducing organism is not so much an irremediably selfish self, defending the whole organism that it alone constitutes, and defending slivers of a self in others; rather here are values, instantiated in the self and conserved by distributing them, by defending them wherever they are present as a result of the various cellular and genetic divisions, and thereby replicating and multiplying them. This results in the transgenerational contributing of genetic values, the only kind of values that the organism has.

When used in ethics, "share" has a positive moral tone, and our point in using it biologically here, additionally to describing what is going on, is to neutralize, to unbias, the negative moral tones left by "selfish." "Share" is difficult to interpret selfishly. When genetic information is passed on to a next generation, when that information overleaps death, it would seem as appropriate to say that it has been "shared" (distributed) as that it has been "selfishly" kept. *Genes are no more capable of "sharing" than of being "selfish" – it must at once be said – where "sharing" and "selfish" have their deliberated, moral meanings.* Since genes are not moral agents, they cannot be selfish, and, equally, they cannot be altruistic. But they can transmit information, and, if one is going to stretch a word sometimes employed in the moral world and make it serve in this amoral, though axiological realm, then "share" is as descriptive as "selfish" and without the pejorative overtones. Sometimes one has to lean into the wind to stand up straight. "Dividers" and "multipliers" too find it hard to be selfish. The survival of the fittest turns out to be the survival of the sharers.

We do need to choose our words carefully – "distribute," "disperse," "allocate," "proliferate," "divide," "multiply," "transmit," "recycle," or "share" in "portions." We want a nonhumanistic, nonanthropocentric account, one unbiased by our morals, either for worse or for better. The distributive account is a much more descriptive paradigm, because there is no good reason to think that genes are selfish; there are no moral agents in wild nature even at the organismic level, much less the genetic one. But there is good reason to think that there are objective, nonanthropocentric values in nature, on which survival and flourishing depend, and that these are defended and distributed by wild creatures in their pursuit of life. Only humans are moral agents, but myriads of living things defend and reproduce their lives.

This is value vocabulary, but the point here is that in the genetic world *value*-based vocabulary is more accurate descriptively than is *morally*-derived vocabulary, for genes essentially are information, and information is of value. A gene is an information fragment, a puzzle piece in a picture of how to make a way through the world, and such a fragmentary piece can be of value to survival. That is not a selfish thing; that is a valuable thing. We are first describing what is the case when we model the phenomena so, and, after that, we may also value such value, often prescribing that such value not only *is* present, but *ought* to be conserved in the world. What kindred organisms have is a set of shared values, more or less.

From this point of view one can worry that the "selfish gene" perspective is driving a humanly biased value-laden interpretation of nature, one that has become a kind of paradigm. The jaundiced view is not coming from nature, but from the lens through which the sociobiologist or behavioral ecologist promoting such views is looking. Looked at through the lens of biologically based values, the system contains intrinsic values (such as the somatic lives of individuals, defended for what they are in themselves, transmitted to others); it also contains instrumental values (such as one organism's depending on another, or parenting that contributes to the welfare of offspring, or food chains with organisms eating and being eaten). Every such value is networked interactively into ecological systems, of systemic value. Increasing complexity and diversity require both logically and empirically increasing specialization of parts and roles, which requires increasing coaction, cooperation, and interdependence of these evolving selves. The evolutionary and ecosystemic arrangements require for these values initially to be generated and then regenerated, and subsequently distributed and shared over many millennia. The means to this end is genes.

8. STORIED NATURAL HISTORY

Earth is the planet with genetic natural history, several billion years worth, and that genesis is stored in genes. There are no genes on the moon, nor Jupiter, nor Venus. Physics and chemistry are, scientists think, true all over the universe; so astronomers and cosmologists confidently spin their theories extrapolating across the vast reaches of space and backward and forward in time. Elsewhere in space, where conditions permit, one expects to find the mineral classes and rock types that have been discovered on Earth. One expects chemical reactions to conform to the atomic table. But biology is Earth-bound. Even if there is life elsewhere (as we may hope, but for which we have only modest evidence), we do not expect to find trilobites, or dinosaurs, or tigers, or Neanderthals. Physics, chemistry, mineralogy, geology are nomothetic sciences, but biological history is idiographic to Earth. We do not expect elsewhere these historically derived genes and their Earthly generated products. They are more particular than universal, more story than law.

Max Delbrück, trained first in physics and turning to biology, reflected:

> A mature physicist, acquainting himself for the first time with the problems of biology, is puzzled by the circumstance that there are no "absolute phenomena" in biology. Everything is time bound and space bound. The animal or plant or microorganism he is working with is but a link in an evolutionary chain of changing forms, none of which has any permanent validity. Even the molecular species and the chemical reactions which he encounters are the fashions of today to be replaced as evolution goes on.... Every biological phenomenon is essentially an historical one, one unique situation in the infinite total complex of life. (1966, pp. 9–10)

And yet, he continues, it is in just this historical character that the most impressive genesis occurs, this enormous complexity in life. He marvels at what seems almost "magic":

> how the same matter, which in physics and chemistry displays orderly and reproducible and relatively simple properties, arranges itself in the most astounding fashions as soon as it is drawn into the orbit of the living organism. The closer one looks at these performances of matter in living organisms the more impressive the show becomes.... Any living cell carries with it the experiences of a billion years of experimentation by its ancestors. (1966, pp. 10–11)

This Earth story is not simply coded in genes; and we shall, in due course, give ample space to cultural history with its novelties. Even the genes, as they spin a natural history that they also record, are placed in larger events of climate, geomorphology, or marine hydrology. The story takes place at multiple levels, of which the microscopic genes are only one, the level of smallest scale. There are, on smaller scales still, atoms, electrons, quarks, on which the story motifs are superimposed, but we do not know of any cybernetically transmitted, accumulating historical coding that is registered in structures and processes at these lower levels. There are, on larger scales, the native range events with which the phenotype must reckon, the blooming, buzzing confusion of life on land and in the sea.

Once upon a time on Earth, there was no biology, only geophysics and geochemistry, and these materials organized themselves into biological molecules, into organisms. So the creativity does not begin in biology; it is already latent in the precursor materials. We are dealing with self-organizing in nature (autopoiesis), a spontaneous nature that on Earth organizes selves, but whose processes transcend those selves to increase their diversity and complexity. In a world that coheres through connections, one must put into place the assertive individualism epitomized in the selfish gene theory, find the appropriate place for the gene in a world where what is of value is widely shared, distributed, reproduced.

The story becomes memorable – able to employ a memory – only with genes (or comparable predecessor molecules). That means that the story can becomes cumulative and transmissible, that is, historic. Acetylcholine molecules and their transmembrane receptor channels are distinctively Earth-bound and historically derived; they are not intrinsic to physics and chemistry, not universal laws that one can expect to find repeatedly expressed when we explore outer space. If there is life elsewhere, one can expect levels of coding and coping, mutating and mating, and perhaps there too the best adapted survive. Wherever there is life, it will have to be defended somehow. But no biologist will predict ribosomes and Golgi apparatus in the Andromeda galaxy. This memory is loaded into the Earthen genes; these are events peculiar to Earthen biological genesis.

The production and defense of natural kinds are what is ultimately involved in the alleged "selfishness" of these genes. The historical evolution and reenactment of individuals instantiating the diverse natural kinds cannot be evil. After all – anticipating a monotheist view of the matter (Chapter 6) – God created Earth as the home (the ecosystem) that could produce all those myriad kinds. "Let the earth bring forth living creatures according to their kinds" (Genesis 1.24). There is nothing ungodly about a world in which every living thing defends its intrinsic value, those brought forth from its own perspective, at the same time that it shares, or distributes, these to offspring, to others, in the ongoing evolutionary narrative. There might be something godly about a human kind that, made in the image of God, could oversee this panorama of natural history, find (again) that it is "very good," and rejoice in it.

Chapter 2

Genetic Identity: Conserved and Integrated Values

Evolutionary genesis requires information to create the display of diverse and complex forms of life on Earth, achieved by locating value in organismic "selves," with these selves integrated into families, populations, and species lines; integrated as well into ecosystems in which organisms must have their niche, habitat, and adapted fitness. The genesis thus requires encapsulated and localized identity, both inside and in place. The "self" question, much discussed in biology, is, philosophically speaking, an "identity" question, which proves also to be an "integration" question. The question is of "belonging," what are the gene's and the self's proper and suitable role and place.

Although any analysis of identity in organisms must locate this significantly in the genes, this analysis soon reveals complex and multifaceted dimensions. Identity is a diverse, composite, developing, and far-reaching phenomenon and need not have a single a priori meaning for rocks, mountains, rivers, continents, genes, cells, organisms, families, populations, species, ecosystems, or persons and their societies.[1] One must work out what identity means a posteriori to events. There is no particular cause to suppose that all identity is genetic, although we next examine that kind. With living things, questions of level mingle with questions of identity, which mingle with questions of persisting and perishing. Genetic identity, the focus of this chapter, is paired with genetic distribution and sharing, with

---

[1] When we reach humans in their cultures, identity becomes yet more complex. In some cultures persons identify most with family, in others with tribe or clan, with nation state or companions who share their ideology, with the causes they serve, or the landscapes they inhabit. Religious identity may be with other believers, with the church, or God, Brahman, or the Tao.

54

which we concluded the previous chapter. Genetic identity must be located in "selves," but this is only half the story; it cannot be confined there but must be copied and portioned out. Such copy portions must have value, making a contribution to both identity and integration, to the generation and regeneration of life.

1. GENETIC IDENTITY

Biological identity is itself complex – organismic, neural, genetic, familial, species-specific, ecosystemic, Earthen. At the genetic level some genes are unique to oneself; others more or less identical to near kin, others shared within the breeding population, others with kindred species. Biological identity seems to attach most readily to an organismic individual, to a somatic "self." Individuals can be counted only when I can tell where one leaves off and another begins, either where there are several in front of me (cows in a field) or when there is replacement over time (father, son, grandson). We have to identify what is where, what dies, and what lives on (those cows are all Holsteins; the son is "just like" his father). We can start by working from the concept of a "self," and here it is logically essential to the concept of "survival" that one be able to identify what entity survives.

(1) Material Identity

A rock yesterday is a rock today, tomorrow, and a century hence (though not forever). Material identity persists over time. The same silicon, oxygen, aluminum, and magnesium atoms remain even if the rock is crushed. At the levels of subatomic physics, where particles are annihilated, converted into energy, and retransformed into new particles, or where probabilistic wave packets collapse into discretely located electrons, involving uncertainty about momentum, one can become troubled even about material identity. Setting aside such submicroscopic problems, material identity in macroscopic-level chemistry, mineralogy, and geology is often unproblematic. Inorganic identity is sometimes material, though not always. A mountain today contains the same material next week; the same river next week does not. Nearly everything with form can persist despite the replacement of some of its material. Material identity cannot be reproduced; no rock can be cloned.

Turning to biology, even a clone has new material identity, different atoms from those in the yet-continuing original. Organisms cannot reproduce their exact selves at all; a clone is a different self, inside different skin or bark. From the skin in, even the same organism, maintaining its organismic identity over time, with its continuing metabolism, does not require material identity; to the contrary it requires replacement – resource input and waste output. Genetic identity involves material identity over short ranges; a chromosome this morning contains largely the same atoms and molecules this afternoon. Over longer ranges there are material replacements. To make any sense of a gene's surviving, one needs some concept of genetic identity superimposed on a material identity that shifts over time.

(2) Somatic Identity

A cat yesterday is the same cat today and again tomorrow. But the cat was not here a decade ago and will not likely be here a decade hence. The biological identity of an individual begins at conception, is brought into the world at birth, and continues until death, during which time there is repeated material turnover. Most complex things can survive the replacement of their lower-level components. In living organisms, such identity is organismic, a biographical or career identity. Spatial boundaries, the edges of the organism, are reasonably clear, though identity questions arise with clones, when an amoeba divides by fission, in colonial organisms when slime mold cells aggregate for reproduction. Grass plants (ramets), formed by stolons, at first remain connected and later disconnect, all having the same genetic set (genet).

Maintaining somatic identity is not possible for long. Given aging and death, one must reproduce. In sexually reproducing organisms, in an offspring one's genome is halved; in the third generation it is quartered; in the fourth it is one-eighth. Soon the persisting amount of an individual's genome collection in any descendant individual reaches negligible proportions. The idea of having 1/8 of a self in a great granddaughter does not make much sense, if 7/8 are other 1/8 selves copresent there, and if the granddaughter also enjoys an 8/8 self of her own. If any biological identity is to survive very long, it will have to be identity at the genetic, familial, populational, or species level. At the organismic level, self-preservation is a game

soon lost. All that an organism can really transmit to future generations are genetic elements of itself, slivers of a self, or, in the vocabulary we prefer, "portions" or "shares" of itself.

To put this in the first person, the first scope of "my" in "my
genes" is somatic. There is a copy in most cells in my body (in all the
nucleated cells, though not in some short-lived cells, such as erythrocytes). No one cell uses more than a fraction of this information
copy found in each cell nucleus. Most is switched off, though, combining what this cell uses, and that one and that one, "I" use, overall,
much of the information there. Somatic cells divide and reproduce,
regenerating both the whole genetic set and particular kinds of cells.
In mitosis, when an aging liver cell makes a replacement and shortly
dies, with its replacement continuing on, this is self-maintenance.
The capacity for faithful self-maintenance of somatic identity is the
first principle of life, vitally important. There can be no advancing
biological genesis without it. Yet somatic identity, however necessary, is not sufficient for ongoing life.

(3) Kinship Identity

In meiosis, at reproduction, genes divide and get redistributed. In
result, when we ask about any self-genes one has to enlarge the
scope of that "self," to be more "inclusive" (Hamilton 1964, and
Chapter 1, Section 6). An organism must reproduce: make others,
who will be kin, that is, more or less like it. The immediate such
group is the family – we could speak of familial identity – but families soon grow extended. Kin result, and the more comprehensive
term is "kinship identity." The self belongs with its kin. Inclusive
though this is, such identity also, as we noted, divides rapidly in the
1/2, 1/4, 1/8, 1/16 progression. In sexual reproduction, which nature requires for survival in most fauna and flora, these offspring
will be half-different even in the first generation – "half-different" at
least from the perspective of the diploid–haploid–diploid recombination of genes that takes place at meiosis. Sexually reproducing
organisms cannot make identicals; offspring must be others (*alteri*),
and in this sense sexual reproduction is by necessity "altruistic" in
an others-unlike-self sense.

Organisms can only make similars with differences, and such variations over evolutionary time are as critical as the similarities. It is
not possible, of course, for an organism to make other-very-

differents; it can only breed after its kind. There is another perspective from which a great deal in the mate with which an organism breeds is quite similar. From this perspective the fifty–fifty split is a misperception. Those genes in the mate that seem other when meiosis couples up unrelated or "alien" partners are in many respects quite similar. After all, my wife too has hemoglobin in her veins and an opposable thumb on her hands, as do all "alien" humans around the globe. She has "alleles" ("allelomorphs," other, alternative forms) of many genes, but the idea of similarity is as much a part of the concept of an "allele" as is difference. An allele is another form, more or less, of the same gene, a variant.

Alleles are valuable because genetic identity can only be conserved if it is replicated, subjected to repeated testing and retesting, and modified as may be required in a world that is dynamic and changing. Alleles are genetic diversity, needed for ongoing genesis. The organism has a "share" of the alleles in a population, but all the alleles available cannot be included within the singular somatic organism;[2] they must be included in others, located outside the self in kin or in other selves out there in the population and species. Genetic identity, we are now saying, is mixed with genetic variation. Without such alleles in the population ("others" similar but different), genetic identity, again, is a game soon lost. By mating, with meiosis and division, and producing kin (offspring who are relatives, only relatively like oneself), the variation arises in which there can be both the conservation of past genes that remain functional and the development of modified genes that are better adapted fits in a changing environment.

When an organism is faced with defending similarities against differences, in competition with others of its species, with different alleles, each organism will defend its similarities in offspring. That way, if its alleles have a survival advantage, the fittest will survive. If faced with a "choice" of benefiting oneself versus benefiting more than two siblings or more than eight first cousins at cost to oneself, the self-defending thing to do is to benefit the relatives. That costs the somatic self of the acting organismic individual, but it benefits the reproductive self of the genes. Individuals and their selves do not really divide up like this, of course; one cannot meaningfully speak

---

[2] Recessive genes can be included in an individual in which the dominant gene is expressed.

of 1/2 of a self, 1/8 of a self, as though a whole self here at cost could benefit 1/8 of itself there, 1/8 of itself yonder, 1/8 of itself elsewhere. But we can compute like this if we revert to the genetic level. Meanwhile, when the self spills over and fragments to benefit many family members and further kin, one is still pursuing a kinship identity.

Animals have no capacity to defer to other alleles elsewhere in the genetic pool, that is, "altruistically" in a defending-alleles-not-possessed sense. They can only defend their kinship. An animal can only defend the genes it possesses, not alleles that it does not have. As a general phenomenon, the defense of kinship identity seems inevitable as a subroutine in the defense of species identity. Of course when an animal is defending what is kin in offspring, it is also defending what is different in them, what is "other" than its "own" self, but, in their genome, connected with what is "their" self. In that still further sense, animals routinely behave "altruistically," defending kin who are other than itself. Those other selves, too, are tested for their survival capacities, assisted when one kin defends its self, or, more accurately, its portion of its self in this other. A full "self" here defends a half or a quarter of itself there, but the same act defends the other half or three-quarters that differ from self, not-self in the kin.

Already, genetic identity is getting mixed up, or distributed, or shared. We hardly know whether to say that some helping behavior, directed at a relative, who partially contains a copy of one's "self," is a "self-sacrificing" or a "self-interested" act. It depends on where one posts the boundaries of "self." Richard Alexander sums this up: "We are evidently evolved not only to aid the genetic materials in our own bodies, by creating and assisting descendants, but also to assist, by nepotism, copies of our genes that reside in collateral (non-descendant) relatives" (1987, p. 3). Assistance to a relative will be favored if the benefit to the relative, proportioned to the degree of relationship, exceeds the cost to the donor.

Any such "inclusive" self clouds the seeming clarity of having located a "self" that can be identified, much less one that can be (as we later worry) "selfish." It is not just the organismic (somatic) self that counts; it is the reproductive (genetic) self. Genetically speaking, the "self" is more and less present in all conspecifics, or, we could say, "divided out" among them, or "shared" with others in the family. In relatives, a self's act preserves a self's genes even if the self is not the one to perpetuate them. One can feature in one's theory, if one likes, the sameness, but the variants are always just as much there, just as vital in the maintenance and development of life.

(4) Species Identity

Sexual species must mate with their kind, and the offspring, with only half of the familial identity, fully represent the species. The off-spring of a mating pair of *Panthera tigris* is a third tiger, but still the same species. All three share a species identity. Though species are not fixed over evolutionary time, *Panthera tigris* changes little enough over native-range experience. The biological identity that organisms defend in reproduction is not somatic but species identity, or at least the representative portion of species identity that they instantiate, together with that representative portion in their mates. Species identity cannot be had without somatic identity, but these are different identities. A female tiger may live for a decade (preserving somatic identity), never breed, and die (without contributing to species identity). Alewives (*Alosa pseudoharengus*) swim upstream each year to spawn, then die, losing somatic identity to contribute to species identity.

From a biological point of view it is troublesome to decide whether somatic identity trumps species identity, or vice versa. Nothing can reproduce, of course, unless it has survived somatically itself; that would seem to place somatic survival uppermost. The dead cannot breed. On the other hand, nothing can live unless it instantiates an ongoing species line; it inherits life from the previous generation. The living have been bred. This life lineage instantiated in the present generation must also be passed on to the next generation. Nor will it do simply to have many offspring simply in the *next* generation; those offspring must be positioned to reproduce success-fully in the *next + 1* generation. An organism must leave reproduci-ble offspring with an open prospect of indefinite reproducability, *next + n* generations. This reproductive urge seems to place species survival first. The self belongs with its ongoing species.

No one member of a species contains all the species identity, how-ever: a particular organism carries only one genetic set, this allele here, that allele there, drawn from a species population in which there are many other alleles, which this individual does not carry.

One individual contains a sufficiently representative set to re-present the species. Given the geography of breeding, mating takes place in populations, which are subgroups of the species, but, over time, there is typically gene flow between populations.

Reproduction, as we earlier noted (Chapter 1, Section 6) is a matter of the conservation and transfer of intrinsic value. We can now interpret that from the perspective of identity. The reproducing individual is keeping up its own kind, that is, its species identity. The female's reproductive system is not for her health, or identity maintained, but to preserve her species identity. This preserves her genes, if you like, but these genes of hers flow in reproduction into the populational and species pools. The gene flow is at one level the species flow at another scale. So a mother tiger, giving birth, is *Panthera tigris*, recreating itself.

She plays a part, or has a role or a "share," in the ongoing line. Her behavior may be "self-propagating" or even "self-interested" from the point of view of species identity, although the vocabulary of "self" or "interest" applied to a species line seems awkward. But what she does is not "self-interested" from the point of view of somatic identity. Richard Alexander says, "In a sense somatic effort is personally or phenotypically selfish, while reproductive effort is self-sacrificing or phenotypically altruistic but genetically selfish" (Alexander 1987, p. 41). Well, perhaps, but it is proving difficult to figure out what "genetically selfish" means, because the "self" is "inclusive" of others who are kin and instantiated in and representative of a species line. Meanwhile, genetic survival often requires somatic sacrifice.

The organism will be selected what is called an "evolutionarily stable strategy," that is, a behavior pattern such that, if all the members of the species adopt such behavior, no alternative behavior pattern can invade the population under the influence of natural selection (Maynard Smith, 1982a). In that sense, behavior patterns will be selected that can be also adopted by, or shared with, other members of the species. In the game theory metaphor, they will play by the same rules. Such behavior patterns will be subject to statistical distribution, to probabilities, and to the variation on which natural selection works, but the tendency toward uniformity is there. This behavior will be genetically based, and, in this respect, the genetic identity of one individual will parallel that of others in its species.

(5) Genetic Type Identity

Eventually, with the death of the organism, all the gene tokens rot or are digested and incorporated into something else. The molecules fall apart. Any particular gene is quite mortal. All that can survive in any long-term sense is the gene type. A few genetic tokens are passed intergenerationally, one copy for each ancestor–descendant crossing. As the zygote develops, that one gene token thereafter makes myriad other gene tokens. If there is any identity preserved, it is an identity of replicas.

Somatic cells produce new organisms in vegetative reproduction, but in most of this discussion a self's genes means such a self's gametes, transmitted during meiosis. Gametes are not permanent, not even persistent; they are recurrent. Looking to the past, any particular "self" inherited its genes from parents, via gametes, and grandparents, via other gametes, and on back in an exponentially widening of ancestral family into community. The self does not originate its own genes at all; any self-set is unique only as a kaleidoscopic recombination of a gene pool, perhaps with a few novel mutations. The self is a transient carrier of a historical line, receiving copies (in shuffled, mutated set) from predecessors and, looking to the future, passing copies on (in shuffled, mutated set) to descendants.

There is a short-range viewpoint from which an organismic self has its own genes, owns its genes; there is a long-range viewpoint from which any such "own genes" are "owned by" or "belong to" the genetic line that a particular self instantiates. Better, since any particular self has only some of the alleles that are also being passed down the line, a particular self's "own genes" means, This is the genetic network that any self samples as an example, drawn from the populational pool. That sample, drawn from the past and bequeathed to the future, is all the identity that any particular organism has inherited or that it can transmit. This is its genetic type, and survival of the gene type is what "counts."

Animals and plants do not just reproduce when survival of the species is at risk; they maximize the number of tokens of their type. What seems to "count," to dandelions at least, is not just leaving some of your kind to count, but leaving the most to count. On the other hand, species must have an adaptive fit, and, within the constraints of most ecosystems, the average population growth is more or less zero, as birth rates equal death rates. Other species (called k-selected rather than r-selected) are not programmed so much to max-imize their counts as to replace themselves dependably. Pelicans are an example.

Moving from the level of the organism down to the level of the gene is remarkably similar to moving up to the level of the species. The passage of genes is the passage of species. The genes are the species writ small, the macroscopic species in microscopic code; the species is the genes writ large, the microscopic code in macroscopic species. A genome is a map of a life-form. Intergenerationally, the transmission phase occurs microscopically (sperm, egg, spore). The interaction-with-environment phase (for metazoans) occurs macro-scopically at the skin/world boundary as the whole organism con-fronts its world.[3] The survival of the genes is the survival of the kind, and vice versa, since genes code kind and kind expresses genes. No one organism expresses all of the alleles available in a species, but every member organism (unless defective) represents the species as a token of the type. Its genes code the kind, representatively, and the organism, an expression of the kind, presents and re-presents the kind in the world.

Whole organisms are ephemeral. The genes have more of an eye on the species (so to speak) than on the individual. The solitary or-ganism, living in the present, is born to lose; all that can be transmit-ted from past to future is its kind. Though selection operates on individuals, since it is always an individual that copes, selection is for the kind of coping that succeeds in copying, that is re-producing the kind, distributing the information coded in the gene more widely. Survival is through making *others* (altruism, again), who

---

[3] David Hull terms the genes the replicators and organisms the interactors (Hull 1980). But whole organisms replicate themselves, tigers make baby tigers, as evidently as do genes replicate themselves. Replication takes places at multiple levels; it is the coding that is genetic.

William Wimsatt suggests an analogy: the business of life takes place at the organismic level; the bookkeeping is at the genetic level. No business, no book-keeping; no bookkeeping, no business. The levels are entwined (Wimsatt 1982, p. 172).

Such an analogy, however, makes the books kept rather too passive, merely recording events in the business. These gene books actively code for the organ-ism produced, and they introduce the variance in coding on which selection can work – disanalogies with business bookkeeping. Gene books are more like policy than account books. They are a problem-solving, cybernetic process that is a precursor of rationality (Chapter 4).

share the same valuable information. Survival is of the better sender of whatever is of genetic value in self into others. Survival of the fittest turns out to be survival of the senders.

"In a Darwinist sense the organism does not live for itself. Its primary function is not even to reproduce other organisms; it reproduces genes and serves as their temporary carrier. . . . The individual organism is only their vehicle, part of an elaborate device to preserve and spread them. . . . The organism is only DNA's way of making more DNA" (Wilson 1975a, p. 3). Depending on perspective, of course, one could as well say that DNA, transmitted in genes, is the organism's way of making another organism. Either way, genes get "spread" around, in Wilson's vocabulary, or "distributed" by organisms who do not live to benefit their "selves" but to spread what they know to other selves.

When we shift from the microscopic genetic perspective (the DNA) to the macroscopic species perspective (tigers making tigers), the genes are what they are, and survive as they survive, only as shaped to fit a species for tracking its way through the world. In this fuller perspective we see better what fitness is. Fitness is not measured by an individual's own survival, long life, or welfare. Fitness is measured by what any individual can "contribute to" the next generation in its environment. Such fitness is not individualistic, not "selfish" at all; it is fitness in the flow of life, fitness to pass life on, to give something to others who come after.

It is possible to insist on a gestalt in which the gene is said to be protecting itself selfishly in the next generation. "The ultimate benefit is clear enough: genes help themselves by being nice to themselves, even if they are enclosed in different bodies" (Barash 1979, p. 153). The trouble with that kind of claim is that the "self" essential to the claim has no firm identity – whether somatic, material, genetic token, genetic type, or cybernetic. And when one clarifies that identity in terms of a cybernetic flow of information – familial, populational, species – the phenomenon under discussion is more appropriately viewed in another gestalt. Fitness is the ability to contribute more to the welfare of later-coming others of one's kind, more relative to one's "competitors." The organism contributes all that it has to contribute, its own proper form of life, what it has achieved that is of value. The system facilitates congruence between generations. In view of the larger religious horizons in which we are eventually interested, one could even employ a religious metaphor: fitness is dying to self for newness of life in a generation to come.

(6) Genetic Symbiotic Identity

Sometimes two life lines, once independent, have fused into a single identity. Two of the most important processes energizing life on Earth use endosymbionts (Margulis 1993). One, involving mitochondria, powers animals; the other, with chloroplasts, powers plants; and, of course, plant power is the basis of animal power. Mitochondria, which anciently had a free-living identity, have been incorporated into the organisms they now empower. Similarly with the chloroplasts. Multicellular organisms may have formed by joining up of one-celled organisms, as well as by their differentiation. In the differentiated multicellular organism, even if all cells have the same genes, different cells express those genes differently, and each cell takes its organismic identity from its association with those other cells. If there are "jumping genes," each time they are incorporated into the DNA of a new species, or into another population within the same species, they revise the context of their identity.

Symbionts, as they are found at present, such as lichens, have genetic identities that are complemented by the genetic identities of the other species with which they are mutually entwined. Many ungulates depend for their survival on cellulolytic ciliates and bacteria in the rumen. Certain luminous fish have light organs that are formed of groups of symbiotic bacteria, producing photons that light the way for the fish inside of which they reside. In the full drama of natural history, identity can be a multileveled, dynamic phenomenon. Biological identity mingles with biological solidarity. Where genes belong, that is, have a proper or suitable place, can and does change depending on symbiotic relationships (Sapp 1994). With increasing recognition of this, says John J. Lee, "the time is right to recognize that symbiotic association is a central theme of life on Earth and that symbiosis is a major contributor to evolutionary change."[4]

---

[4] Quoted in *Science* 276(25 April 1997):539.

(7) Genetic Identity in Ecosystems

Fitness is not something a gene, or even an organism, as such has. A coyote in a zoo has yet its liver, teeth, and skin, but its fitness has vanished, because native-range, real-world fitness occurs in the coyote–wildlands relationship, on a tall-grass prairie or in a ponderosa pine savanna. Morphology, metabolism, and behavior result from expression of the inherited DNA, but tomorrow's DNA is determined by trial fitness today. Neither gene nor organism can be intrinsically adapted; both can only be adapted in an environment. Adaptation, the central word in Darwinian theory, is an *ecological* word, not a *genetic* one. One does not know the fitness when one knows the output of a gene, not even when one knows how this output integrates hierarchically in the whole organism. We know fitness only when we know how this output operates in the environmental niche that the organism inhabits. The value intrinsic to the gene is value that must be instrumental to survival in an ecology. Identity is identity in an environment.

From one perspective it can seem that the gene's-eye view takes life down to the molecular level. Life is reduced to genetic biochemistry. But what determines the shape of these genes? Their information content about behavior and morphology that results in reproduction in the big-scale world. The genes have been selected for – not at the microscopic level but at the level of organisms in ecosystems. The genotype can seem to determine what the phenotype is, but, then again, the form of life, the needs, the environmental niche of the organism determines what genotype, what biochemistry is selected and maintained. So the conformation, the information at the molecular level, is thrown back up to the macroscopic level and the confrontations in its community of life there. That is where its value is evaluated.

Selection operates on the whole organism, not the gene, and so the information stored in the molecular shapes and codings is a story about what is going on at the native-range level – something like the way in which a book with its small print contains a story of the big world (though with disanalogies due to the activism of these genes). Genes are being selected not so much because they are being selfishly successful in replicating themselves but because they function well in the organism in which they serve, an organism that occupies a niche in an ecosystem. Genes take their molecular shapes in response to this life history at native ranges. The causal power is not all in the genes; ecosystem changes shift selection forces and cause genetic changes in result. Molecular and organismic biology tracks big-scale ecosystemic and evolutionary biology.

No sooner is one tempted to say that the genes are in control and that life is nothing but biochemistry than one realizes that the biomolecules, selected to provide survival of the better-adapted organisms in ecosystems, are nothing but the recorded and continuing evolutionary story, set to push through the next generation. The reduction can seem complete in biochemistry at the molecular level only in a momentary cross section of what is a dynamic historical process, and the extended process is not merely molecular, but molar, indeed regional and planetary. The ecosystem determines the biochemistry as much as the other way round. Although the mutants bubble up "from below," the shape that the microscopic molecules take is controlled "from above," as the information stored there is what has been discovered about how to make a way through the macroscopic, terrestrial-range world. Sometimes it is hard to say which level is prior and which is subordinate; perhaps it is better to say that there are vital processes at multiple levels. Biological identity is multileveled.

The organism does not live alone; any "self" is embedded in an environment. Only those organisms survive that find a fitness in an ecosystemic community. A grass plant survives with other plants, more and less kin, as well as other species, embedded in the same soil, capturing nutrients released by fungal and microbial decomposers. Plants depend on the carbon dioxide released by animals, who depend on the oxygen released by plants. An animal must eat the grass, or eat what has eaten the grass, and so the trophic pyramid builds up. Energy and materials cycle and recycle through the system. In this system, the only capacity that the individual organism has is to be "self-interested," to defend its self and its kind, but the truth is that the system requires the organism to coact, to operate within the dependencies, resources, and constraints of its situation, and in that sense to cooperate, to operate together with what else is around it. Any evolutionary stable strategy will need to work generally in the repeated environment of seasons, decades, centuries, and also to be flexible enough to track changing environments, perhaps as climates alter or as new arrivals change the shape of the niche space the organism occupies. *Ecosystem* is as ultimate a truth as is *gene*.

(8) Genetic Cybernetic Identity

Physics and chemistry, as noted earlier, feature *matter* and *energy*, with both conserved through transformations, which occur as a law of nature. Biology, by contrast, features *information* by which matter and energy are informed, formed into structures and metabolisms, encoded in the genotype, and expressed functionally in the organism, the phenotype, making a living in the world. A gene is an information fragment. "A gene is neither an object nor a property but a weightless package of information that plays an instructional role in development" (Williams 1985, p. 121). A gene is a bit of replaceable material; it functions by using a bit of energy to code for and maintain vital morphologies and metabolisms. Neither the matter nor the energy counts so much as the information. What is conserved is not the matter, not the organism, not the somatic self, not even the genes, but a message that can only be conserved if and only if it is "distributed," "disseminated," "portioned out," "divided," "multiplied," "shared."

The inclusive gene's-eye view (so to speak) sees its own information over there in offspring and cousins, and what matters is not the matter but the message that gets through. A gene is a cybernetic unit; that is what makes biochemistry more than chemistry, biophysics more than physics. In the tiger-organism genetics, from the viewpoint of chemistry and physics there are only causes, but there is, from the viewpoint of the gene and of the species, information superposed on these causes. We do not try to move from the gene's-eye view further down to a carbon- or nitrogen-atom's-eye view, because we have no concept of information storage and transmission below the genetic level.

The gene makes no sense except as coding a discovery about life. By a serial "reading" of the DNA, a polypeptide chain is synthesized, such that its sequential structure determines the bioform into which it will fold. Ever-lengthening chains (like ever-longer sentences) are organized into genes (like paragraphs and chapters). Diverse proteins, lipids, carbohydrates, enzymes – all the life structures are "written into" the genetic library. The DNA is thus a logical set, not

68

less than a biological set, informed as well as formed. Organisms use a sort of symbolic logic, use these molecular shapes as symbols of life. The novel resourcefulness lies in the epistemic content conserved, developed, and thrown forward to make biological resources out of the physicochemical sources. This executive steering core is cybernetic – partly a special kind of cause and effect system and partly something more: partly a historical information system discovering and evaluating ends so as to map and make a way through the world, partly a system of significances attached to operations, pursuits, resources.

The dominant/recessive phenomenon has evolved as a way of carrying potential in a gene line that is unexpressed, "recessive," variation present in the genotype but infrequently actualized in the phenotype, usually not used but available when environmental circumstances alter. That transmits a cybernetic resilience, coupled with intergenerational stability (the dominants are more stable in the typical present environment). "Large numbers of alleles are stored in populations even though they are not maximally adaptive for that time or place; instead they are maintained at low frequency in the heterozygous state until the environment changes and they suddenly become adaptive, at which point their frequency gradually increases under the influence of natural selection until they become the dominant genetic type" (Ayala 1978, p. 63). In the metaphor we will presently examine, it would seem curious to call these many genes-in-waiting "selfish" genes. They are not aggressive about their self-expression, although they are duplicated and kept in reserve, should a need arise.

If one is born and bred to transmit information, then one must transmit the information one has, and not some other information one does not have. One will transmit that information as it both contests and complements other information simultaneously transmitted by others. Anticipating our discussion in later chapters, compare the genetic cybernetic struggle in natural history with the ideational struggle among philosophers or scientists in cultural history. Karl Popper notes, "The tentative solutions which animals and plants incorporate into their anatomy and their behavior are biological analogues of theories; and vice versa" (Popper 1972, p. 145). Is a scientist, an ethicist, or a saint pushing his or her own ideas and arguments in the world acting selfishly? The DNA code makes reprints, generation after generation, but we do not complain that only the books that sell (influence others) remain in print. Philosophers and biologists should no more object to DNA's replicating itself than they do to their books' remaining in print in constant reedition. Nor, in the larger debate that surrounds these books as they influence others, should scholars wish to be exactly copied. In both academic and ecological systems one wants developing knowledge produced as variants are tested whether they improve epistemic power for living better in the world. Fruitfulness is essential in this testing for survival power.

In nature, biology is an epistemic process, and that puts us in a position to look more directly at the claims that the conservation and distribution of this biological value demand "selfish genes."

2. GENES IN ORGANISMS

Richard Dawkins opens his influential *The Selfish Gene*, "We are survival machines – robot vehicles blindly programmed to preserve the selfish molecules known as genes" (1989, p. v). He claims that, over the last decade and a half, since he first set forth this claim, "its central message has become textbook orthodoxy" (1989, p. viii, 1976). Edward O. Wilson opened his *Sociobiology* with a claim about "the morality of the gene" (1975a, p. 3), since repeated over several decades (recounted in 1994). George C. Williams concludes, "Evolution is guided by a force that maximizes genetic selfishness." "The evolutionary process is immensely powerful, . . . it can reliably maximize current selfishness at the level of the gene" (1988, pp. 391 and 400; 1993).

Dawkins calls his selfish gene thesis, now textbook orthodoxy, "an astonishing" claim (1989, p. v). But a seemingly bold hypothesis, on closer examination, sometimes dies the death of a thousand qualifications. What happens to the bold hypothesis of selfish genes, we shall argue, is that these genes live the life of ten thousand interconnections. By the time the allegedly selfish gene has made these myriad connections it has been so transformed that a paradigm of integrated parts, of shares in a whole, is more adequately descriptive. Genetic identity is too multileveled to be so simply selfish.

(1) Integrated Organisms Versus Selfish Genes

The first astonishing claim is to meet genes, microscopic entities, labeled with a word borrowed not only from several orders of magnitude up the biological scale, that of a "self," but also with a word taken from the cultural phenomenon of morality, "selfish." L. E. Orgel and F. H. C. Crick report the discovery of "Selfish DNA: The Ultimate Parasite" (1980). Genes are a startling phenomenon on Earth, and here is a startling account of them. No one believes that there is any selfishness on the moon, on Jupiter or Mars, but the phenomenon of life on Earth is a phenomenon of selfishness.

We must take a closer look at this form of speech.[5] One would be amused if these references were to the morality of the liver or of the endoplasmic reticulum, for organs and organelles cannot be moral agents. But genes do code for life (for livers, cells, and also organismic behaviors as a whole), and perhaps there can be a morality of genes. Genes govern the process; they are not simply products, and maybe there is some selfishness in the executive program.

The definite articles are particular (*the* morality of *the* gene; *the* selfish gene; *the* ultimate parasite), as though one gene could be moral, but it must be immediately clear that rhetoric and analogy are present here. There are also striking disanalogies and the rhetoric needs to be brought under logical control. It is essential to any morally censurable selfishness that the agent has an option otherwise. *Ought not* implies *can do otherwise*. Genes have no such options. We can only be dealing with a compulsive selfishness, governed by the genes as they determine (but do not choose) behaviors. Already we need to be circumspect. If selfishness in genes exists, it will be a quite different phenomenon from human selfishness.

Those who speak of selfish genes say at once that the words "selfish" and "altruistic," as they use them here, have nothing to do with motivation, and so that is not yet an issue (though it will become so in Chapter 5 on ethics, where we find an unavoidable carryover). Genes have no intentions to consider; genes are not conscious. The issue at the genetic level is consequences, and genetically governed behaviors do have consequences. Now the question becomes whether one can make sense of a gene's acting with selfish consequences, in view of the fact that these consequences are quite intertwined with behaviors resulting from other genes. Perhaps any genetic value is more corporate.

---

[5] Pace the linguists, our analysis too is a form of speech, but within the Western tradition of linguistic discourse, one can debate the meaning and adequacy of truth claims as these are couched in various forms of speech.

Further, it is logically essential to the concept of selfishness that some entity (a "self") act or behave in its own interests in an arena where one can identify peer entities (other "selves") that have interests that can be acted for or against. Else the analogy will fail; the rhetoric will collapse. In this "behavioral" rather than "subjective" definition of selfishness, "An entity . . . is said to be altruistic if it behaves in such a way as to increase another such entity's welfare at the expense of its own. Selfish behaviour has exactly the opposite effect" (Dawkins 1989, p. 4).

What is the contrasting class of entities? If there are selfish genes, this must be other genes located within or without a particular individual organism. Gene A benefits; gene B loses. There are sometimes nonrival benefits, of course. Gene C benefits; gene D benefits; no gene loses. But unless, on some occasions, one gene can behave in its interests and against the interests of others, the possibility of selfish behavior lapses.

An initial question is whether one gene can act against the interests of others that cohabit the same organism. No gene is fit by itself; if it has fitness at all, it has fitness only in the company of other genes in the organism. "What is good for one gene is good for all" (Dawkins 1989, p. 235). Dawkins insists, though, that any gene will, if it can, act selfishly against the interests of other genes in the same body (1989, pp. 235–237). The power of his interpretive gestalt disposes him to look hard for such genes, and, indeed, he finds that some genes succeed in being what he calls "outlaws" (1983, p. 133). Some curious genes, called B chromosomes, are rare but present in thousands of species, including humans. They do not seem to contribute to the functioning of the cells and do not divide up in reproduction; they just seem to reproduce themselves and may slow down rates of growth and give reduced health and fertility in the organisms who have them. They are rather like parasites (Bell and Burt 1990). Genes such as these are, notes Trivers, "truly selfish genes, genes that lower the success of other genes in the individual but are selected because they increase their representation among offspring" (1985, p. 137).

But genes are not favored because they replicate and nothing more. They are favored because they do something for the organism (thereby including all the other genes in the organism). Genes are favored if and only if they "make a contribution" or have a part, a "share," in the integrated coping capacities of the whole organism.

They have to "convey an advantage," which in the integrated organism is distributed to all the other genes, which then have a part, a "share," in this advantage. Any particular gene has to translate itself into a "benefit" given to the organism. Genes, that is, are favored because they have conservation value for the organism. Replication of the coding is not enough; the gene must generate some survival value for better coping, done by the organism as a whole.[6]

The name "outlaw" gene, colorfully borrowed from human affairs, gives away the problem. Organisms must be well ordered (with "law and order") to maintain the metabolism, physiology, and reproductive success needed to manage successfully in the world. "Outlaw" genes that do not fit in with the larger needs of the organism, to which these other genes contribute, are not going to increase adapted fit and will be selected against. Natural selection will encourage well-orchestrated genes. It is hard for a gene to be a repeated winner inside organisms that lose when that gene wins. So this has to be self-limiting, as Trivers at once realizes. "To the extent that these selfish genes do lower reproductive success, they set up selection pressures on the rest of the genome to modify or extinguish their effects; this is presumably why such genes are not more common" (1985, p. 137).

Egbert Leigh puts it rather anthropomorphically: "It is as if we had to do with a parliament of genes: each acts in its own self-interest, but if its acts hurt the others, they will combine together to suppress it" (1971, p. 249). The metaphor is rather misleading, however, since no gene has, all by itself, more than a bit fragment of the larger know-how that an organism must have to make its living in the world. It takes many genes to have enough of a self to have an interest, enough self to know how to defend a life. Such know-how is distributed among many genes, not so much aggregated into a parliament as integrated into an organism, in which any possible "self-interest" of a gene is inseparably identified with the collective interest of them all. A gene has to collaborate to get anywhere, indeed to be anything alive. So any such "outlaw" genes are anomalous, of disvalue; "truly" successful genes have to integrate "unself-

---

[6] Primevally, in the earliest quasi-biotic chemistry, there might have been molecules that only replicated themselves. But mere-replicator molecules cannot advance far. Such molecules had to be integrated with other molecules, each doing something beneficial for the whole assembly, packaged in a cell.

ishly" with the rest of the genome. Most functional genes, as Dawkins realizes, are in-law genes.

Still, Dawkins insists, what genes do is replicate themselves, without regard for any functional contribution, and that is the explanation for all the so-called junk DNA, which is genes that are not functional, but not harmful either (else they would be selected against). "The simplest way to explain the surplus DNA is to suppose that it is a parasite, or at least a harmless but useless passenger, hitching a ride in the survival machines created by the other DNA" (1989, p. 45). But if such DNA is a sort of virus, then it does reduce functional efficiency. If it is harmless or useless, it is hardly selfish, since, being neutral, it does not increase its own genetic well-being at the expense of the well-being of other genes. The function of such DNA is not yet well understood; another interpretation is that it is important for regulation and plays a role in the reshuffling of genetic structural modules (exons), important in the recombination process producing the variation over which natural selection acts.

Any functional gene, then, plays a part in a whole. This is elementary biology, which the advocates of "selfish genes" fully understand. "Selection has favored genes that cooperate with others." The result is "the intricate mutual co-evolution of genes" (Dawkins 1989, p. 47). "To survive in the long run, a gene must be a good companion" (1995 p. 5). So much for outlaw genes. Nevertheless, the best gestalt in which to interpret these genetic activities is under debate, so let us see how "gene sharing" is as descriptive as "gene selfishness," certainly inside the organism.

Genes code at one level for morphology and for behavior at another level, with structures and metabolisms also at various in-between levels. The genotypic level is multiplexed, or crosswired, to the phenotypic level. One gene may affect numerous phenotypic traits (pleiotropy); a single morphological or behavioral trait may depend on the contribution of many genes (polygeny) (Fig. 2.1). If there is any such thing as the fitness of a gene, for most genes this is an incremental contribution wired into a mesh. A pleotropic gene might increase fitness at phenotypic characteristic$_1$ (the main locus of its expression), but it may also increase fitness a smaller amount at phenotypic characteristic$_2$ (serendipity at another locus) and even decrease fitness slightly at phenotypic characteristic$_3$ (a benefit typically has its cost). Most genes are also epistatic; they affect one another's effects.

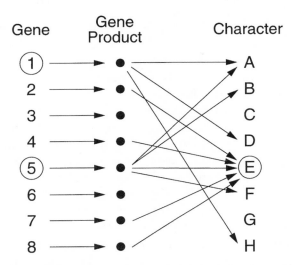

Gene	Gene Product	Character

Figure 2.1. Pleiotropy and polygeny. From *Animal Species and Evolution* by Ernst Mayr, p. 265. Copyright © 1963 by the President and Fellows of Harvard College. Reprinted by permission of Harvard University Press.

Seldom does any one gene produce one phenotypic product that confronts the environment independently. Perhaps there is a single gene that produces the melanin that colors moths darker for better camouflage in a smoke-filled countryside. Such a gene might be selected for rather directly. Another might produce an enzyme that makes an animal more aggressive in defending its territory. But most enzymes are polygenic and most behavioral traits polyenzymatic. Few genes will operate so simply and directly.

Large numbers of genes code for and control the assembly of enzymes, mitochondria, organelles, and various cellular components that never face the outside world directly. A mutation that codes for better mitochondria (resulting in more efficient energy use in the citric acid cycle) can be "seen" by the forces of natural selection only if emplaced in an organism with the ten thousand other proteins it needs for living.

Protein assembly is complex. Coded on a stretch of DNA in the nucleus, protein assembly takes place, during the first stages, on the ribosomes, complex organelles outside the nucleus that serve as protein factories. These ribosomes are themselves a combination of three ribosomal DNA (rDNA) molecules plus some fifty proteins, in two subunits, each ribosome component made by the information coded on several different genes (Lake 1981; Nomura 1984). The ribosomes make what proteins they make when instructed by messenger ribonucleic acid (mRNA) molecules from specific genes. No gene can manufacture the protein for which it codes, except as it does this in the ribosome factory previously assembled by the concerted action of proteins built from the information in many other genes.

Nor are the proteins finished here. After initial assembly on the ribosomes they are sent to the Golgi apparatus, where they are modified in at least three stages, by subunits of the Golgi apparatus, each of which is made with the aid of yet other genes (Rothman 1985). That done, most of the proteins, finished and now ready for service, are sent (under the control of still further enzymes, made with the information on yet other genes) to the sites of their functioning within the cell (Farquhar 1983; Stryer 1995, pp. 875–948).

Other proteins, though modified in the Golgi apparatus, are still unfinished and afterward sent to the lysosomes (digestive and recycling organelles) for further processing. Here the input from new protein assembly is coordinated with proteins that are being recycled. The lysosomes also receive material from other parts of the cells, including ingested material, and their output is a controlled mix of newly assembled and recycled materials. These lysosomes contain up to fifty enzymes, which are coded on different genes, often made from subunits separately coded and later joined (Bainton 1981). Each lysosome enzyme breaks down different nutrients and recycled materials, and the "benefits" of this digestion and recycling are mixed with the input of new proteins. The lysosome output is dispersed throughout the cell and often secreted for functions elsewhere in the organism, benefiting organelles, organs, tissues, and metabolisms coded elsewhere on other genes.

A monomeric protein may be built on the information contained in one gene, but in multisubunit proteins various units that are coded on several different genes come together. Control enzymes (for example, translocases, required in protein synthesis) and motility proteins (myofibril, used in muscle contraction) are constructed when half a dozen genes coact to produce complex proteins that do not themselves meet the outside world but regulate and motivate biochemistries in cells whose structural proteins are produced by yet other genes. The work of the cell, of the body, is distributed to diverse parts, "shared" by them and by the genes that produce such parts. Only the integrated organism confronts the outside environment, and any genetic identity can only be understood in such integration.

Mitochondria supply the energy to power the products made on the ribosomes. They do not make anything structural; they are only dynamos. These mitochondria arise by their own self-duplicating fission, coded first by their own DNA, which is made from genes located outside the nucleus, and coded second by nuclear DNA. They therefore arise by coaction of quite different genes (genes that may in fact have diverse ancient historical origins in quite different organisms, subsequently symbiotically assembled). These multiple genes coact to produce power for the rest of the cell, having themselves no control over where and how this power will be used. The control of that power belongs to enzymes produced on yet other genes.

When acetylcholine functions as a neurotransmitter, crossing synaptic junctions in the cybernetic network of animal brains, the acetylcholine released by one neuron has to be received by another. This reception takes place across a channel through the cell membrane of the receiving neuron. This channel is a complex protein that is embedded in and spans the lipid bilayer that forms the cell membrane. One such acetylcholine receptor channel is constructed of five subunits, each with a relatively heavy molecular weight. Each subunit is encoded on a different gene (Stevens 1985; 1987; Brisson and Unwin 1985).

A gene that forms a product that functions only when coacting with the product of four other genes, which even then serves only as a channel in a cell membrane through which neural signals pass that are controlled by many other factors – enzymes, feedback loops circulating through morphological structures and using metabolic processes coded for on thousands of other genes, and powered by still other genes – is indeed a vital gene. But that such a gene might be acting selfishly makes no biological sense.

There are a great many regulatory genes, which switch on and off the structure-producing and enzyme-producing genes. So a "selfish" regulatory gene, if there were such a thing, could only be expressed in the phenotype if it switched on and off appropriately some structural or other gene, presumably too a "selfish" gene, but one that does not express itself automatically or spontaneously, but that can, in turn, be selfish only subject to the operation of a regulatory gene.

Similarly with structural genes for the main bodily organs. The abductor pollicis longus, one of the muscles that operate the thumb, is buried in the human forearm. It will be hard for the genes that code for this muscle to code for much selfishness specific to these genes; they must operate in concert with all the genes for over six hundred muscles in the human body, indeed with all the metabolic genes as well. The human skull has about eighty-five named openings (foramina, canals, fissures) that provide passages for the spinal cord and the major blood vessels serving the brain. Each of these openings has to be controlled genetically, but does this mean that one bit of DNA selfishly reproduces itself without regard for whole brain functionality? Can the genes that produce the supraorbital foramen, for passage of the supraorbital artery and nerve, find competitors elsewhere in the body?

One might claim that although each genetic locus is a long way from the full phenotypic expression of a beneficial structure or trait, still each locus affects the phenotypic expression incrementally. A mutation at a single locus produces some phenotypic adaptiveness that is just a tiny bit better, and selective forces can detect very small increments of fitness. By this account there is a one-gene-per-incremental-survival-benefit connection, one that can be analyzed as an independent variable/dependent variable relationship. Single genes, though they seldom face the environment directly, still produce results that have tight, though incremental, contact with the environment; thus they can be effectively selected for. But since this benefit accrues not simply to the one gene producing it, but equally to all the hundreds of genes with which it is inseparably integrated along the causal chain of product delivery and functioning effectiveness, it is hard to think of that nonrival, shared benefit as being selfishly obtained by the single gene.

Though Wilson believes in selfish genes, he also concedes, "Real selection, however, is not directed at genes but at individual organisms, containing on the order of ten thousands of genes or more" (1975a, p. 70). Ayala cautions: "It must be remembered that each locus is not subject to selection separate from the others, so that thousands of selective processes would be summed as if they were individual events. The entire individual organism, not the chromosomal locus, is the unit of selection, and the alleles at different loci interact in complex ways to yield the final product" (1978, p. 64). There are 100,000 or so gene loci in humans (Ayala 1978; McKusick and Ruddle 1977).[7] Heredity may be particulate, but survival is corporate to the whole organism.

Especially with regard to behavior, which involves complex neural, cognitive, and muscular activities, the whole organism is involved. In a study of the role of inheritance in animal and human behavior, Robert Plomin concludes, "Most behavioral traits appear to be influenced by many genes, each with small effects. . . . Genetic influence on behavior appears to involve multiple genes rather than one or two major genes, and nongenetic [acquired] sources of variance are at least as important as genetic factors" (Plomin 1990).

The genetic reductionist approach falls into thinking that the organism is nothing but an aggregation of genes and their outputs, each gene being individually "selfish," a kind of bottom-up approach. Dawkins finds himself unable to think any other way:

> What *are* DNA molecules for? The question takes us aback. In my case it touches off an almost audible alarm siren in the mind. If we accept the view of life that I wish to espouse, it is the forbidden question. DNA is not "for" anything. If we wish to speak teleologically, all adaptations are for the preservation of DNA; DNA itself just *is*. (Dawkins 1982, p. 45)

But for those able to entertain a more comprehensive view, the truer picture is a top-down approach, where the organism is a whole, a synthesis, and codes its morphologies and behaviors in the genes, which are analytic units of that synthesis, each gene a cybernetic bit of the program that codes the specific form of life. A gene exists in the microworld of coding, though its output functions in the ecological macroworld of coping. It is a long way up to that big world, about which a single gene, strictly speaking, "knows" nothing. A gene only "knows" how to code "for" a protein; everything else going on is "over its head." It hasn't the slightest hint what a "predator" or a "mate" is, even if the protein it codes is used in fleeing or copulating. These bitsy "knowings" in ensemble are integrated into what the organism "knows." The single gene is of no value on its

---

[7] Ayala estimates 100,000 effective loci in humans. McKusick and Ruddle estimate 50,000 structural genes, with an amount of DNA sufficient to code 50 to 100 times that amount of genetic information. The remaining DNA is used for regulation and other functions; some may be idle, and some is repetitive.

own; it is evaluated as it coacts "for" conserving life in the integrated organism. That is what the genes are "for."

In the language of computer science, the integrated knowledge of the whole organism is "discharged" level by level as one goes down through organ, cell, organelle, enzyme, protein molecule, DNA coding, similarly to the way in which an expert computer system, at the executive level, is "discharged" into subroutines, translated into assembly language, machine language, and so on, down until the ultimate coding units only "know" on from off, flip from flop, and nothing more. No one gene "knows" enough to be selfish, any more than does a computer byte. On its own, a gene is only a tiny bit of knowledge fragment.

"Genes" thinks of them as though they were books in a library, each a discrete unit with its covers. Genes are as cybernetic as they are somatic; they are information bits as much as biochemistry. Genes may be more like thoughts in my mind, or sentences in a book, than like neurons in my brain. Thoughts are somewhat particular but often joined to other thoughts without sharp edges; thoughts frequently are what they are in their interconnections with other thoughts. Sentences in a book can be isolated, but they mean what they mean only in a narrative context. If we must use analogies from our conscious and moral life, a gene may be more like a word in a sentence in a paragraph than like a selfish moral agent.

Locating a gene in such a fishnet of fishnets of fishnets, it is difficult to think what it would mean for a single gene to operate "selfishly" in any biological sense (even after one has set aside whether such an idea makes moral sense). The benefits and costs accrue at a level different from that at which the gene immediately acts. The know-how of the organism is "distributed in parts," or "shared," by the many genes; each gene has only a part, a "share," of this knowledge, which is "divided out" among them. Such a knowledge fragment is functional only if a gene "makes its contribution" to the metabolism and behavior of the organism as a whole. The feature of any gene is not so much that it must be "selfish" as that it "contributes" to other genes, supplying information that they lack, and they do likewise. This kind of participatory vocabulary seems much more descriptive of what is going on than any effort to interpret each gene as "self-interested," much less "selfish."

In the integrated organism, cells, all with the same DNA, do not simply replicate. They differentiate, each taking a share of the work of the organism, each expressing only a portion of the information in its DNA, each depending on myriad other cells, which express other portions of that information. The cells divide out the life process in symbiosis. None of this is "selfish" at all, not even biologically or behaviorally, could we stabilize what that means, much less morally or philosophically.

If one insists on the word, the process is sociobiological in the fuller sense, where all biology is the coaction of multiple parts that exist only in *social* or organic corporation. No one gene is in any position to act in its "own" interests separably from the conservation interests of other genes. Adaptive fit within living things organized as and in hierarchical structures is not always competitive struggle; it can be coaction and organic integration. Reflecting over the evolution of complexity, especially the evolution of wholes made of parts, John Maynard Smith concludes that if adaptive evolution is to occur among any biological entities, beyond their capacity for multiplication, variation, and heredity, there are further properties: "Of these, the most important is that they should not themselves be composed of smaller entities, between which selection is acting" (Maynard Smith 1991). The evolution of competition at one level requires the evolution of cooperation at another. Whole organisms may compete successfully only if their parts work together, and that precludes selfish genes. Any particular gene shares in an organismic whole.

(2) Self-Actualizing Versus Selfish Organisms

Turning to skin-out biology, it can first seem that although the biological individual, skin-in, is quite organismically coordinated, nevertheless, facing outward, life is lived as a singular biological individual. The organism is on its own. Within its environment, the organism has some capacity of individual fitness, and this quantity of fitness must compete with other organisms who likewise themselves individually have more or less fitness. At this point there is natural selection operating to select the best adapted fits.

Now "selfish" behavior becomes more plausible. Behavior is a molar characteristic of the whole organism, not a distributed characteristic of this or that gene. Again one asks, as logically essential to the concept of "selfish," whether there is an identifiable entity (a "self") that can act or behave in its own interests in an arena where peer entities (other "selves") have interests that can be acted for or against. In the case of a selfish organism, the contrasting class will be other organisms, either of the same or of other species. Still reserving for later the question whether there is any moral component here, one can begin to make sense of biological selfishness. An individual organism, encased by skin or bark, is little trouble to identify.[8] Such organisms can and frequently do behave so as to benefit themselves at cost to others. A bird grabs a seed, and others foraging nearby do not get it. A bird eats a worm and benefits; the worm loses. Genotypes cannot be selfish, but the phenotypes they produce can.

Organisms include plants as well as animals, and one does not usually think of plants as "behaving" this way or that; they cannot "act" to move around and do things. Nevertheless some things that plants do as a result of their genetic programming benefit one plant at cost to other plants with which it competes – for water, sunlight, or nutrients. Neither plants nor animals have intentions in the reflective sense. Except possibly for some higher animals, it is not possible for them to do otherwise. So it is still not clear that selfishness is an appropriate category to apply to genetically based performances, where there are no options. But at least we can see how one organism can gain while other organisms lose, and so perhaps we have a precursor, or a biological analogue, of selfishness.

The "selfishness" alleged here is one of many characterizations that biologists use that are loaded with moralistic and pejorative overtones. There is "aggression" in ants, honeybees, hamsters, crustaceans, birds, carnivores, primates (Wilson 1975a; see the index). Gorillas and wrens "lie to one another" and get "cheated" (Wilson 1975a, pp. 119 and 326). Lions, guppies, salamander larvae, even termites are "cannibals"; not just langurs but even wasps practice "infanticide" (Wilson 1975a, pp. 84–85 and 246). "Hyenas are truly murderous" (Wilson, 1975a, p. 246). There are "warfare" and "slavery" among ants (Wilson, 1975a, pp. 50, 244–245, and 368–371). A mallard duck commits "rape" (Barash 1977; 1979, p. 54, defending the term when challenged). There are "adultery" in the mountain bluebird and "prostitution" in the tropical hummingbird (Barash 1979, p. 78; 1977, pp. 159–160; Wolf 1975). A wren is caught in "cuckoldry"

[8] There are anomalies, such as slime molds and colonial organisms. Plants are often modular as much as unitary organisms. An aspen stand may be one genetic set (genet), the individual trees (ramets) once or still connected by underground roots.

(Wilson 1975a, p. 327). "Promiscuous" male primates are "fickle" and "desert" females after breeding, though other males "guard" their mates (Maynard Smith 1982a, p. 27). Animals can be "jealous." "Spite" may exist in caterpillars (Wilson 1975a, 118–119).[9] There is "homosexual rape in acanthocephalan worms" (Abele and Gilchrist 1977). Williams lists various "other sins" (1988, p. 389).

Shades of the big bad wolf! Despite the reservations that these are behavioral and not moral descriptions, the overtones are clear. If one takes this model seriously, the natural world is being negatively judged. Williams insists: "The process and products of evolution are morally unacceptable . . . and justify an . . . extreme condemnation of nature" (1988, p. 383). Animal behavior is "patently pernicious" (1988, p. 392). He can put this with derisive rhetorical flourish: "Mother Nature is a wicked old witch!" (1993). Dawkins's most fundamental biological truth is "the gene's law of universal ruthless selfishness" (1989, p. 3). But before we reach such conclusions, we should ask whether the seemingly pejorative picture is theory-laden because the "selfishness" is in the eye of the beholder. Such a beholder might be viewing wild nature through a particular kind of human prism, and, though this is said to be objective hard science, it could really just be a subjective way of framing the problem.

Other prisms might be equally plausible, indeed more so; we humans cannot escape using some prism, but they are not all equally clear in what is reflected compared with what is projected. Using this lens, sometimes it seems as though sour morality is being disguised as hard science. There is almost an echo of animism – talk of "selfish genes," "adulterous bluebirds," and "spiteful caterpillars" – even though such talk will be checked by repeated cautions that the scientist can strip off the metaphor and retranslate the whole behavioral pattern as cause and effect. If it is only cause and effect, however, all these moralistic overtones package up the theory in a pejorative terminology, but the theory is not revealing anything about values in nature; it is just confusing us.

Many sociobiologists still approach biology from a perspective where the fittest survive by violent combat. "I think 'nature red in tooth and claw' sums up our modern understanding of natural selection admirably," Dawkins concludes in *The Selfish Gene*. "The argument of this book is that we, and all other animals, are . . . like suc-

---

[9] "Spite" is harming others without benefit to self.

cessful Chicago gangsters" (Dawkins 1989, p. 2). More metaphor, more rhetoric, but what is the reality? A less pejorative theory will avoid reading back objectional features from culture into nature, avoid speaking as though animals and genes were ethical agents in conditions of only superficial similarity. Theories are like suits of clothes; they do have to fit the data more or less, but a great deal depends on how you want to dress things up (such as nature in a witch costume). We are sewing together here such a new suit of clothes.

A seemingly scientific hypothesis, on closer examination, sometimes interconnects with metaphysical roots and lives by these underground roots as well as by the more evident scientific evidence. There is a current fashion to dress up biological phenomena by interpreting behavior in a framework that has been borrowed and reduced from morality, even to the point of taking "selfish" language down to the genetic level, regardless of the fact that it is quite problematic there. But this is revealing, and we must watch for bias. Perhaps at the organismic level also, though the "selfish" language at that level can seem more plausible, we need to be on guard. Claims that all organisms are "selfish" may depend not so much on empirical evidence as on the choice of a general interpretive framework within which to view the phenomena. Such biologists could be committing what Alfred North Whitehead called "the fallacy of misplaced concreteness," whereby, selecting out one feature of a situation, one forgets the degree of abstraction involved from the real world and mistakenly portrays the whole by overenlarging a factor of only limited relevance. A careful analysis must evaluate the metaphors with which scientists evaluate nature.

In less pejorative language, one can more simply say that an organism is "self-actualizing." It pursues its integrated, encapsulated identity; it conserves its own intrinsic value, defends its life. The organism does this in both competition and symbiosis with other organisms of the same and other species in its biotic community. This involves "self-defense," without which life is not possible. An organism must make claims on its environment, for food, mates, territory. It must use, instrumentally, other organisms, for example as prey. It must resist being made use of by other organisms, where this is detrimental to its interests, for example, again, as prey. An organism is "self-constituting," "self-realizing," "self-developing," "self-conserving," "self-generating"; an organism acts "for its own sake" – all these things can be said in a descriptive language that stops short of framing organisms with the "selfish" overtones of the "selfish genes" theory. Self-maintenance and self-propagation are not evils; both are necessary and good: without them no other values can be achieved or preserved. In the metaphorical language of game theory, an organism "plays to win," and this too is without negative connotations.

An organism can only conserve what identity, or value, it has. This genetically based knowledge will be tested in the trials of life. Any particular organism has a "good-of-its-kind," that is, a species identity, but it does not have all of the good-of-its-kind, since other alleles, which it does not have and which are not expressed in its structure and behavior, are not present. They are elsewhere in the population. So the organism expresses as much of the good-of-its-kind as it possesses, both that conserved from its inherited past and that ventured in novel recombinations and mutants. Others, conspecifics, do likewise. Some reproduce better than others. In the contests of life, natural selection operates to optimize the good-of-that-kind in the niche in which that species resides. The outcome is species-actualizing – self-actualizing, so to speak, at the level of the species – whose members are later more fit in their environments than they were before.

This places this organismic self-actualizing in a more inclusive context. Insisting on seeing everything from the perspective of either individual genes or individual organisms could be a metaphysical atomism that fails to appreciate how these self-units, these "atoms," are structured into a community, parts within larger wholes, a network that constitutes their identity quite as much as does anything internal to their genetic or organismic "selves." The truth could be far more social, or ecological, than this so-called "socio"-biology envisages. Life must be encapsulated in selves, and such selves must reproduce and spread in an environment in which they both play a part and have an integrated fit. They must have a part, a "share," of the resources in their environment, and they themselves will, sooner or later, enter that resource chain and become parts claimed, or shared, by others. The self-actualizing takes place, as we have been arguing, in the context of a shared identity.

The organism can only conserve what value it has, and none other. But the biological system, in which the individual self-actualizing and self-reproducing organism plays its role, is more selective. Individuals are evaluated for their increased fitness, for what geneticists call their "adaptive value" (Tamarin 1996, p. 558; Ayala 1982, p. 88). Individuals that have more of this survive, but when they survive they make their contribution to survival value in the species line, in the ongoing environment. What is conserved is what any individual "knows" that is better than its less well informed competitors, the losers. Such vital information gets distributed, portioned out, increased in frequency in the next generation, increasingly actualized, and thereby "shared." The cumulative result is the genesis of diversity and complexity in natural history.

3. ORGANISMS IN COMMUNITIES

Every organism, plant or animal, lives in a biotic community. Nothing lives alone. There are insects and fishes that simply hatch and live on their own; still they are more or less together with many siblings. Plants live together with other plants. Animals, undeniably, are often social and cooperate. They mate in pairs and rear their offspring, hunt in packs, nest in colonies, give alarm calls, lead each other to food sources. After we look more closely at these patterns of life together, of "symbiosis" in this larger sense, the organismic self will turn out to be as much entwined with its community as was the single gene entwined within its organism. We continue expanding the interconnections.

(1) Organisms in Families: Genes and Their Kin

Biologists are rather insistent that selection occurs at the level of the individual, not the group, and yet when they come to define the individual, these biologists go down to the genes and spread these out in the kinship group. What gets defended and selected is not just the genes of any particular individual but some set of genes in relatives, wherever they are in the kinship group. The behaviors selected are not atomistic and individualistic after all; they are diffused in the kin, in the nearby kind. Some biologists insist also on calling this "selfish" behavior, but it is a groupy selfishness. Dawkins's first book is *The Selfish Gene*, but his second is *The Extended Phenotype*; perhaps the more that phenotypes have to be extended, the less plausible it is to think of any of their genes as being very selfish.

For most higher animals, the first group, or community, is parents, siblings, cousins, offspring; each is a member of a family. Further, the individual is a member of a breeding population, and of a species; it reproduces itself, but it is also and willy-nilly reproducing its kind, and so one needs an account of community at population and species levels (Section 3[2]). Individuals in species inhabit niches in ecosystems, the larger community (Section 3[3]). Virtually all higher animals reproduce sexually, and so the account of identity in family, population, and species will also require an account of sexuality (Section 4). Finally, our account must also reckon with interspecifically shared genes (Section 5). The organismic self is located at the center of series of widening circles.

To put the self-actualizing of the organismic individual in its place in the bigger picture, we have to recall how it is located in a family line, where the self has a more inclusive fitness (Hamilton 1964; also see Chapter 1, Section 6). As already noted, many of a particular self's genes are copresent in relatives, copies within kin in a different skin; indeed all of a particular self's genes are somewhere carried also by others, save for those rare mutants it might possess.

It is impossible for the organism to be occupied all of the time in helping such relatives and the shared genes they proportionately carry, since most of its behavior must of necessity involve maintaining its own somatic metabolism and reproduction. The animal must itself eat, sleep, metabolize, move around, find shelter, protect itself, reproduce, care for its own young. Any activity devoted to preserving its own self somatically will be time, energy, and resources spent not helping (and therefore by omission hurting) copies of its genes in relatives. Some of its behavior directly hurts those copies in relatives (when it eats food that a cousin might otherwise have eaten) and is negative or costly to them. If it helps one cousin, perhaps it will not have time or resources to help another, so helping one who has copies of its genes is failing to help another who has copies of the same genes. So one must figure in fractionalized costs to others as well as fractionalized benefits.

Nevertheless, the organism, in addition to its own self-actualizing, assists in the self-actualizing of its kin. Such an expanded self-actualizing takes place along what can also be thought of as a kind of wave front moving through time (sketched in Fig. 2.2). What constitutes the particular self is not so much unique genes as a particular recombinatorial package of them, and that idiosyncratic mosaic the self can and must actualize and defend, though (save for clones) it

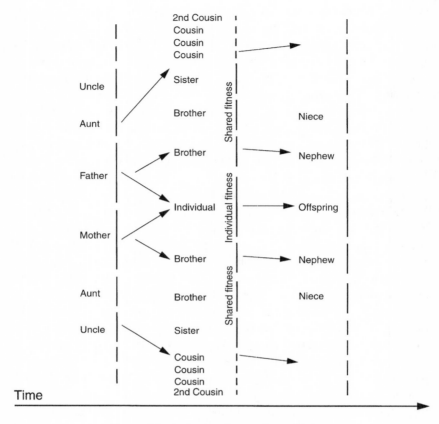

Figure 2.2. Inclusive fitness through time.

cannot reproduce this exact combination. Its offspring but likewise its cousins, aunts, uncles, and other relatives share some of its genes; they also contain alleles that it does not.

Consider animal siblings and relatives. Sister and brother share many alleles, and both share some with a cousin. Brother and cousin both have alleles that sister does not have; sister has some that they do not. If sister carries some one allele (allele B6 in Individual I, Figure 2.3), the behavior resulting from its output will benefit either sister or allele B6 in brother (shown by arrow 1). But the genes are so deeply embedded in organisms that, in sister, the output of B6 is interlocked with benefits to alleles A8, C7, D11, E5, all those other alleles that sister carries, of which brother shares some but not others. All these other alleles within sister have their beneficial outputs too,

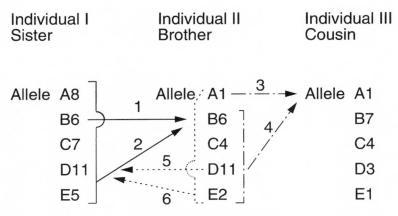

Figure 2.3. Shared alleles.

interlocked with the output of B6, and these outputs are therefore through sister's behavior delivered to B6 in brother (arrow 2), whom sister is helping because of B6 in herself. Allele B6 in brother benefits from allele B6 in sister, but, because of the interlocking, B6 also benefits from sister's A8, C7, and E5, alleles different from the A1, C4, E2, which in fact brother carries, though, luckily, identical to D11, which brother happens also to carry. But if B6 in brother is at once benefited by C7 in sister and by brother's own C4, it is difficult to put C7 in competition with C4.

Meanwhile allele A1 in brother through behavioral output is benefiting allele A1 in cousin (arrow 3), with simultaneous delivery to such cousin of all the benefits of B6 (arrow 4), interlocked with his C4, D11, and E2 (which package is meanwhile benefiting from sister's behavior). Allele A1 in cousin is interlocked with benefits to cousin's genome (B7, C4, D3, E1), which means that B6 in brother via this interlocking with A1 is benefiting B7 in cousin, rather than competing with it. Meanwhile, back in brother, the behavioral output of D11 is benefiting D11 in sister (arrow 5), which requires the delivery of the interlocked package of benefits in brother's genome, A1, B6, C4, E2 (arrow 6). Thus E2 in brother, willy-nilly, benefits E5 in sister.

Figure back in how the output of most of the genes just discussed (B6 in brother, D11 in sister, and so on) is pleiotropic, affecting several behavioral traits, and how most behavioral traits are polygenic, requiring the output of several genes. Sister's behavioral trait X requires C7, D11, and E5; in brother, parallel behavior requires C4,

D11, and E2. Cross-connect the pleiotropy and polygeny in brother, sister, cousin. And on and on. There is a mutual interweaving of each other's fitnesses.

The concept of inclusive fitness posits that one genome can locate its interests partially located elsewhere among kin, buried though these genes are in the networked genomes they inhabit. Perhaps that is true up to a point, but it seems equally true that any such set of interests is interlocked with (distributed among) the interests represented by various other sets of genomes, including other alleles. So the organism, placed among kin, is disposed to help partial copies of itself, found in kin, and that disposition is inseparable from helping dissimilar genes, more or less, found in relatives. Genetic identity gets mingled with kinship identity, which gets mingled with kinship otherness. The organism will actualize itself, but it also will assist in the self-actualizing of others. There is an ensemble of linked fitnesses, carrying forward the family line.

The fractions of relationship divide and diminish quickly past cousins, though the number of such relatives multiplies and increases rapidly also, could one add them up. An organism has millions of relatives, were it possible to identify and count all of them, but the theory deals only with near relatives, both near genetically and near enough geographically to be helped or hurt by a particular organism's behavior. In this sense, relationship is not a clear-edged affair, despite the fact that who breeds with whom and who is descended from whom is (or could be) precisely identifiable, as are the probabilities of gene distribution in mating. Defense of inclusive fitness in nearby kin would seem to be irrelevant for any genes that an individual shares widely with all or nearly all the members of its breeding population, or its species.

In any particular circumstances, better information sometimes appears, coded in some genes, which can result in improved fitness. What is being selected, along this wave front of fitness, is an inclusive fitness, distributed in multiple bodies, in which some one individual participates (shares); that is a better way of framing it than to see selection as operating to favor the isolated (and selfish) individual. It is not simply that these kin are included in its self-interest (selfishness); it is included, partially, in their selves; all its parts (alleles) are included out there among its kin here and there, though it packages them uniquely. The unit or target of selection is such genetic coding, dispersed in the family line. Genetics is a question of defending and sharing better information.

(2) Organisms in Species Populations: Genes and Their Kind

The actual reproductive group is the breeding population. That is where the individual, in a sexually reproducing species, finds mates (as well as competitors), and so that is the mixing pot, from which the most successful genes are selected. The individual can, biologically speaking, breed with any member of the opposite sex within that species anywhere on Earth; geographically speaking, the mate must be within its range of mobility. There will also be some relevant territorial boundaries and perhaps other constraints, such as dominance hierarchies. Depending on powers of mobility (wings, legs, fins) and on habitat through which one can move, as well as on contingencies of weather, fire, disease, and so on, there is genetic migration into and out from breeding populations. In plant species, apart from clones and self-pollination, the pollen for fertilization, carried by wind or insect, comes from outside the "family" line, again subject to geography and mobility.

Genes do not stay within individuals; they are spread around families, and, beyond that, they cannot stay within families either. In mating, outbreeding, they must be mingled with those of others in the population, and those of populations can (as mobility and opportunity permit) be shared throughout the species. The mutant or novel combination is preserved, if it conveys survival advantage, or perhaps, in neutral cases, if only there is no selection against it. Such a gene is also passed along, distributed to descendants, some of whom may make more of it.

The organism has a good of its own; it also instantiates a good of its kind. Does the organism act for the good of its kind? One must be careful about phrasing an answer. The individual organism acts out its genetically programmed information; it can do no other. There is no question of intentions; neither genes nor organisms such as plants and insects can entertain such intentions. Behaviorally, does the organism act for the good of the species of which it is a member? There is an important sense, often insisted on by biologists, in which the answer is no. The organism can only actualize whatever good it contains in itself. It cannot take the viewpoint of the whole species line. Nor has it, behaviorally, any means of evaluating conspecifics in its species line, some of whom will be relatively more fit, and deferring to those more fit others. In that sense, the organism cannot be an "altruist"; it can only be self-actualizing. It can only, so to speak, defend its case.

But there is another sense, sometimes overlooked by biologists, in which the answer at a higher level is yes. The unit of selection is the individual, but the evolving unit is the population, with its common gene pool, dynamically exchanged, recombined, mutated, gradually improving the adapted fit of a species. When the organism actualizes itself, and when that genotype expressed as a phenotype is tested for its relative fitness, that behavior is a good thing for the species line. The more fit organisms will succeed, relatively; the less fit will fail, relatively. In result, the species line will have members who are relatively more fit. In that sense, the organism that actualizes itself, and relatively fails, leaving fewer offspring, is still a good thing for the species line, since that possibility has been tested and found wanting. Likewise, the organism that actualizes itself, and relatively succeeds, leaving more offspring, is also a good thing for the species line, since that possibility has been tested and selected for. Individuals, winners and losers, have their roles in that genesis.

What is to be explained is a shifting toward increased fitness of the average behavior of the population, the group with which any particular individual is identified, and such an individual, even though not a winner with the particular genetic information (variation) it ventures, is nevertheless a vital part of that information-searching process. Only by such generation and testing of trial variants can the species line become more fit. The advantage is genetic diversification, with the resulting capacity to adapt to a variety of environments. At this point one can say that both the relatively more fit and the relatively less fit have a part, a share, in this searching process (Chapter 1, Section 3).

The genetic information is divided out in the population, various alleles here and there, various recombinatory and mutant trials, and the good of the species vitally depends on such distributed and shared genetic values. Though the individual organism does not act for the good of the species – it is incapable of doing so – it is good for the species that the individual organism act so. The losers, used in the genetic search, get sacrificed, relatively, for the good of the species, but that does not mean they have no share in the generative process. Though their alleles are less frequent in the next generation, the species line in which such organisms also have their identity continues for the better. Losers in one sense can be winners in another – rather as those who lose an argument win if, in the discipline or tradition with which they identify, better arguments prevail. Most ball teams are losers, but the champions require the testing that the losers provide, and the sport that the losers love is a better sport in result.

Evolutionary genesis depends on such individuals, both winners and losers, to comprise the variation over which natural selection can act. In that sense the individual organism, self-actualizing as it is, is a player in a bigger drama that is going on, so to speak, "over its head" or that is "bigger than itself." The uniqueness of a particular genetic makeup is a one-off event, temporary, instantiated in an organism, tested for its fitness, and thereby it has a role in a recombinatorial process by which the species survives, making possible the myriad other lives that ensue in that species lineage.

The confusion resulting when the word "selfish" is applied to genes and organisms suggests that they can be "unkind." Both can only defend their kin and kind. To make the point by playing on etymologies, in the only sense available to them they cannot be unkind (against their kind). The fauna and flora do not so much "love" propagating themselves as propagating their know-how. One is not so much dealing with a narcissistic biology, nor nepotistic biology. Rather, genetic transmission and organismic reproduction are epistemic biology: the propagation of kinds and the information required for their generation and regeneration in populations and species lines.

(3) Organisms in Ecosystems: Genes in Their Places

At a still more comprehensive level, organisms, families, populations, and species live in ecosystems. The organism is webbed into a life support system, and, in this sense, both inherits and is dependent on very much more than its genes. An organism inherits genes, but it also inherits the system of life into which it is sprouted, hatched, or born. Genes are located within ecosystems that are developmental (evolutionary) systems, and without their location they are powerless to create anything at all. They themselves are created as a result of this context in which they emerge. Any genecentrism needs to be complemented with environmental location.

A phenotype is generated as genes interact with an environment. In this construction, an appropriate environment is as necessary as appropriate genes. In the resulting phenotype and its performance, there are not two kinds of behavior (genetic and acquired), but various behaviors with various mixtures of genetic and acquired components interacting. Many genes do not express themselves spontaneously but are expressed only if there is an environment that stimulates their expression. The environment produces *this* particular phenotypic expression (turning on that region of the genotype), rather than *that* one (turning off another region of the genotype). The phenotype is a product of both genes and experience, or, better, genes and experience in environment. There are extragenetic inheritances such as food in the embryo or nest or den, immune molecules in the mother's milk, home territories, hunting strategies, migration routes, birdsong dialects, or ecological niches.

The genes within an organism are dependent on all the genes in all those other species with which it significantly interacts. One can think of this as value capture and contest, as it is, but it is value dependency as well. Any particular organism, with its genes, must live "together with" those on whom it is dependent, more and less. And in turn, others will be dependent on it. That genes "cooperate," operate together, evidently true from the skin in, does not cease to be part of the truth from the skin out, although the character of the cooperation shifts from the organismic to the ecosystemic. Each species is a node in a network, and genes elsewhere in that network are quite vital to it, "alien" or "other" genes in the somatic sense, but genes with which it is quite "at home" in the ecological sense.

Animals occupy niches in a trophic pyramid; they eat and will be eaten. Animals have no genes for photosynthesis; such genes, in plants, are quite vital to them. Ungulates cannot digest cellulose without the bacteria in their rumen. Carnivores eat herbivores. Higher animals may lose enzymes, rather than gain them, because they depend on the lost enzymes' remaining in species on which they depend. Natural selection shapes animal behavior according to such dependency, which may involve several trophic levels. The raptors eat the warblers that eat the insects that eat the leaves. That makes raptors dependent on the successes of all the genes with survival value in warblers, insects, and the plants they eat. That these creatures are in contest and competition cannot be denied; nor can it be denied that they are bonded together in interdependencies.

Vertebrate genes code, at the microscopic level, for coping at the macroscopic level, and so they will be assisted and hindered by behaviors that are coded for in the genes of myriad other organisms. A gene that enables a rabbit to run faster, for instance, costs something to the genes in coyotes, who find the rabbit more difficult to catch, at the same time that it benefits genes in the fleas that inhabit the rabbit's fur. What about a gene that enables a rabbit to suckle its young more nutritiously? Will it benefit the rabbits? Or only provide more young for the coyotes to catch? When the climate is growing drier, the fate of a grazing species may depend not so much on the adaptive capacities of its own genes as on those in the plants it eats, whether they can adapt to the reduced rainfall. All species participate in energy loops, nutrient cycles, hierarchies, and depend on the genes in others that make this possible.

Genes are cross-wired not only within individuals, within families, within populations, within species; they are cross-wired within ecosystems.[10] Another way of phrasing this is that a gene is an information bit in the story of natural history. Still another model is that any self, with its integrated genes from the skin in, distributed genes round about, and web-worked connections from the skin out, is a kind of holon, a genuine whole but one in which also its environment is fully reflected.

The question is how an organism behaves, but now this must also include the context, constraints, and consequences of that behavior in its environment. Since a species is what it is where it is, genes have an ecology. Biological phenomena take place at multiple interconnected levels, from the microscopic genetic through the organismic to the ecosystemic and bioregional levels. Bigger networks are superposed on smaller networks, and these on lesser networks still; there is descent from continental and global scales to those in nanometer ranges. Genes have what identity they have only as they play a part in this larger biotic community in which they code a role. For these reasons, the name of the branch of biology dealing with genet-

---

[10] As recognized in Dawkins's "extended phenotype," though here still, he argues, each gene is selfishly computing its interests in all its effects in the world (1983).

ically based behavior has shifted from "sociobiology," which has been stigmatized with its "selfish genes," to "behavioral ecology" with more emphasis on how any such genes must produce behavior that is an adapted fit in an ecosystem.[11]

4. SEXUALITY, SELFISHNESS, AND COMMUNITY

A paramount force that keeps enlarging the family is sexuality, combined with the pressure toward outbreeding. In sexually reproducing species, when the genes go through just that phase of the life cycle where the fully selfish genetic set might be expected to construct a faithful copy of itself, a cloned organism with identical genes, there are chopping up and reshuffling, as though to bar genetic fidelity as the only rule in the game. The system insists on variation. It is hard to be selfish, if one is a genome and must be split in half at every reproduction. "In sexual reproduction only half [the genes] are identical; the organism, in other words, has thrown away half its investment" (Wilson 1975a, p. 315). This is "the paradox of the cost of meiosis" (Dawkins 1983, p. 160).

Sex is ubiquitous in all groups of organisms living today. Less than one in a thousand animal species reproduces asexually; there are rare examples among lizards, fishes, grasshoppers, all thought to be hybrids of sexual species (White 1978, p. 287). Plants too reproduce sexually.[12] David Hull comments, "The prevalence of sex re-

---

[11] For example in the journal *Behavioral Ecology* (Oxford University Press).

[12] It is doubtful that plants are amenable to kin selection theory or to sexual selection theory. They do not have behaviors by which they can help other plants, though they may live in close association with other plants of the same (or other) species, which may contribute to their welfare. Plants are not unitary organisms (highly integrated with centered neural control) but are modular organisms, with a meristem that can repeatedly and indefinitely produce new vegetative modules (additional stem nodes and leaves when there is available space and resources) or new reproductive modules (flowers and fruits). Using iterated modules of growth and reproduction, plants do not separate somatic and gametic lines as do higher animals, and a somatic mutation in a developing meristem can be subsequently reproduced in flowers and fruits on that stem.

Male and female are often included in the same plant (monoecious, with perfect flowers), though some species have male plants and female plants (dioecious, with staminate and pistillate flowers on separate plants). In the perfect plants, there is typically an arrangement to promote outbreeding and to dis-

mains the major roadblock to an entirely 'individualistic' interpretation of evolution" (1980, p. 329). George Williams, the most thoroughgoing advocate of selection at the individual level, is much frustrated by sexuality, an anomaly of the first magnitude in his conceptual scheme. "There is a kind of crisis at hand in evolutionary biology. . . . The prevalence of sexual reproduction in higher plants and animals is inconsistent with current evolutionary theory." In higher animals, birds, and many insects, "sexuality is a maladaptive feature, dating from a piscine or even protochordate ancestor, for which they lack preadaptations for ridding themselves" (1975, pp. v and 102–103). But another evolutionary biologist, Graham Bell, not so determined to cast every process, sexuality included, into an individualist framework, calls sexuality "the masterpiece of nature" (1982).

Niles Eldredge says, "The details of the ultra-Darwinian internal wranglings on the subject are labyrinthine, with the main point being that they cannot agree." The problem, he thinks, is an overly "dogmatic" assumption "that organisms are locked, willy-nilly, into an eternal combative struggle to leave more copies of their genetic information to the next generation" (1995b, pp. 217–219). John Maynard Smith, after a determined attempt to explain sexuality from an individualist genetic perspective, finds that he cannot do so. "One is left with the feeling that some essential feature of the situation is being

courage selfing. Some plants have outbreeding flowers as well as self-pollinating flowers (cleistogamy). An individual seed is composed of parental sporophyte tissues in the seed coat, remnants of separate, but related gametophytes, the descendant sporophyte embryo, and a triploid endosperm. Figuring the interests of some one gene amid these others, calculating kin ratios for genetic self-interest becomes quite complex, perhaps implausible (Burnham and Stout 1983; Queller 1983; Willson and Burley 1983).

Many plants also reproduce vegetatively, and one genome (the genet individual, a strawberry plant originating from seed) can clone many organisms (the ramet individuals originating by stolons) which are genetically identical, and, if they compete with each other at all, cannot compete in ways by which natural selection differentiates genotypes. Ramets still attached to each other through stolons and root systems may sometimes also share resources (water, nutrients, photosynthetic sugars). Predation (grazing) on plants usually does not kill the organism, which may recreate itself with new modules of growth and reproduce thereafter; predation on animals destroys the animal and terminates reproduction.

overlooked" (1976, p. 257; 1978). Though the adaptive function of sex remains disputed in evolutionary theory, there is general agreement that long-term adaptive potential is maintained by recombination. Species that reproduce by cloning lose out to those that reproduce sexually. Groups that diversify extensively are made up of sexual, not clonal species.

It would indeed be odd if sexuality were an evolutionary mistake, persisting a billion years, and prevailing in 999 out of 1,000 faunal species, as well as overwhelmingly in floral species. If sexuality embarrasses the theory, perhaps the theory is incomplete. Sexuality seems rather to be at the center of biological creativity. How is one to frame this phenomenon, so widespread in nature? We could say that, in sexual reproduction, the genes "throw away half their investment," or that they get "corrupted" ("adulterated"!) with every reproduction. We would prefer to say "distributed," or "divided out" or "shared" – the metaphors are important; they color how we perceive what is going on. "Sex," says Michael Ghiselin, is "synonymous with 'mixis' – literally 'mingling' " (1974, pp. 52–53).

Sexuality is so pervasive that it must convey some survival advantage. The most plausible account is that it allows the interchanging of genetic discoveries, which permits faster evolution, especially in rapidly changing environments or in coevolutionary contests (Stanley 1979, pp. 213–227; Maynard Smith 1978). In species that usually reproduce asexually, individual organisms begin reproducing sexually under adverse conditions (Bell 1982). Only in asexual reproduction can an organism make identicals, clones, but asexuals are disadvantaged over evolutionary time. There is not enough variation, and no way to crossbreed discoveries. Steven Stanley holds that sexuality enables whole new species to arise relatively rapidly; it permits quantum rather than incremental innovation. From the viewpoint of a "selfish" gene, dramatic speciation is even more upsetting than incremental innovation.

If an animal must mate, then mating with siblings would more nearly preserve the particular set of genes that an organism has. Given the necessity to breed sexually, it might be thought advantageous to breed with near kin. That way the organism can transmit its own genes somatically coupled with its genes that are also in relatives. This sometimes happens, but the system discourages close inbreeding. Breed an organism must with its own kind; breed it often does within its tribe, perhaps even its larger family; but breed it should not with immediate relatives. There are selective pressures toward outbreeding, where an animal mates with kind, not kin.

Inbreeding costs, known also as inbreeding depression, include reduced viability and fecundity of offspring and susceptibility to disease and genetic deformities (as breeders discover), so that close inbreeding is selected against rather strongly and virtually absent in natural populations of animals (Ralls, Ballou, and Templeton 1988). These detrimental effects have also resulted in suppressed self-pollination in plants. In humans, if there is any sexual behavior imprinted in the genes, the most likely candidate is a disinclination to incest. The system discourages kin selection in sexual pairing; it forces outbreeding, against the "selfish" tendencies of the genes.[13] It requires spreading genes around, mixing them up.

To reproduce themselves selfishly one's genes have to join with an alien set. That interlocks any "selfish" set of genes with those of another line; it must outbreed at a fifty-fifty split to protect its genes within. From the gene's-eye view, this is a curious system in which the chances of transmission are fifty-fifty by required coupling with nonkindred lines. If one still wants to think of it that way, the system limits, or mixes, the permitted "selfishness" with other-directedness. Competitors are forced to be cooperators, the selfish to share. An organism must mate to breed.

So also must others in its family line. So even further still, all those bits and pieces of "my" inclusive fitness in cousins and nephews (to phrase the issue in the first person) must similarly and inseparably be joined in sexual reproduction with the genes of more remote in-laws, strangers that these relatives marry, and there is all the more entwining willy-nilly with much "alien" fitness. The further away these blood relatives are, the more reduced is the proportion of genes that any one of them shares with me, and yet the more of such relatives there are. Summed up, these reduced but numerous proportions remain important. And, via their mating, whatever proportion of "my" fitness is present in them becomes dependent on the coaction of persons with the alien genes with which these relatives intermarry. Every one of these relatives has also to outbreed, as much as do I.

---

[13] With enough unrelatedness, in populations evolved more or less differently in more or less distant environments, there can also be outbreeding depression (Thornhill 1993).

The whole thrust of sexual reproduction is toward bonding the individual into a community exceeding itself, but any selfish gene theory will have to find a way of interpreting this enlarged bonding as mere "selfishness" preserved. It seems equally obvious, however, that any "selfishness" is rather getting diluted or divided out; what one in fact confronts is survival by way of incorporation into, and cooperation with, others. Males, for example, depend on females for their mitochondria, since these descend in the female line only, and thus males are empowered by females. Females (in mammals) depend on males for Y chromosomes, which they do not have or need somatically, but without which in the males, of course, the females cannot mate and reproduce.

Pure replicators, making only identicals, may do well enough in the short term or in little-changing environments, but in the long haul and in complex environments, they go extinct. So an organism arrives in the world as a beneficiary of past variations, and it inhabits a natural system in which it can continue only if it can make variant copies of itself. So far as they are copies, the organismic history is inherited; so far as they are variants, history is generated anew. The organism is itself a product of history, but its "self" cannot continue long somatically: it dies. And it cannot replicate itself except as it also generates otherness, copies with variance.

In such mating an animal defends its kin over its kind and is unable to take a specieswide view. The organism can only venture itself as a trial in the survival game, actualizing itself on the basis of its genome. It can only defend its genes in kin, although in so doing it must also defend all the other genes, other alleles, in those relatives in whom its own alleles are mixed. There is no particular cause to think this defense of one's familial genes is selfishness, any more than to think that somatic self-defense is deplorable.

Nevertheless, there are cases that seem to involve ruthless genetic selfishness. Hanuman langurs (*Presbytis entellus*) are lanky long-tailed monkeys of Asia. When a new male takes over a harem, it kills (it is claimed) the young sired by the previous male, so as not to invest effort caring for another's young. Further, the lactating females cannot breed until they are no longer nursing; killing the young returns the females to estrus and they will breed the more quickly with the new male (Hrdy 1977a; 1977b; Vogel and Loch 1984; Hausfater 1984). It is to the female's genetic advantage to nurse her young a few weeks more, since she already has considerable investment in them, and they can soon be weaned, later to reproduce themselves. The new male's genetic interest is served by infanticide. So he kills the young. Such behavior would be favored by natural selection because it defends kinship identity over the identity of the species.

These killer males have infanticidal genes. Such genes, the moralist might say, are deplorable. Labeling such behavior "infanticide" is problematic, however, since the word is borrowed from the realm of moral agents, who can reflect on such behavior and choose to do otherwise, as langurs are unable to do. Nevertheless, such behavior, instinctive though it is, might be judged an evil in the system, even if we did not fault the male langurs, who are merely acting out their instincts.

Ethologists are not agreed whether such behavior has been correctly reported or interpreted. It has been directly observed only three or four times, suspected and inferred some fifty times, but not infrequently the paternity of the slain infant is difficult to assess. Also, in some cases where extensive field studies have been carried out, such behavior has never been observed (Vogel and Loch 1984; Boggess 1984; Eibl-Eibesfeldt 1989, pp. 93–95). A first question is whether this behavior is species typical or pathological. There are recurrent genetic errors (like Down's syndrome) that have no survival advantage. Atypical behaviors can be produced by unusual crowding, and these langurs are often stressed by human-caused habitat degradation. Aggressive tendencies often get exaggerated in crowded conditions. It would be a mistake to try to interpret aberrant behavior as though it is a long-standing reproductive strategy. Even if it is locally adaptive within some populations, it may not be species typical. On various counts, the adaptive significance of infanticide is problematic.

Langurs live in small groups, extended family lines, and these compete for territory and resources with such groups nearby. Imagine that in some family line these infanticidal genes become quite widespread, since they are a successful reproductive strategy. Such genes are used and selected for often. Compare a family line with frequent infanticidal takeovers with one in which there are none. It is difficult to think that a population that slays, say, half its infants can compete advantageously with a population that cares for them all. The losses would soon strain the reproductive capacity of such a population. Will there not be selection pressures to keep any such infanticidal genes a minority in the population? The usefulness of the gene will be frequency dependent; the more of them there are, the less useful it is.

The usurper is, after all, killing large numbers of his own genes, since he is in the species and in the population with the infants he slays. He is probably killing large numbers of his familial genes, since (in typical breeding populations) he is not very distantly related to these infants. His infanticidal genes, passed along to the young he will sire, will sow seeds of destruction in the offspring that descend from him. If one grandson takes over the harem of another grandson, his great grandchildren will be slain. If a grandson takes over a harem in which two granddaughters are present with young, his great grandchildren will be slain.

Are the females always the losers? Can they evolve no means of protecting their genetic interests? According to the prevailing theory females make more investment in offspring than do males, since they have to gestate and nurse them, and this should result in more capacity to protect their investment. Females often hold a good deal of power in animal family groupings, especially as regards caring for the young. A female has a good deal of control over which male will fertilize her eggs, and, especially if she has previously experienced an infant-killing male, perhaps she can select against those with infanticidal genes. Most of this male's genes only happen to be present in this generation in a male; 50 percent of the time they are carried by females. Do females carry these genes, disadvantageously, for the males; or are these genes peculiarly on the male chromosome? The infanticidal genes are to the advantage of this killer male's other genes in this generation, embodied in this male, but they will be, over generations, as often embodied in females, to their disadvantage half the time. It is complex to figure out how such genes can really convey survival advantage over the recurrent generations. Above some threshold level, they are likely to prove counterproductive and to be selected against or suppressed in the population.

Nevertheless, there is little doubt that some animals kill the young of competitors, as when lions kill the cubs of cheetahs. Something like this might be present in langur infanticide. If so, it will have to be interpreted as behavior that results in the selection of the most fit genes, which requires diversity as well as identity in any surviving population. Meanwhile, the larger constraints of sexuality remain. This male, breeding with these newly acquired females, will have to mingle his genes with those of these others. That is the only way he can distribute any genes that he owns.

There is a great deal of well-documented cooperative animal behavior that will also have to be fitted into any complete picture (Dugatkin 1997a). In African wild dogs, a ranking female regularly gives birth to many more pups than she can nurse, and other females in the pack begin lactating and nurse her pups. Or if other females bear pups, the dominant female may take them over as her own, with the result that she has more than she can nurse, and the nursing becomes communal again. Any lactating female can nurse any of the young. As the pups grow, various adults return from the hunt and regurgitate food for various pups, as well as for other adults who have remained behind to guard the pups (Bueler 1973).

Among the wolves reintroduced to Yellowstone National Park in the Rose Creek Pack in 1996, a male and female gave birth to nine pups but, unfortunately, established a den site outside the park, where the male was shot. Later, an unrelated male in another pack left his sibling males, joined the female, and adopted her pups, helping to feed them, even regurgitating food. Later still, although he recognized his siblings and greeted them on boundary encounters, he was defending his new pack's territory against incursions from these siblings in his former pack.[14] In another case in December 1997, the alpha male of the Druid Peak Pack was killed, and five days later an unrelated male, dispersing from another pack, joined the Druid pack, adopting five yearling pups. Elephant matriarchs often adopt lost or orphaned calves not their own. So infanticidal genes, if such there are, are complemented by more social genes.

5. INTERSPECIFICALLY AND INTRASPECIFICALLY SHARED GENES

Each genetic set is unique, except for clones and twins. Philosophers may rejoice in this uniqueness, personally expressed in humans, noticed in proper names, and founded on distinct biological identities. Brother differs from brother, child from parents: humans are endlessly variable in their traits. There are over five billion humans, yet the immune system in any one human body will recognize as foreign

---

[14] *Survival of the Wolves*, National Wildlife Foundation documentary on Yellowstone wolves, Turner Broadcasting System, November 1996.

a bit of tissue from any other person or other living individual. In natural populations of all organisms, there is great variation. This idiographic distinctiveness is made possible by genetic variation, both in the kaleidoscopic assortment, the set of alleles that one individual inherits from the genetic pool, and in the diverse alleles available in that pool.

From what is known about meiosis and independent assortment, dominant and recessive genes, and so on, one might first think that the animal offspring have, on average, 50 percent of the father's and 50 percent of the mother's genes. This is so, from the perspective of the particular set assortment in some "self." But since the animal or plant self belongs in a family, in a population, and in a species where others have, less and more, the same or similar genes (alleles), there is an equally valid perspective from which there are both many shared and many alternative genes (other alleles) found in an ever widening circle.

Genetic variation is a vital part of genetic creativity, but so also is genetic similarity. Genetic variation is itself complex. How many loci vary? How often does a locus vary from individual to individual? How many different alleles are available at a particular locus? Do the alleles make much or little difference in the phenotypic expression? Do they interact equally with other genes? A frequently used measure is the heterozygosity, which estimates the probability that two alleles at any particular locus will be different when taken at random from members of the population. The average heterozygosity is 4.6 percent for plants, 13.4 percent for invertebrates, 6.0 percent for vertebrates, 6.7 percent for humans. In a human with 30,000 structural genes, 2,010 will be heterozygous, and the resulting possible variation is dramatic (Ayala 1982, pp. 45–55). Meanwhile it remains true that 93.3 percent of the human loci are homozygous, a measure of genetic similarity.

Humans are "all of one blood" in the species sense. The man and the woman, like any mating pair, must have enough in common to interbreed; they share far more in biochemistries than do they differ in their idiosyncrasies. If one is thinking about the genes that make ribosomes, Golgi apparatus, erythrocytes, acetylcholine molecules and their receptor channels, lipid molecules, and microtubules, whatever distinctive mix there is of these alleles in the husband's body, these genes are quite similar to, if also somewhat different from, the genes that do those things in the wife's body. Most of my genes are nonrival with most genes in most other humans. We share enough to make blood transfusions possible, or to mate, or to form bonds of attraction between male and female.

Genetic studies show a remarkable uniformity from one human population to the other. In human blood types, only 15 percent of the variation exists between groups, whereas 85 percent of the variation is shared across groups. There are only four blood types as far as transfusion is concerned. "Based on randomly chosen genetic differences, human races and populations are remarkably similar to each other, with the largest part by far of human variation being accounted for by the differences between individuals" (Lewontin 1972, p. 397; 1982). "In man's case, curiously, genetic studies have shown a remarkable uniformity from one population to another. Differences between populations, for most genes studied, are of frequency only; the genes themselves tend to be the same in population after population from the equator to the arctic circle. Despite years of intensive study, the number of human genes whose selective advantage is actually understood can be numbered on the fingers of one hand" (Reynolds and Tanner 1983, p. 4).

Genes for dark skin provide protection in sunny climates. The sickle-cell gene gives resistance to malaria. But hundreds of human blood group variations do not make any known difference to reproduction rates. Some geneticists believe that most of these differences are due to genetic drift and are neutral to selection. Several forms of a gene or a trait may be present in a population (polymorphism); many of these alleles can be equally functional, that is, similar so far as natural selection is concerned. It is difficult to think that such genes could defend themselves "selfishly," since selection does not act on them. Some of these differences (as with dark skin or strong muscles) may formerly have made more difference than they do now; they may be relict genes.

All five billion humans have copies of genes more or less like the copies that "my" self shares with them, and the differences between us, if we must compete about these, all turn on a few percentage points and a different turn of the genetic kaleidoscope. It is really only the relatively idiosyncratic genes about which we are quarreling, and, from another perspective, this variety is the spice of life.

Humans share widely many, even most, of their genes not only with conspecifics but even, as earlier noted, with other species. "At the biochemical level we are today closer relatives of the chimpanzees than the chimpanzees are of gorillas" (Ruse and Wilson 1986, p. 176). If one were to translate such genetic similarity into the vocabulary of kin selection theory, when Jane Goodall devotes her life to saving chimpanzees, this is really 99 percent "selfish" and only 1 percent "altruistic," at least for structural genes. Dian Fossey is only being 2.3 percent charitable to the gorillas, sharing 97.7 percent of their DNA. Likewise with the Siberian tigers to whom George Schaller is perhaps 95 percent related. And so on down the evolutionary lineage. There is a much expanded circle of relationship, several orders of magnitude past the usual parameters of siblings and cousins, aunts and uncles. Any account of genetic "selfishness" seems unlikely to be plausible in explaining such behavior.

Many genes produce products that are biologically rather like cultural artifacts such as nuts and bolts, light bulbs, zippers, resistors, capacitors, transistors. Once invented, molecules – acetylcholine, actin, myosin, and so on – get conserved and show up as biological universals in all kinds of organisms in all kinds of places. It makes rather little sense to think of the genes that construct them as always and only acting in their particular self-interest, since they are so widespread and incorporated in one way or another in so many different species. There are variations on such genes, of course. But the similarities are as important as the variations.

Such genes are employed in the genesis of biotic diversity and complexity, valuable for the information they contain. "Selfish genes" is a doubtful category for interpreting genes used repeatedly in such constructions. What unbiased beholders behold, when they overview the long evolutionary struggle, is shared and distributed values. Insist if you like that this view is only a different bias, using intrinsic value rather than selfishness, but some kinds of filters help us to see better what is there, and other filters distort. What is actually there, and in need of explanation, is genetic identities, which persist in the midst of their perpetual perishing and have their intrinsic and instrumental biological values, conserved as they are integrated into diverse biological communities to which these organisms also belong. Genes play their valuable roles in organisms in communities caught up in a creative evolutionary epic. Genes are "the fine print in the book of life" (de Duve 1995, p. 2).

Organisms with their genes are cognitive systems, and this requires an axiological, rather than a moralistic, account of natural history. This, in biology, is a precursor of more dramatic events to come, when human minds arise, forming their cultures, in which morality does become possible, as well as, more recently, reflection on the role of genetics in human life. To that we next turn.

Chapter 3

Culture: Genes and the Genesis of Human Culture

Animals do not form cultures, not, at least, cumulative transmissible cultures. Culture, in Clifford Geertz's memorable definition, "denotes an historically transmitted pattern of meanings embodied in symbols, a system of inherited conceptions expressed in symbolic forms, by means of which men communicate, perpetuate, and develop their knowledge about and attitudes toward life" (1973, p. 89). Culture, according to Edward B. Tylor, is "that complex whole which includes knowledge, belief, art, morals, law, custom, and any other capabilities and habits acquired by man as a member of society" (1903, p. 1). Robert Boyd and Peter J. Richerson write, "Culture is information capable of affecting individuals' phenotypes which they acquire from other conspecifics by teaching or imitation" (1985, p. 33). Culture, Margaret Mead often said, is "the systematic body of learned behavior which is transmitted from parents to children" (1989, p. xi).[1] Unlike coyotes or bats, humans come into the world by nature rather unfinished and become what they become by culture.

1. NATURE AND CULTURE

Culture is a contrast class to nature, at least to physical and chemical nature, and, especially of interest here, to biological nature.[2] Infor-

---

[1] For 164 definitions of culture see Kroeber and Kluckhohn (1963).

[2] "Nature" has multiple layers of meaning, not addressed here. If one is a metaphysical naturalist, nature is all that there is, and so everything in culture – computers, artificial limbs, or presidential elections – is natural. Nature has no contrast class. Metaphysical naturalists may complain that in what follows nature, in the sense of wild nature, and culture are too dichotomized. But we do need to be adequately discriminating about the real differences between them.

mation in wild nature travels intergenerationally largely on genes;[3] information in culture travels neurally as persons are educated into transmissible cultures. The determinants of animal and plant behavior are never anthropological, political, economic, technological, scientific, philosophical, ethical, or religious. The intellectual and social heritage of past generations, lived out in the present, re-formed and transmitted to the next generation, is regularly decisive in culture.

Cumulative transmissible cultures are made possible by the distinctive human capacities for language. Language "comes naturally" to us, in the sense that humans everywhere have it; the child picks up speech during normal development with marvelous rapidity; language acquisition is only more or less intentional. The child mind is innately prepared for such learning (Chomsky 1986). Human language, when it comes, is elevated remarkably above anything known in nonhuman nature; the capacities for symbolization, abstraction, vocabulary development, teaching, literary expression, argument are quite advanced; they do not arise naturally as an inheritance from the other primates, whatever may otherwise be our genetic similarity with them. Though language comes naturally to humans, what is learned has been culturally transmitted, this or that specific language, and the content carried during childhood education is that of an acquired, nongenetic culture. On this language capacity the development, transmission, and criticism of culture depend.

Sometimes the term "culture" is used of animals. Opening an anthology on *Chimpanzee Cultures*, the authors doubt, interestingly, whether there is much of such a thing: "Cultural transmission among chimpanzees is, at best, inefficient, and possibly absent" (Wrangham et al. 1994, p. 2). There is scant and in some cases negative evidence for active imitation or teaching of the likeliest features to be transmitted, such as tool-using techniques. Chimpanzees clearly influence

---

[3] Animals imitate the behaviors of parents and conspecifics, as when birds, with a genetic tendency to migrate, learn the route by following others. Through various "maternal effects," parents influence their young nongenetically. Animal behavior is not always genetically stereotyped; it may be labile, subject to development only if environmental circumstances require or permit it, including parental behavior. Even in single-celled organisms, acquired characteristics can be transmitted to offspring (Landman 1991). See Chapter 2, Section 3(3). But genetics remains the dominant mode of intergenerational information transfer, and none of these nongenetic hereditary factors resembles a cumulative transmissible culture.

each other's behavior and seem to intend to do that; they copy the behavior of others. But there is no clear evidence that they attribute mental states to others. They seem, conclude these authors, "restricted to private conceptual worlds." Without some concept of teaching, of ideas moving from mind to mind, from parent to child, from teacher to pupil, a cumulative transmissible culture is impossible.

Animals are variously socialized and become what they become interactively with their surroundings, which include the groups in which they live. But there is little or no evidence for any higher-order intentionality, even among primates that are highly social. Organisms with zero-order intentionality have no beliefs or desires at all. Animals, such as vervet monkeys, clearly intend to change the behavior of other animals, first-order intentionality. Second-order intentionality would involve intent to change the mind, as distinguished from the behavior (though perhaps the behavior as well), of another animal, that is, to teach by passing ideas from mind to mind. Third-order intentionality would involve knowledge that another, a teacher, is intending to change one's mind (Dennett 1987). Primates do not seem to realize that there are minds there to teach in others, although they often imitate each other's behavior, as when adults are imitated by their offspring.

In this higher-order sense of communication, conclude Dorothy L. Cheney and Robert M. Seyfarth, "signaler and recipient take into account each others' states of mind. By this criterion, it is highly doubtful that *any* animal signals could ever be described as truly communicative" (1990, pp. 142–143):

> It is far from clear whether any nonhuman primates ever communicate with the intent to inform in the sense that they recognize that they have information that others do not possess. . . . There is as yet little evidence of any higher-order intentionality among nonhuman species. . . . Teaching would seem to demand some ability to attribute states of mind to others. . . . Even in the most well documented cases, however, active instruction by adults seems to be absent. . . . The social environment in most primate species is probably too simple to require higher-order intentionality. (1990, pp. 209, 223, and 252)

Richard Byrne finds that chimpanzees may have glimmerings of other minds, but little evidence of intentional teaching (1995, pp. 141, 146, and 154).

What is missing in the primates is precisely what makes a human cumulative transmissible culture possible. The central idea is that acquired knowledge and behavior are learned and transmitted from person to person, by one generation teaching another, ideas passing from mind to mind, in large part through the medium of language, with such knowledge and behavior resulting in a greatly rebuilt, or cultured, environment. Humans still have genes, of course, but humans live under what Boyd and Richerson call "a dual inheritance system" (1985; Durham 1991).

Richard Lewontin puts it this way:

> Our DNA is a powerful influence on our anatomies and physiologies. In particular, it makes possible the complex brain that characterizes human beings. But having made that brain possible, the genes have made possible human nature, a social nature whose limitations and possible shapes we do not know except insofar as we know what human consciousness has already made possible. . . . History far transcends any narrow limitations that are claimed for either the power of the genes or the power of the environment to circumscribe us. . . . The genes, in making possible the development of human consciousness, have surrendered their power both to determine the individual and its environment. They have been replaced by an entirely new level of causation, that of social interaction with its own laws and its own nature. (1991, p. 123)

Theodosius Dobzhansky, a pivotal figure in modern genetics, reflects:

> Human genes have accomplished what no other genes succeeded in doing. They formed the biological basis for a superorganic culture, which proved to be the most powerful method of adaptation to the environment ever developed by any species. . . . The development of culture shows regularities *sui generis*, not found in biological nature, just as biological phenomena are subject to biological laws which are different from, without being contrary to, the laws of inorganic nature. (1956, pp. 121–122)

The critical issue is whether and how far this cultural inheritance system is nongenetic, or transgenetic, and moves past (and perhaps is contrary to) natural selection. We pursue that question here regarding culture in general, and, in chapters to follow, more specifically regarding three especially revealing cultural phenomena: science, ethics, and religion. The genesis of culture is as remarkable as the genesis in nature; it is nature's most remarkable genesis. The genes outdo themselves.

2. GENE–MIND COEVOLUTION

Fins evolved, wings evolved, hands evolved. Human brains evolved once upon a time, developing from precedent animal brains. The cognitive endowments achieved in *Homo sapiens*, including the objective brain, the empirical neurology, the capacities for perception, and the experiential, psychological subject, have an evolutionary history. The results of this ancient history are now delivered biologically at birth to (all normal) members of *Homo sapiens*. These past evolutionary events (phylogenesis) are recapitulated (more or less) and generate a contemporary brain (ontogenesis), sponsoring a mind. What was achieved in millions of years (even billions if one includes all the biochemistries) is, via DNA, coded and copied, reenacted in the few natal/childhood months and years. In this sense, evolutionary history accumulates, is repeated, and is conserved in the human line with each new generation with their brains. Some reflection of this evolutionary history in the way the mind works is to be expected.

Does evolution repeatedly produce intelligence in other species lines? Increasing diversity and complexity appear repeatedly in evolutionary history. In the animal world, eyes evolved many different times, and similarly with muscles, with organs of hearing, taste, smell. Legs, fins, and wings evolved several times. Genetically based skills, we have argued, are widely distributed and shared. Much of this increased complexity depends on neural development, allowing, from the skin in, centered identity and integrated control of animal life, and, from the skin out, cognitive powers for information perception and processing important for survival. Many animal species have brains. On the one hand, such mental powers evidently have survival value; on the other, most species (plants, insects, crustaceans) survive quite well with little intelligence and develop no more over the millennia.

So one cannot claim that all animals, much less organisms in general, evolve steadily toward higher intelligence. Only some do. But perhaps it is highly likely that some will. Christian de Duve, a Nobel laureate, concludes that neural power, where it luckily arises, has such "decisive selective advantage" that there is high probability of its increase:

> The direction leading toward polyneuronal circuit formation is likely to be specially privileged in this respect, so great are the advantages linked with it. Let something like a neuron once emerge, and neuronal networks of increasing complexity are almost bound to arise. The drive toward larger brains and, therefore, toward more consciousness, intelligence, and communication ability dominates the animal limb of the tree of life on Earth. (1995, p. 297)

Perhaps that is so with certain kinds of intelligence, but still it is rather surprising that, of the five to ten million species on Earth at present, of the perhaps five to ten billion species that have come and gone over evolutionary time, only one has reached self-conscious personality sufficient to build cumulative transmissible cultures. Ernst Mayr, despite finding progress undeniable in the evolutionary record, reflects on the evolution of intelligence with conclusions opposite from those of de Duve:

> We know that the particular kind of life (system of macromolecules) that exists on Earth can produce intelligence. . . . We can now ask what was the probability of this system producing intelligence (remembering that the same system was able to produce eyes no less than 40 times). We have two large superkingdoms of life on Earth, the prokaryote evolutionary lines each of which could lead theoretically to intelligence. In actual fact none of the thousands of lines among the prokaryotes came anywhere near it.
>
> There are 4 kingdoms among the eukaryotes, each again with thousands or ten thousands of evolutionary lineages. But in three of these kingdoms, the protists, fungi, and plants, no trace of intelligence evolved. This leaves the kingdom of Animalia to which we belong. It consists of about 25 major branches, the so-called phyla, indeed if we include extinct phyla, more than 30 of them. Again, only one of them developed real intelligence, the chordates. There are numerous Classes in the chordates, I would guess more than 50 of them, but only one of them (the mammals) developed real intelligence, as in Man. The mammals consist of 20-odd orders, only one of them, the primates, acquiring intelligence, and among the well over 100 species of primates only one,

Man, has the kind of intelligence that would permit [the development of advanced culture]. Hence, in contrast to eyes, an evolution of intelligence is not probable. (Mayr, quoted in Barrow and Tipler 1986, pp. 132–133)

Repeatedly, Mayr concludes, "An evolutionist is impressed by the incredible improbability of intelligent life ever to have evolved" (1988, p. 69; see also 1985a; 1994). Mind of the human kind is unusual, even on this unusual Earth.

Perhaps only one line leads to persons, but in that line at least the steady growth of cranial capacity (Fig. 3.1) makes it difficult to think that intelligence is not being selected for and conserved when it is achieved. The brain gets bigger. "No organ in the history of life has grown faster" (Wilson 1978, p. 87).[4] This know-how for building bigger brains is genetically coded, of course, but perhaps also here, if anywhere on Earth, genetic history transcends itself and passes over into something else. We ought not to forget that the human brain is a consummate product of the millennia-long evolutionary genetic sharing and searching for cybernetic achievements; we ought not also to forget its startling complexity. In the 1,500 cubic centimeters of neural networks, there is more operational organization than anywhere else on Earth, or in the universe so far as is known. The number of possible associations among the 10^9 neurons in the brain, where each cell can "talk" to as many as a thousand other cells, may exceed the number of atoms in the universe.

What is surprising in humans is not so much that they have intelligence generically, for many other animals have specific forms of a generic intelligence. Nor is it that humans have intelligence with subjectivity, for there are precursors of this too in the primates. The surprise is that this intelligence builds cumulative transmissible cultures. *Homo sapiens*, as we have named ourselves, is the "wise" species, and some of this is "wisdom" programmed into our genes, universal to all. Still, the specific reference largely denotes the "wisdom" achieved during human historical careers and passed on culturally to generations to come. The "wisdom" peculiar to humans lies in their cumulatively transmissible cultures.

Humans have lived in cultures for perhaps a million years, during

---

[4] Some early humans had slightly larger brains than modern humans, though a smaller brain to body ratio, and modern brains are more convoluted and complex. Brain size is only an approximate index of intelligence.

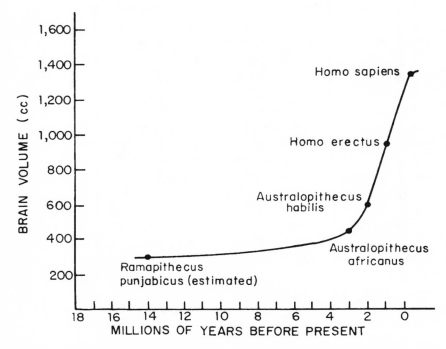

Figure 3.1. Increasing cranial capacity in the hominoid line. From *Sociobiology: The New Synthesis* by Edward O. Wilson. Copyright © 1975 by the President and Fellows of Harvard College. Reprinted by permission of Harvard University Press. This graph is redrawn from David Pilbeam, *The Ascent of Man.* Copyright © 1972 by Allyn & Bacon. Adapted by permission.

which time they have reproduced across thousands of generations. There is every reason to expect that over these millennia, those humans will do best reproductively who do best culturally, and, vice versa, that a genotype will be selected to produce a culturally congenial phenotype. If, as Aristotle put it, man is a political animal (*Politics* I, 2, 1253[a]), then human nature will be adapted for social life. Humans will have a range of motivations, inclinations, propensities, emotions that equip them for living in culture. No one denies that, just as evolution produced our brains and hands, evolution produced humans with brains inclined to culture.[5] Any constraints of genes on the mind come in the context of such genetic enabling.

[5] For efforts to reconstruct the historical stages in the developing human mind, see Mithen 1996; Donald 1991.

What sort of mind is especially adapted for culture? One impossibility is complete genetic control of behavior, so that humans would be like ants. Even higher animal behavior is not that stereotyped. Equally impossible is complete cultural independence of genetic capacities. Culture must be superposed on human biology, and any culture that does not accommodate human biology, minimally the requirements of feeding and breeding, will soon fail. Every Earth-bound culture must provide for persons to be washed, be sheltered, go to the toilet, mate, and so on. Every culture must express and control the human emotions – love, fear, joy, grief, guilt, anxiety – and allow artistic, musical, religious expression; protect property and privacy; and provide for various activities to which humans are "by nature" inclined.

A more plausible possibility for gene–mind coevolution is that certain sorts of minds, produced with certain sets of genomes, will be disposed to certain sorts of cultural practices. Other cultures will go better with other genes. In this strategy, a genetically prejudiced mind presets behaviors choosing certain cultural options, with more and more such persons in each next generation. There might be a channeling specific to persons in selected cultures. Certain behavioral dispositions will coevolve with particular cultures. Since gene frequencies change much more slowly than cultural practices, this would require long-stable cultures. Also, relative isolation of such a gene-culture coevolving population is needed, else both inflow of genes and ideas from foreigners will upset the process. Such conditions might have obtained in the remote past, but they are not characteristic of recorded history.

If the character of human cultures changes over time, then perhaps differing genotypes producing differing phenotypes will be favored according to novel needs. Humans need native abilities for whatever they do in the home or at work, and if these demands shift, perhaps abilities will shift. Although there is no particular evidence for this in surviving peoples, there is nothing initially implausible about supposing that the genotype best suited for hunter–gatherer Pleistocene cultures differed from the genotype best for Malaysian agriculture, introduced into Africa about 2,000 years ago, and that this differs from the genotype best for a high-technology society today. If there were found, somewhere on Earth, an isolated hunter–gatherer culture, they might retain, on statistical average in the population, such an archaic genotype.

As cultures become more fluid and complex, however, another strategy would be more open-ended. The genes would produce a keen, critical, open mind that can evaluate cultural options for their functional usefulness and for their contribution to a meaningful life. The direction of selection in humans, as evidenced by their enormous potential for diverse cultures (from those of the Neanderthals to a high-tech computer age), all of which require intelligence in various roles, would then select for an unspecialized intellect with open educable capacity. The best strategy for slow-paced genes that need to succeed in fast-paced culture is not to build a relatively inflexible mind whose pace and preferences are genetically biased for this or that culture, but to build a flexible mind that can make preferences independently of genetic bias, since these could misdirect persons in the rapidly shifting vicissitudes of culture. The number of neurons and their possible connections is far vaster than the number of genes coding for the neural system, and so it is impossible for the genes to specify all these connections.

When there emerges a later-evolved method of communication at the neural past the genetic level, the genes will need subsequently to develop so as to favor *teachability* above all. What will get selected is not so much specific gene traits coevolving lockstep with matching cultural behaviors, as open teachability, which is to say that the genes will have to abandon tight control of behavior and cast their luck with launching a human organism whose behavior results from an education beyond their control. As more and more knowledge is loaded into the tradition (fire building, agriculture, writing, weaponry, industrial processes, ethical codes, electronic technology, legal history), the genome selected will be that set that is maximally instructible by the increasingly knowledgeable tradition, and this will require that the genes produce a flexible and open intellect, generalized and unspecialized, able to accommodate lots of learning and to do so speedily, able to adopt behaviors that are functional in, or conform to, whatever cultures they find themselves in. Perhaps the owners of these genes may choose another culture and migrate there. Perhaps soldiers or traders from a variant culture will invade their territory and force their culture upon them.

"A genetically fixed capacity to acquire only a certain culture, or only a certain role within a culture, would however be perilous; cultures and roles change too rapidly. . . . Human genes insure that a culture can be acquired, they do not ordain which particular culture

this will be" (Dobzhansky 1963, p. 146). "Genetic differentiation between human populations for determinants of biases is unlikely"
(Boyd and Richerson 1985, pp. 284–285). It is better to be able to learn
any of the myriad human languages than to be genetically pre-
disposed to learn French, better to eat a cosmopolitan fare than to
like only Italian food, better to be able to use any of various cultural
ideas than to be genetically inclined to use only Polynesian-
originated ones.

Intelligence, based on neurology, allows an organism to make an
appropriate, rapid response to an environmental opportunity or
threat, protecting the organism against the necessity of making
slower, less reversible responses at the genetic level. If the genes
supply intelligence in sufficient amounts, they need not themselves
be closely tuned to directing behavior that can track environmental
changes; they turn this over to the general intelligence they have
created. Robert Plomin, in an analysis of human development and
genetics, concludes, "There is no evidence for major-gene effects on
normal variation in general or specific cognitive abilities" (Plomin
1990).

This idea of a "global learning capacity" can be exaggerated. The
genes do not build a tabula rasa mind; humans do need behavioral
dispositions of some kinds (to fear snakes or spiders, to seek mates,
to avoid incest, to protect their children, to reciprocate for mutual
benefits, to obey parents or follow leaders). Boyd and Richerson further suggest that humans could be genetically disposed toward religious beliefs or toward ethical practices, because of cultural group
selection; those in such cultures prosper (1985, pp. 175–177). So a
genetic bias toward ideas useful in various cultures can be expected,
and welcomed. We return later to a behavioral psychology model of
such an adapted mind, proposed by John Tooby and Leda Cosmides
(1992).

After that, all sorts of cultures demand all sorts of capacities and
skills, and nearly all humans have sufficiently rich talents to find a
niche in their culture. If so, there might not be any differential selection pressures when cultural patterns differ across place and time.
On statistical average, different human populations in different cultures might not be detectably different genetically so far as their capacities for either culture in general or this or that culture are involved. S. L. Washburn, surveying the archaeological record,

concludes that "there has been no important change in human abilities in the last 30,000 years" (1978, p. 57). If so, then all the changes are technological, historical, political, religious, or some other form of cultural change.

In present human populations, it seems that a baby taken from any race on Earth, appropriately reared, can receive almost any sort of general education. This does not mean that any baby can become a mathematician, or musician, or professional basketball player. But different babies can be found in any particular race that can do all these things well, and any normal baby can learn enough of these things to function more or less normally in any culture. As we earlier saw, the vastest part of human variation is within populations.[6]

Culture is a quite diverse affair, and it might be that culture reinforcing genetic disposition works for some practices (incest avoidance), but not for others (learning nuclear physics), sometimes with interaction and sometimes with independence. Whether or not adults have enzymes for digesting fresh milk will determine their pastoral practices. But, the differences, say, between the Druids of ancient Britain and the Maoists in modern China, would be nongenetic and to be sought in the historical courses peculiar to these cultures. Such cultures catch their member humans up into an ongoing tradition, give them their identity, and radically differentiate persons historically, even though Druids and Chinese have a biochemistry and a biological nature largely held in common (though there can be differences in skin color or in blood groups).

All organisms are cybernetic systems. Their know-how to solve problems evolved biologically. This is true in natural history; coyotes know how to hunt for ground squirrels. It is true in cultural history; humans evolved brains that could figure out how to make tools to hunt. Natural selection is typically thought to be the key determinant of these events; better knowledge gave better survival power. This would be evolutionary epistemology in the genetic sense; what one can know (Greek: *episteme*, knowledge) is linked with adaptation (leaving more reproducible offspring). Humans have flourished around the globe, and it seems evident that their knowledge has given them survival power. So we must press further the question of genetic determinants in culture (Section 3). After that, we will be in

---

[6] Chapter 2, Section 5.

a position to ask whether the evolution of ideas always functionally serves the production of more offspring in the next generation (Section 4).

3. GENETIC DETERMINANTS IN CULTURE

Sociobiologists insist that in the dual inheritance system biology is dominant. Wilson puts this in a bold, if somewhat loose, metaphor:

> The genes hold culture on a leash. The leash is very long, but inevitably values will be constrained in accordance with their effects on the human gene pool. The brain is a product of evolution. Human behavior – like the deepest capacities for emotional response which drive and guide it – is the circuitous technique by which human genetic material has been and will be kept intact. (1978, p. 167; Lumsden and Wilson 1981, pp. 13 and 179)

Earlier, confronted with "the selfish gene," we found the metaphor problematic; here again, microscopic genes cannot hold any leashes. There needs to be rigorous analysis of the analogy before we know whether this is science, philosophy, ethics, or poetry, or what the truth claims are.

More scientifically put, the claim is that, although there are many options in culture (a long leash), genetic constraints always circumscribe and overrule human behavior. "The central tenet of human sociobiology is that social behaviors are shaped by natural selection" (Lumsden and Wilson 1981, p. 99). Wilson continues:

> The essence of the argument, then, is that the brain exists because it promotes the survival and multiplication of the genes that direct its assembly. The human mind is a device for survival and reproduction. . . . The intellect was not constructed to understand atoms or even to understand itself but to promote the survival of human genes. (1978, pp. 2–3)

This is "the general sociobiological view of human nature, namely that the most diagnostic features of human behavior evolved by natural selection and are today constrained throughout the species by particular sets of genes" (1978, p. 43). A quite simple biological force – producing the most offspring in the next generation – pervades and is the most basic determinant in all human affairs.

Michael Ruse agrees: "I argue that Darwinian factors inform and infuse the whole of human experience, most particularly our cultural dimension. . . . Human culture, meaning human thought and action, is informed and structured by biological factors. Natural selection and adaptive advantage reach through to the very core of our being" (1986, pp. 140 and 147; 1994). Irvin DeVore insists: "We are no less subject to natural selection than any other species" (DeVore and Morris 1977, p. 88).

Biological survival and reproduction are valuable achievements warranting all due respect, and if the intellect is put to work supporting survival and reproduction, well and good. But a problem arises if the intellect can do nothing more than support survival and reproduction. That will limit intellectual capacities, so that there are no higher levels of genesis to be found in culture. The life of the mind in culture will need to be underdetermined by the genes in nature for enough cultural freedom to think about anything else besides survival and reproduction. The genes need to support and permit, rather than to preempt or constrain, the genesis of culture.

Jerome H. Barkow realizes that there is a "complex psychology" in humans, with genes and culture interacting, sometimes working together, sometimes pulling in opposite directions. Nevertheless, he concludes, it remains basically correct "to speak of the genes anchoring the psychological predispositions that tend to pull our cultures back to fitness-enhancing orbits" (1989, p. 8). How correct that is will depend on what counts as fitness-enhancing; it is doubtful whether all cultures pull all their members in all their behaviors toward maximizing their numbers of offspring, a sort of gravity-like social force pervading everything. But cultures that do not reproduce themselves and their members go extinct.

Richard Alexander says, with emphasis, *"It is always true that the cumulative history of natural selection continues to influence our actions by the set of genes it has provided humanity"* (1987, p. 23). His emphasis is undone by the weak word "influence," which no one will contest. We can well expect that humans are, to a considerable extent, the kinds of creatures they are because they once upon a time evolved. Everyone knows that the most cultured humans still digest their dinners with biochemistry; they think on circuits that employ neurons in their brains; they can only reproduce with genes in their gonads; all of their biology once evolved under natural selection. Biology is a prerequisite for culture and therefore "influences" it. Likewise, chemistry and physics are prerequisite for biology and influence it.

In that sense, foundational sciences always trump later-coming ones superposed on them. But we already knew that.

Sociobiology – so most of its adherents claim – is going to reveal springs of our action about which we did not previously know. Freudian psychoanalysis claims that, although humans think they know what is in their minds, much more is going on in their unconscious minds, which are the real determinants of behavior. Now, sociobiology will go deeper, down to the genes that determine the conscious and the unconscious mind. We think we also know that, influenced by biology though culture is, many things happen in human affairs that overleap genetics. In fact, this is not so – say these sociobiologists – or not importantly so; behavior is genetically controlled. "The question of interest is no longer whether human social behavior is genetically determined; it is to what extent. The accumulated evidence for a large hereditary component is more detailed and compelling than most persons, including even geneticists, realize. I will go further: it already is decisive" (Wilson 1978, p. 19).[7]

Biologists distinguish between proximate and ultimate explanations (Mayr 1988, p. 28). Why does a plant turn toward the light? Cells on the darker side of a stem elongate faster than cells on the brighter side because of an asymmetric distribution of auxin moving down from the shoot tip. But the ultimate explanation is that, over evolutionary time, in the competition for sunlight, there were suitable mutations, and such phototropism increases photosynthesis. Analogously, whatever the proximate explanations of how the mind shapes behavior, the ultimate explanation is evolutionary success, natural selection for maximum offspring.

Possibly the content of some but not all beliefs is genetically fixed or favored. We shouldn't be surprised that it is easy to teach children that snakes are dangerous, hard to teach them that milk is poisonous. In a favorite example, almost all cultures teach that a person should avoid incest. Indeed (some say), parents hardly have to teach this at all; brothers are not inclined to be sexually attracted to sisters, nor

---

[7] Sometimes Wilson is less decisive about these genetic determinants: "Culture is not just a passive entity. It is a force so powerful in its own right that it drags the genes along. Working as a rapid mutator, it throws new variations into the teeth of natural selection." Nevertheless, "the genes continue to hold culture on a leash" (Lumsden and Wilson 1983, p. 154).

sisters to brothers. Close inbreeding tends to produce handicapped offspring.[8] Over the millennia, humans who had genes that inclined them to avoid incest (by reducing sexual attraction to those with whom they were closely reared, by propensity to adopt an ethic prohibiting incest) left more reproducible offspring.

So now humans are genetically disposed both unconsciously to avoid incest and in conscience to believe that incest is morally wrong. This ethical belief is more or less blueprinted into our genetics. Notice that there is nothing undesirable about such biases of conduct. To the contrary, there would be a certain wisdom in such dispositions, and one should be thankful to have them built into human nature. These all involve beliefs relevant to our common, Earthbound biology (toxins in snakes, nutrients in milk, deleterious genes).

Cultural practices cannot be indifferent to genetic transmission, because every generation dies and must be replaced by a subsequent generation. Beliefs, crafts, skills, mores, and other cultural features figure into successful replacement. Biological replacement is absolutely necessary, vital to any ongoing culture. There is the logical possibility that such a culture could be transmitted so rapidly to new converts that it could survive by proselytizing, with each new generation of proselytes failing to reproduce, the culture meanwhile having spread to yet a newer generation of recruits. This would be something like an order of celibate priests, whose members never reproduce, but who recruit new priests each generation from those who do.

Such priests, however, belong to a religion, such as Roman Catholicism, that amply encourages biological reproduction among its lay members. In a culture as a whole, cultural survival by evangelical recruitment without biological reproduction seems unlikely. On statistical average, we can well expect that the genetic dispositions of potential recruits will leave them indisposed to a culture that educates them into nonreproduction. Such genetic dispositions would be, we might say, unnatural; there are strong reproductive urges in all animal species and in *Homo sapiens* as well. Nor do we have any historical examples of such cultures.

But it does not follow that all cultural practices will be determined

---

[8] See Chapter 2, Section 4.

by this need for reproduction across generations. One simply needs adequate provision for such reproduction, which will keep population levels above some threshold of flourishing, not some law by which cultural practices are always and only tested for their power to maximize the number of offspring in the next generation. There might well arise within a culture some convictions about the optimum size of families, or of the population as a whole, which differed from the maximums biologically possible. And such convictions might well spread to new recruits in other cultures, so that numerous cultures adopted this belief. In fact, there is good reason to believe that exactly this has happened in modern times (as we note later).

Nor does it follow that the only kinds of beliefs and practices that can be sustained within a culture are subject to this must-produce-a-new-generation constraint. As we will demonstrate in chapters to follow, persons are quite capable of entertaining scientific theories, or ethical principles, or religious convictions; evaluating these; and acting upon them, independently of whether these beliefs and practices optimize their numbers of offspring in the next generation. The same is true in art, or literature, or politics, or economics. Natural selection, that is, is relaxed for great areas of cultural activity. So there may be numerous nonbiological beliefs (whether democracy is the best form of government, or whether one ought to share possessions fairly, or conserve endangered species) that are not coded into our Earth-bound genes, not even dispositionally, and have to be discovered some other way.

We said that smart genes will have to build, and then default to, a smart mind. Any such smart mind will be constrained by its own mortality and must produce in its culture an ongoing generation of minds, embodied in biological bodies. Minds are "leashed" to bodies, to phrase it that way, and this is inescapable. But if genes have to default to the minds they build, that also means, in the metaphor of "leashing," that smart genes must trust a smart mind that they unleash, because the leashed mind will be disadvantaged in the marketplace of cultural options, where others are more free to choose the smart ideas.

(1) Epigenetic Rules and Culturgens

One model of this leashing involves epigenetic rules and culturgens. Earlier versions of sociobiology supposed that the genetic shaping of beliefs was rather direct and one-way, as liking milk, hating snakes, and avoiding incest suggest. In later versions, more attention is given to gene–culture coevolution. Genetic variations launch some cultural variations, and successful cultural variations (whether genetically or nongenetically launched) get tracked by the genes. The genes are still in control, however; cultural variations are selected and persist only when the genes can use them the better to reproduce. Subsequent generations of humans will be genetically disposed to these kinds of cultural variations, although the detail of such innovative practices will be transmitted to the next generation culturally and nongenetically.

The genes build an epigenetic mind. "Epigenesis" conveys the idea of a secondary genesis, ancillary to the primary genetic determinants, a sort of epiphenomenon. Ruse and Wilson put it this way:

> Human thinking is under the influence of "epigenetic rules,"
> genetically based processes of development that predispose the
> individual to adopt one or a few forms of behaviours as opposed
> to others. The rules are rooted in the physiological processes
> leading from the genes to thought and action. (1986, p. 180)

The claim that there is genetically based, channeled development in anatomy and physiology is uncontroversial; such development is termed primary rules. There are also secondary epigenetic rules, "innate mental dispositions" (Ruse 1995, p. 97); these control cognition and behavior. An example is a genetic disposition to see four primary colors, blue, green, yellow, and red, and this might convey survival advantage (though many mammals are not especially sensitive to color) (Lumsden and Wilson 1981, pp. 45–46 and 370).

The debate starts with the claim that the content of what humans can learn and what cultural practices humans can take up is blueprinted in the genes. Humans do not all have the same beliefs; their beliefs can differ radically. The question is whether such cognitive development is under genetic control. American Indians believed that the tree is as green as ever when no perceiver is there; modern scientists believe that reflected electromagnetic waves enter the eyes, and that the experience of green is in the eye of the beholder. All humans have a conscience, but the ancient Scythian nomads in southern Siberia believed that when chieftains die, their concubines should be buried with them, along with their horses and other necessities for the next life; modern Americans believe in women's rights and doubt that horses ought to be treated this way. Which of these various beliefs one comes to hold depends more on education than on genes. So the claim that genes control the content of belief, and resulting behavior, seems often to fail.

No, say Lumsden and Wilson; the leash is long, but the limit is reached. The secondary epigenetic rules operate on "an array of transmissible behaviors, mentifacts, and artifacts which we propose to call *culturgens*"[9] (1981, p. 7). "In cognitive development, the epigenetic rules . . . influence the form of learning and the transmission of culturgens" (1981, p. 370). A culturgen is a cultural trait that an individual can form an attitude about, choose, and use for success in living – a marriage custom, a religious belief, a dietary preference, a clean shaven face. There are three kinds: artifacts, behaviors, and mentifacts (1981, pp. 27 and 317). Individual persons (phenotypes) have differing propensities, resulting from their differing genotypes, to adopt differing culturgens, and so the genes determine these behavioral tendencies to adopt this or that culturgen. Which culturgen is adopted depends on our genetic bias toward it (our epigenetic rule) in combination with how many others have also adopted it, since its value or disvalue may depend on the extent of its use. Which one is adopted makes a difference in reproductive survival, and so in the next generation there will be more of those genotypes with the fitter propensities that favor the productive culturgens. The hardware coevolves with the software.

Culture is fully recognized, but fully leashed, though there is a long leash. The unit of interest is the culture-*gen*, the culture that serves *gen*etics. "The rules comprise the restraints that the genes place on development (hence the expression 'epigenetic'), and they affect the probability of using one culturgen as opposed to another" (Lumdsen and Wilson 1981, p. 7). In the coevolutionary circuit, "culture is generated and shaped by biological imperatives while biological traits are simultaneously altered by genetic evolution in response to cultural innovation" (1981, p. 1).

"Can culture have a life of its own?" According to this view, not

[9] "A culturgen is a relatively homogeneous set of artifacts, behaviors, or mentifacts (mental constructs having little or no direct correspondence with reality) that either share without exception one or more attribute states selected for their functional importance or at least share a consistently recurrent range of such attribute states within a given polythetic set" (Lumsden and Wilson 1981, p. 27).

really. Culture is "the product of a myriad of personal cognitive acts that are channeled by the innate epigenetic rules . . . translating them upward to the social level through the procedures of statistical mechanics" (1981, p. 176). Culture is a statistical summary of an equilibrating population of persons. The microscopic genes produce the macroscopic personal acts, and the megascopic society results. "There are in fact three steps: from genes to epigenesis, from epigenesis to individual behavior, and from individual behavior to culture" (p. 343). "Culture is in fact the product of vast numbers of choices by individual members of the society. Their decisions are constrained and biased in every principal category of cognition and behavior" (p. 177). This view from below, where culture bubbles up from the genes, is counter to the prevailing understanding of social scientists, "which views culture as a virtually independent entity that grows, proliferates, and bends the members of the society to its own imperatives" (p. 176), a view from above superimposing culture on genetic potential.

One problem is that any set of propensities (epigenetic rules, resulting from a genotype) is presumably fixed across an individual's lifetime, though perhaps different rules could mature at different life stages. Fixed propensities could be maladaptive. A set of propensities that is an adaptive fit for life in rural Nebraska might result in reduced fitness if the individual chances to have a college roommate who persuades him to look for work in Boston. What one learns depends on opportunities to learn (the culturgens available), not just on propensity to learn. The options available depend on politics, economics, and religion, often just on chance, and which is the right one to take can be different in town and country. Encountering such diverse problems and opportunities, as much hangs on social history as hangs on genetics.

Further, although adopting this and not that culturgen seems rather atomistic, culturgens are no more likely to be singular than genes. A mutant in genetics (a gene, a bit of DNA) is a different item from a mutant in culture (an idea, a new fashion). There could be variation and selective retention in both, but it does not follow that natural selection can reach through to detect variant culturgens and retain them selectively. To find out whether the genes–culturgen theory is true one would have to analyze clusters of traits (culturgens) that can be counted and subjected to statistical analysis and matched against human genomes. This measurement problem for human scientists further reveals a selection problem in gene–culture coevolution. Culturgens must be distinct, particulate enough for natural selection to operate on them. That seems doubtful. Whether adopting just one of them yields any benefits will depend on the package of other culturgens simultaneously adopted, the culturgen in gestalt, and whether those others in the gestalt are adopted will depend on other propensities that the individual may or may not have. Culturgens will blend and piggyback on each other; traits will be variously correlated, or not, for better and worse reasons and causes. Also (as Lumsden and Wilson recognize), there needs to be still another feedback loop, since whether or not a particular culturgen is the fittest in particular circumstances may depend on whether everybody, somebody, or nobody else is using it.

There will be a fishnet of propensities operating within any individual, interacting with a fishnet of culturgens encountered outside, interacting with the fishnets of propensities within hundreds of persons in society. It will be difficult for natural selection genetically to code for epigenetic rules that can generalize successfully enough to compute the probable result in offspring and their likelihood of survival. It will be difficult to calculate (or blindly to select for) one's inclusive fitness in a mobile society, and so on. Even if the inbuilt genetic dispositions can do this computing in today's circumstances, cultural change can occur rapidly. New culturgens come up for choice, and yesterday's disposition may not serve tomorrow's opportunity. Flexibility is more important than programmed reaction patterns.

Linking genes and culturgens hooks together two processes that occur at speeds differing by several orders of magnitude. Genetic evolution usually occurs at a snail's pace; cultural evolution can occur rapidly. When information is transmitted genetically, since nothing changes during the lifetime of an individual human, it takes thirty years (a generation) to venture an incremental difference, and the incremental variations that bubble up from the genetic shuffle during reproduction, since they are "blind" to the needs of the oncoming generation,[10] can only be tested over the lifetime of that next generation (another thirty to sixty years). Information can only be

---

[10] Though we have doubted whether the genetic search strategies are as blind as orthodox theory insists. These are smart genes with searching algorithms (Chapter 1, Sections 3–5).

transmitted to offspring; that means that it disseminates slowly through a population.

The biological form of information transfer and modification works quite well at the pace at which ecosystems evolve, or climates change, or continents drift. Except for anomalous, geologically short eras of catastrophic extinctions, the pace of genetic change and the pace of environmental change are of the same order of magnitude, nearly enough that life has flourished. The pace of cultural change is enormously faster. That makes it unlikely that specific genes co-evolve channeled with specific culturgens.

Lumsden and Wilson try to meet this criticism by claiming that genetics and culture can be colinked in a thousand years, thirty generations (1981, pp. 295–297). Behavioral differences between contemporary human cultures that have been isolated for a thousand years are likely to be caused in part by genetic differences between those groups, as different genes have arisen to track different culturgens. But Maynard Smith and Warren (1982) find that the conclusion is simply a result of presumed strong selection and high heritability that have been built into the premises of their mathematical model. Individuals adopting the selectively favored culturgen accumulate resources five times as fast as those adopting the less favored culturgen, an extremely unlikely event. The thousand-year rule is an assumption, not a discovery.

Even if there were a thousand-year detectable difference, the differential pace is still orders of magnitude apart. Significant cultural changes can occur within a century, even within a decade. Entire cultures rise and fall in less than a thousand years; culturgens come and go far faster than that, and the millenarian genes will find it impossible to track the ephemeral culturgens.

When information begins to be transmitted neurally, the individual can learn, gaining new information constantly throughout its lifetime (this much happens even in animals whose behavior can be conditioned). With enough neural power to sponsor consciousness, new information can be deliberately sought, and ideas can be tested in imagination or in experiment; humans can think up new ideas faster than they can breed children. Using language, we can communicate these ideas, not just to our children but to anyone we tell about them. When oral cultures evolve to become literate cultures, people can transmit ideas to thousands who read books a thousand miles away or a thousand years later. Cultural practices get borrowed, traded, adapted, and so forth; they intermingle across genetic lines. One man can only intermingle his genes with those of one woman (a monogamist) or a few (a polygamist), but a person can intermingle his or her ideas with those of hundreds of persons with whom one never copulates. The genetic process can neither be hurried nor directed nor disconnected from sexuality. All this accelerates the pace of cultural information transfer by orders of magnitude over that of genetic information transfer. It is difficult to yoke horses and jet planes in coevolution and have them travel anywhere together.

One might suppose that one particular set of epigenetic rules can cover the span of many centuries of culturgen variation, but the longer that span is, the more culture can develop, vary, progress, regress, all within the scope of one set of epigenetic rules (the longer the leash is), and the less genetics is either tracking or determining what is going on (the less leash there really is). Which cultural option is being adopted is not really being genetically determined, since all these centuries are covered by the same set of epigenetic rules. Since there are few or no observable differences in basic ability and behavior in the many existing races of *Homo sapiens,* whose lines of descent parted many tens of thousands of years ago (native American tribes and Scottish clans, for instance), either humans are all on the same leash or there is no leash. There is little or no evidence of coevolutionary tracking of particular genetic sets in particular populations, channeling particular epigenetic rules and biased toward particular culturgens, nor of such genetic sets, as a result, outcompeting other genetic sets favoring other culturgens.

Of course any cultural options that are adopted for long have to be more or less successful, getting themselves reproduced over generations, but that can operate through a cultural transmission and selection that work without modifying the genetics. The real problem is that sociobiology reverts to a causal view of cultural success. Persons succeed to some extent because of their minds, and yet, deeper down, almost despite their minds, because their minds are not free for objective, rational, critical evaluation but are prejudiced by their genes.

(2) A Dual Inheritance System

Another account finds a "dual inheritance system" (Boyd and Richerson 1985; Richerson and Boyd 1989), whereby humans have some dispositions to which they are genetically disposed and other dispositions into which they are culturally educated. Their actual behavior is an interactive resultant of the strengths of these independent sources. Human behaviors in many areas are culturally determined without significant genetic bias (what color automobiles they prefer); human behaviors in other areas can have steady genetic bias (beliefs and practices with regard to incest or snakes). Human behaviors fall within an ellipse with two foci, one genetic and one cultural, and, depending on where one is within the ellipse, behaviors may be dominantly under the pull of genes, or culture, or various hybrids with components of both. In the "leashing" analogy, the leashing can be of culture by nature, or nature by culture, or each restricting the other with various lengths of leash.

But if so, we are fully prepared to recognize that how individuals behave in fact is often determined by their learning experiences and by social trends and contingencies. Choices depend on parents, teachers, peers, advertising pressures, fads and fashions, social policies and institutions. Even in behaviors regarding biological reproduction, cultural beliefs can override genetic dispositions to maximize offspring, if indeed these remain in humans from their evolutionary heritage. L. L. Cavalli-Sforza and M. W. Feldman (1981) show that fertility has declined in Europe in the last century, that Italian women, for example, do not maximize their offspring, differing in their beliefs and behavior from those of their mothers and grandmothers. In modern Western societies, parents have fewer children than they could successfully raise with their resources, a counterexample to what sociobiology predicts (Boyd and Richerson 1985). The average fertility rate per woman in the United States fell from 7.0 in 1800 to 2.1 in 1990, in a period in which resources rose at a rate matching the fall in fertility. The reasons for the changes must be cultural, not genetic, and there does not seem to be any favored genotype producing phenotypes inclined by epigenetic rules to reject these novel beliefs about desirable family size.[11]

A sociobiologist might reply that such reduced reproducers will inevitably be self-eliminating; they will be replaced in subsequent generations by those who do maximize their offspring. The behavior

---

[11] With genetic reproductive technologies as projected, the next generation of parents will even have possibilities for designing the genes of their children (Robertson 1994).

and the idea are doomed to disappear. But this does not follow, because the idea that one ought to have fewer children is not itself genetically transmitted, as is proved by the fact that the couples now restraining reproduction are only a generation or two removed from parents and grandparents who had other ideas. If the idea is contagious enough culturally, and if it is appealing for good reasons, it can spread indefinitely through the population, jumping genetic lines and at a speed of transmission many orders of magnitude faster than any behavioral tendencies transmitted genetically. Those with the new cognitive beliefs convert the oncoming generation to their view.

(3) An Adapted Mind

An evolutionary psychology account finds that humans have not so much an all-purpose or unified mind as what John Tooby and Leda Cosmides call an "adapted mind" made up of "a complex pluralism of mechanisms," "a bag of tricks," a set of "complex adaptations" that, over our evolutionary history, have promoted survival. "What is special about the human mind is not that it gave up 'instinct' in order to become flexible, but that it proliferated 'instincts' – that is, content-specific problem-solving specializations" (Tooby and Cosmides 1992, pp. 61, 91, and 113). "These evolved psychological mechanisms are adaptations, constructed by natural selection over evolutionary time" (Cosmides, Tooby, and Barkow 1992, p. 5). These form a set of behavioral subroutines, selected for coping in culture, by which humans maximize their offspring. The human mind is "an integrated bundle of complex mechanisms (*adaptations*)" (Symons 1992, p. 138). The mind is, says Cosmides, more like a Swiss army knife, tools for this and that, rather than a general purpose learning device.[12]

Humans have needed teachability, yes, but they have also needed channeled reaction patterns. The adapted mind evolved a complex of behavior-disposition "modules," "Darwinian algorithms," each

---

[12] Cosmides started a lecture by holding up a Swiss army knife as a model of the mind at a joint meeting of the Royal Society of London and the British Academy, April 4–6, 1995, London, "Evolution of Social Behavior Patterns in Primates and Man," the proceedings published as Runciman, Maynard Smith, and Dunbar (1996).

dedicated to task-specific functions in this or that dimension of life, such as picking mates, or helping family, or obeying parents, or being suspicious of strangers, or dealing with noncooperators by ostracizing them, or preferring savannah-type landscapes. In picking mates, for example, men are disposed to select younger women, likely to be fertile. Women are disposed to select men of social status, likely to be good providers (Buss 1989; Buss et al. 1990; Symons 1992). Further, these dispositions to behavior, present still in any contemporary culture, are those that meant survival in a Pleistocene environment (such as fear of strangers or desire for many children), and this may mean that they are neither optimal nor altogether desirable dispositions in a modern environment (where people may need to cooperate with strangers, have fewer children, and live in cities) (Cosmides, et al. 1992, p. 5).

The human mind is indeed complex, and various subroutines to which we are genetically programmed (caring for children, obeying parents, and even ostracizing noncooperators or being suspicious of strangers) may indeed be convenient shortcuts to survival, reliable modes of operating whether or not persons have made much rational reflection over these behaviors. It seems plausible that humans are disposed to see colors in certain ways, or to like sweets and fats, or to use nouns and verbs in our languages. Some more or less "automatic" behavior is desirable. It is hardly surprising that males look for a female likely to be a good mother (able to bear children and care for them) and females look for a male likely to be a good father (able and likely to provide resources and to care about his family). It would be surprising if evolution had selected any other dispositions.

It is also possible that selective forces in earlier cultures (for men with strength enough to hunt or plow) differ from those of later cultures (for persons who can read, write, and do arithmetic). We should probably not assume, however, that there was some one kind of Pleistocene environment, either in the various kinds of landscapes on which humans lived or in the various cultures that they developed. The Pleistocene environment too demanded multiple skills and an adaptable mind that could integrate them well. Many of the successful behaviors (recognizing faces, planning for tomorrow, being resolute in difficult times, cooperating, learning from mistakes, exercising appropriate caution, controlling jealousy or lust, or forgiving others) were just as relevant then as they are now. There is much evidence, for example, that humans now taken as infants out of aboriginal cultures can do quite well when educated into a modern European culture.

Nevertheless, though there is something to be said for behavioral modules, the mind is not overly compartmentalized, because behaviors interconnect. Behavioral and genetic psychologists are fond of speaking of mental "mechanisms," and any machinelike function, working instinctively, diminishes the cognitive reflection required. But if women are prone to choose men of status, that requires considerable capacity to make judgments about what counts as status, economically, politically, religiously. They will have to judge which one, from among their suitors, often still relatively young, is most likely to attain it in the decades of their childrearing. If men are to be good providers, that requires judgments about cooperation, and if one is operating in a barter or market culture, judgments will be needed about trading with strangers or ostracizing merchants who renege on their promises. Men need to judge potential mates not just on whether they are likely to be fertile, but on whether they too are probably good providers, able and willing to care for offspring and to educate them successfully into their culture, until they reach childbearing age.

Any such articulated behavioral modes need to be figured back into a more generalized intelligence (Sterelny 1995). Genetically programmed algorithms seem unlikely for the detail of such decisions under changing cultural conditions. Such decisions are difficult even for well-educated persons; they may require insight into character and evaluation based on intuition, additionally to conscious, explicit calculations; decisions at this level take considerable capacity for judgment, not simply mental mechanisms. The strongest finding by far in the cross-cultural study of mate preference is that both sexes from cultures around the globe consistently agree on the most promising characteristics they look for in a mate: kindness, understanding, and intelligence (Buss 1989, p. 13; Buss et al. 1990, pp. 18–20). Capacities to select such a mate are perhaps somewhat "instinctive," but they are unlikely to be an adaptive mechanism isolated from general intelligence and moral sensitivity.

Apparently, the mind is not so compartmentalized that humans – modern ones who read this literature at least – cannot make a critical appraisal of what behavioral subroutines they do inherit by genetic disposition and choose, if they wish, to offset these "Stone Age" dispositions in their evolutionary psychology. Cosmides and Tooby are doing just that – if we may be permitted an ad hominem argument. They themselves illustrate that the human mind is more than a patchwork of naturally selected response routines when they call for "conceptual integration" of the diverse academic disciplines studying humans, their behavior, and their minds. These include "evolutionary biology, cognitive science, behavioral ecology, psychology, hunter-gatherer studies, social anthropology, biological anthropology, primatology, and neurobiology" (Cosmides, et al. 1992, pp. 4 and 23–24). These are not disciplines in which one becomes expert by behavioral mechanisms in a Swiss-army-knife mind. At the least, they and their readers must have quite broadly analytical and synoptic minds.[13] The mind is fully capable of evaluating any such behavioral modules, and of recommending appropriate education so as to reshape these dispositions in result. These psychologists seem to be quite able to readapt by critical thought their own adapted minds; nor is there any reason to think that they and their colleagues in evolutionary psychology are alone in this capacity. In the chapters ahead, the performance of the human mind in the fields of science (looking for a general theory of evolutionary natural history), ethics (evaluating self-interest and altruism), and religion (asking whether natural history provides evidence for the existence of God) seems quite nonmodular.

4. THE EVOLUTION OF IDEAS

The cumulation of a billion years of biological experimenting ends up, in the human species line, with individual humans with a hundred thousand or more genes, coding these discoveries. These genes make a brain with ten billion neurons, each with hundreds and sometimes thousands of possible synaptic connections, providing virtually endless opportunities for encoding ideas. These hookups begin to code cumulative cultural discoveries and to transmit them in new networks of information transfer (language and books, and, more recently, telephones, television, and computer networks). When this has gone on for a hundred thousand years and more, one

---

[13] As Cosmides must have believed speaking at a joint meeting of the Royal Society of London, dealing with the sciences, and the British Academy, dealing with arts, asking the audience to evaluate the model of a Swiss-army-knife mind.

can expect some startling outcomes. Wilson admits, "Clearly, such mechanisms are far more complex than anything else on earth" (1978, p. 77). In evolutionary history, with the coming of humans, there appears the genesis of ideas; in culture thereafter, ideas are perennially generated and regenerated. This phenomenon too has to be incorporated into any unified worldview.

Superposed on genetic endowments more or less common to all members of *Homo sapiens*, humans develop myriad diverse cultures. They think their way through the world in amazingly diverse ways, from Druids to Albert Einstein. In computer imagery, the same "hardware" (biology) supports diverse programs of "software" (culture), even if there are many repeated subroutines. The evolved brain allows many sets of mind. These ideas too "evolve" in the sense that they change and develop. Physics, for instance, has developed from Ptolemaic to Copernican theory. Biology has progressed from belief in the fixity of species to evolutionary theory. Ethics has rejected slavery and the unequal treatment of women, practices once widely accepted. Religion has progressed from polytheism to monotheism, or monism. One culture is reformed into another; cultures sometimes degenerate and go extinct.

These are not claims about the evolution of the hardware that humans inherit natally; they are claims about the evolution of the software. Ideas are discovered and transmitted, and the mechanism of transmission is cultural. One does not have to have Plato's genes to be a Platonist, Darwin's genes to be a Darwinian, or Jesus' genes to be a Christian. The thinkers responsible for shifting physics to a Copernican view, biology to a Darwinian view, and ethics to universal human rights were not from any particular racial or ethnic group. The system of inheritance of ideas is independent of the system of inheritance of genes.

True enough, it may be replied. But still the genes are in control of any such evolution of ideas. The functioning of the brain, with its output in cultures – Greek, Christian, or scientific – is always a matter of cognitive beliefs that are selected for their capacity to survive in the world. Better beliefs in this respect are selected over generations, and they are selected because they better enable one generation to raise a next generation successfully. Epistemology too is evolutionary in the biological sense. So even in the selection of cultural beliefs and behavior, notwithstanding any dual inheritance system, natural selection is still the dominant determinant.

Is it? One possibility is that selection theory transcends both biology and culture, and that natural selection and cultural selection are subsets of a more formal theory of variation and retention. Some things get varied; through differential preservation, some variants are selected to replicate and others eliminated. In fact, selection theory may be something like a constitutive tautology: that is, in a world where there are competing variant x's, where the x's are tested against each other for their capacity to cope with the world, and where x's may and must be copied as the crux of their coping, the best coping x's will be the most copied x's. That says nothing about whether the replicating code for the heritable variation is biological, cultural, scientific, electronic, legal, ethical, religious, extraterrestrial, or whatever. It only says that if one supposes a world in which things come and go; and a process that emits variants, some of which come more quickly and go more slowly; and competition, then those best able to come into being and continue on will. These x's can be genes, protein molecules, organisms, species, or humans, or their scientific theories, ethical codes, legal codes, or religions. They can be chess players in meets, competing in repeated matches with strategies they learn from each other, or housewives whose recipes, replicated meal after meal, are passed from mother to daughter. But they cannot be oxygen atoms, salt crystals, rocks, rivers, or mountains. Though the latter too come and go, they do not cope at all, neither for better nor for worse; nor do they make copies of themselves.

If there are some sorts of persons who make a transmissible and developing culture possible, and other sorts who make it impossible, then, over time in societies where death and replacement of the idea carriers occur, the former sort will be favored and the latter sort will fail. All that is built into the logic of the model and will be true for androids in the Andromeda galaxy. One does not need to know anything about Earth-bound genetics to be sure of that. In this sense, the best adapted survive. Cultural selection and natural selection have been subsumed under the bigger theory, but the formal theory is so high-level, permitting so many diverse applications, that we hardly need to look to see whether it is true.

We do not know before we look that this formal model is the right empirical one, in the sense that there are other logically possible world models. There might be a Platonic model, in which forms are imposed on matter, without any incremental variation and retention. There might be a "Hebrew" fundamentalist model (for the lack of a better name) in which inspired prophets insert divine information periodically into the world, thereafter without variation and with infallible retention. There might be a Hegelian model in which incremental development is always thesis, antithesis, synthesis, so that everything zigzags into other things, partial survival, partial loss, partial transformation. Or a Taoist model with yang and yin in binary opposition. Or a Hindu model with Brahman, *maya* (illusion) arising through *avidya* (ignorance). And so on. One model might be found in biology and another in culture. One might be found in economics and another in theology. So we have to look to see whether the incremental variation–selection–retention–transmission model is the right one.

But if this model is the one applicable, then we know before we look that the best coping will link with the best copying, and will be coded in transmission. If we know that the best teams win, and that winning means scoring the most points, then we know before we look that the best teams are, on average, in the playoffs. But are there any such point-scoring games being played? Are such winning teams advancing to playoffs? That much we have to find out empirically. After that we have to find out how the formal theory is worked out in this or that empirical situation. Biologists must tell us how it operates genetically, since in biology the coding is genetic and the coping is organismic. Anthropologists, sociologists, philosophers, and theologians must tell us how it operates culturally, where the coding is neural-ideational and the coping is social, psychological, ethical. Selection theory wins after all. But the victory is pyrrhic, since the win comes with the loss of genetics as essential to the theory; the genetics is only local to the biological application.

Perhaps in their governing worldviews, whatever may be said about behavioral subroutines, humans have a tabula rasa mind genetically; none of the orienting beliefs that they hold are fixed or even favored by their genes. The content of what they come to believe depends on their education, on nurture not nature. After they come to believe this or that, cultural selection begins to operate. Those who receive the appropriate, culturally functional beliefs survive and have more children. Those who receive inappropriate, culturally dysfunctional beliefs are misfits and go extinct. No genetic alteration occurs, however; the next generation, and the next, are born with ideologically tabula rasa minds all over again. Genes are not determining any beliefs; they are producing open, empty minds that can be taught various beliefs. Cultural selection is determining beliefs. The beliefs that dominantly persist are the culturally functional ones.

Cultural traits can help to produce more offspring, but their transmission need not be genetic. Parents who can build fires stay healthier in winter and have more babies, who also keep warmer and healthier and survive; these children, when grown, having been taught to build fires, do likewise. But such a trait, though it moves from parent to child, does not just run in families. Everyone else who is cold in winter is soon building fires too, and the differential survival advantage is soon lost to particular individuals and their families, although it remains in the culture as a whole.

The human cognitive equipment has what structure it has, like a computer hardware, as a given to work with. Quite diverse software programs can be run on this hardware, and, in terms of selecting among the broad cultural options faced, nothing is hard-wired. All of them have been run on the human hardware, else they would not be encountered by the developing person. A variant software that won't run never gets to be a cultural option. When humans choose between competing options, those who use this and not that software will better succeed in reproducing, and their children will inherit copies of it. This better capacity to survive is copied, but not by hardware rebuilding (not by genetics), rather by software duplication (cultural transmission). There is no reason to call this Darwinian selection. This is Lamarckian selection, since acquired traits are being transmitted. Better still, let us simply call it cultural selection and recognize that it is paralleling and even transcending biology. The evolution of ideas (culture) is a deepening of the plot, launched by the evolution of the brain/mind.

As these ideas come and go, some of them will be conserved (survive) and others will be forgotten (go extinct). Perhaps ideas are competing, and evolving, or developing, on some other selective basis, beyond natural selection. Competition between ideas does not take place in geographical space, but in discursive space, so to speak. Organisms have a life of their own; they metabolize and physiologically function with an organized somatic identity. Ideas have no "life of their own," though they are sometimes metaphorically said to be "alive and well" or to have "died." Yet there is information content in an organism, in the DNA as well as in any acquired information, if the organism is so able. The organism embodies an "idea" in the biological sense; ideas, in the noetic sense, are embodied in humans, who are organisms, and so perhaps there is a cognitive parallel. To this issue we must return, examining the selective forces in science, ethics, and religion. Meanwhile, all this is pointing steadily to a difference in being human, to a complex mind indeed adapted for culture, that is, to a distinctive human genius.

5. THE HUMAN GENIUS (*Geist*)

Genesis is linked with genes, and these genes in "the wise species," *Homo sapiens*, produce a brain that sponsors a mind with what we will call, provocatively, a unique "genius." That reconnects the word "genius" with its etymology, recalling the Latin *genius*, spirit. Other derivatives are found in such words as "ingenuity" and "engineer"; the Latin root, *gigno*, is the same, interestingly, as that of "gene," to generate, involving now the generation of a procreative animating spirit. In German, our choice of words would be *Geist*. At issue is whether the impressive genesis across evolutionary history, which in retrospect has been linked to genes, is now, with culture in prospect, so constrained by these genes that no one can think without survival and reproduction as the bottom-line logic determining the outcome of all thought. Or have humans some further genius that is displayed in the generation of their cultures, with breakthroughs to new achievement and creativity? Earth is the planet where the most complex creativity of which we are aware has taken place; on Earth the most effective and complex creative instrument known to us is the human mind.

Some may caution that although the existence of nature is as certain as can be, the existence of spirit is doubtful. Indeed spirit cannot be found empirically; this is the myth of the ghost (*Geist*) in the machine. But nothing is more evident experientially than that human life has this "genius," or "spirit." "Spirit" in the sense of conscious self-awareness (*Geist*) is a first fact of experience, memorably noted by Descartes's indubitable *Cogito, ergo sum*. That humans are embodied spirits, bodies with self-reflective psychological experience, is really beyond dispute. The act of disputing it, verifies it.

Conscious experience evolves incrementally and is already present in the animal world; the higher animals are "animated" (Latin: *anima*, spirit, in-spirited, in-spired, full of breath). There is "somebody there" behind the fur and feathers, behind those eyes. These animals have psychology beyond biology; they have points of view. But only in the human world does spirit become recompounded through the compounding of transmissible cultures; that is the peculiar genius of the human spirit. Superposed on biology, spirit is nurtured within culture and takes on a life of its own, not free from its world but free in it.

Natural selection is relaxed and superseded. Though culture is superimposed on biology, the leash is not just loose: there is some release from biological determinism (Cavalli-Sforza and Feldman 1981; Boyd and Richerson 1985; Durham 1991), a finding consistent with what anthropologists and sociologists have long maintained. "Biology, while it is an absolutely necessary condition for culture, is equally and absolutely insufficient: it is completely unable to specify the cultural properties of human behavior or their variations from one human group to another" (Sahlins 1976, p. xi; Bock 1980; Breuer 1982; Barnett 1988). Biology determines some outcomes but under-determines many others.

Humans must mate, their genes degenerate unless they outbreed, and so, perhaps, biology shapes marriage customs or what humans think about incest. But consider what educated people think about polygamy, or abortion, or birth control – or disarmament, or evolutionary theory – all done on circuits in the brains that the genes have made. What is happening when a developed nation sends food to those underfed in a developing nation? Such beliefs and events are the result of decisions, perhaps individual, perhaps corporate, but it no longer seems plausible to hold that the principal determinant is producing more offspring in the next generation, or that the decision is some resultant of some complex of still rather instinctive, adaptive behavioral subroutines. It seems that culture relaxes the pressures of natural selection.

The human being is born and develops in some one of thousands of cultures, each historically conditioned, perpetuated by language and tradition, conventionally established, using symbols with locally effective meanings. Nothing in animal society approaches this. If there are animal antecedents to culture, then control by genetics might already be passing over into something higher, and one can well expect such transitional animal skills. Careful analysis will welcome any continuity with animal life, while resisting reductions and rejoicing in the distinctively novel phenomena when humans arrive.

Quantitative differences add into qualitative differences; they recompose an emerging gestalt that exceeds previous evolutionary achievements.

Everyone knows that culture is adaptive in the broad sense. Economic, political, technological, scientific, philosophical, ethical, and religious institutions and artifacts, none of which exists in wild nature, help humans to live well in their world. So there are markets and governments, universities and churches, medicinal drugs and high-yield fertilizers. When humans succeed by using culture, their biochemistries succeed along with them, so in that sense the cultural successes are biologically adaptive. Not all cultural innovations contribute to such success; it is hard to see how, for instance, abortion on demand or the gay rights movement leads to more offspring in the next generation. But, on average, culture helps humans cope, and such culture will be copied from generation to generation. Our ancestors, in whatever cultures they had, behaved in such ways that they had offspring and successfully reared them over many centuries. Here we are, ourselves living proof of that!

All this leaves quite open, however, the question whether our forebears' cultural processes operated under genetic control and were determined by natural selection. Cultures that move past the hunter–gatherer stage rebuild their natural environments; in this artifacted environment, the rules are different. Culture must be consonant with biology, and yet culture repairs biology. Contrast, for example, what goes on in a medically skilled culture with what goes on in wild populations.

When the bighorn sheep in Yellowstone National Park caught pinkeye in 1980–81, many became partially blinded. Often unable to feed properly, more than three hundred bighorns, 60 percent of the herd, perished. By park policy the sheep were left to the forces of natural selection. Only those that were genetically able to contend with the disease survived, and this capacity is now coded in the altered allele frequencies in the survivors. The herd has recovered and the population is genetically more resistant to the *Chlamydia* microbe.

When human children catch pinkeye, their mothers put them to bed and draw the curtains, and their doctors prescribe eyedrops with sodium sulfacetamide. The *Chlamydia* microbes are destroyed, and the children are back outside playing in a few days. They are not genetically any different than before the disease, nor will the next generation be different. When the grandchildren catch pinkeye, they too will get eyedrops. In the sheep the biology is altered by natural selection; in the medically treated children the biology is altered by prescription of sodium sulfacetamide. In both, the result is better dealing with disease. But the similar outcome is entirely different in kind in the human case. Humans make decisions about pinkeye and are able to treat it because of the advances in medicine acquired over the centuries; sheep do not make such decisions and do not have any such cumulation of acquired knowledge.

A determined biologist can reply that any such decisions and knowledge are framed by the epigenetic rules, requiring human minds to adopt whatever "culturgens" – in this case, "medical ideas" – promote more offspring in the next generation. These doctors and parents are operating out of a "care for your children" instinct. Whatever changes culture introduces (looseness in the leash, sodium sulfacetamide for children) do not change the ground rules: genes that code for the best coping survive. The smart doctors and caring mothers have the superior genes. Although the human genetic composition is not different as a result of pinkeye, still, those people who have the best mother/doctor genes will be selected for. These caring-coping subroutines will be coded in the next generation of humans.

This fails, however, to note critical differences. The doctors who discovered sodium sulfacetamide did not do so by using any instinctive behavioral complexes. Further, they have shared this information widely. The information is a discovery in science, carefully researched, and doctors all over the world can learn about it and introduce the drug into their culture. Distribution of the drug is promoted by the World Health Organization and by the United Nations International Children's Emergency Fund, now the United Nations Children's Fund (UNICEF). Some doctors who have shared the discovery were medical missionaries who did so out of ethical and religious convictions. Doctors regularly care about nonoffspring, and the beneficiaries of this sharing help non-genetically related (foreign!) humans to cope. Those foreigners too, when they become good mothers and doctors, will presumably be selected for their caring-coping subroutines. But even if the sorts of people who can share this information are being genetically selected, information is being passed around culturally, overleaping both genetics and innate dispositions. With all this transmission of medical information, there is no genetic transmission or alteration at all. Nature is not taking its course but being interrupted by technology and ethics. More is going on than natural selection of the genetically superior stock.

The "genius" in culture is nongenetic transmission. Humans teach each other how to grow wheat, make fires, bake bread, cure diseases. Some information discovered in the past is transmitted culturally (exogenetically) from parent to child, or, since this information is nongenetic and nonsomatic, as easily from any teacher to any pupil. There is what philosophers have classically called "emergence," the appearance in later evolutionary history of phenomena not present earlier. Once upon a time, out of abiotic nature, living things evolved. Once there were only prokaryotes; eukaryotes evolved. And so with metazoans, neurons, instincts, conditioned behavior, terrestrial life, and psychological experience. Once there was no smelling, swimming, hiding, defending a territory, taking risks, making mistakes, or outsmarting a competitor. All these things appear gradually, also without precedent if one looks further along their developmental lines. In each quantum jump there is a little more of what was not there before, and if one integrates the differentials one gets something in kind where before there was nothing of that kind. In this sense, the evolutionary story regularly produces more out of less.

The new phenomenon is not simply that humans are more versatile in their spontaneous natural environments. Their versatility now extends to rebuilding that wild environment intentionally on the basis of knowledge acquired and transmitted culturally. Humans live from the tropics to the arctic, from the deserts to the rain forest, because they insulate themselves from the environmental extremes by their rebuilt habitations (houses that are air-conditioned and centrally heated; grocery stores with produce from afar). They pass these rebuilt environments on to the next generation. Indeed, modern, medically sophisticated humans are at the point not only of rebuilding their environment but of rebuilding their genotype. Is there more in the cultural consequent than in the lesser biological precedent? For eons, life governed by natural selection was the big story. But in culture the earlier story has become an understory. Climaxing in culture, natural history overtops itself and passes over into something else.

Biologists may celebrate these innovations but believe that natural selection is the principal determinant when any of these phenomena appears incrementally. They are retained because of their survival benefits, even though they are quite complex adaptations. Does that age-old determinant still operate to limit culture? Or does the explanatory power of biology run up against a limit? Is natural selection itself, pervasively present before in botanical and zoological nature, transcended in human nature?

John Maynard Smith, though himself a principal evolutionary theorist, frankly notices that limits appear when evolutionary theory undertakes cultural explanation. Does the sociobiological claim mean that in a society with fixed rules, "the actions of different people in that society – rich and poor, young and old, male and female – are those which would be predicted if each individual is behaving, subject to the rules, in the way which would maximise his or her inclusive fitness" (1982b, p. 3)? He has serious reservations:

> The explanatory power of evolutionary theory rests largely on three assumptions: that mutation is nonadaptive, that acquired characters are not inherited, and that inheritance is Mendelian – that is, it is atomic, and we inherit the atoms, or genes, equally from our two parents, and from no one else. In the cultural analogy, none of these things is true. This must severely limit the ability of a [biological] theory of cultural inheritance to say what can happen and more importantly, what cannot happen. (1986)

But with that, the limiting seems the other way around. Evolutionary theory is not limiting what can happen in culture; to the contrary, it is itself being limited because it does not have the resources to say what can and cannot happen when culture appears. Phenomena arise (the Protestant Reformation, debates on nuclear disarmament, the computer revolution) for which the categories of biology are inadequate as explanations, just as physics was incompetent to explain swimming, hiding, and nursing young. Nor should one be surprised if this involves the deeply historical recompounding of nongenetic ways of information discovery and transmission.

Dawkins also recognizes this. He posits memes that are the cultural analogues of genes. "An 'idea-meme' might be defined as an entity which is capable of being transmitted from one brain to another" (1989, p. 196). "Examples of memes are tunes, ideas, catchphrases, clothes fashions, ways of making pots or of building arches. Just as genes propagate themselves in the gene pool by leaping from body to body via sperms or eggs, so memes propagate themselves in the meme pool by leaping from brain to brain via a process which, in the broad sense, can be called imitation" (p. 192). Memes, although somewhat like genes, are not under genetic control. They are culturally, not genetically selected. "We biologists have assimilated the idea of genetic evolution so deeply that we tend to forget that it is only one of many possible kinds of evolution" (p. 194). "For an understanding of the evolution of modern man, we must begin by throwing out the gene as the sole basis of our ideas of evolution" (p. 191).

Memes transcend genes. Humans can, after all, make some critical evaluation of possibilities, retaining some ideas and throwing out others. "Memes," though it draws an analogy with "genes," is only a nonce word for "ideas," a concept that has been around for millennia. Ideas of various kinds persist so far as they are able to outcompete and outreplicate other ideas, so far as they get "mimicked." Everybody knows that. This perspective will be useful only if it relates idea information to gene information. Curiously, the connection is that Dawkins insists that memes are still "selfish." The selfish theme is invoked once again, now in a guise rather difficult to recognize. "If a meme is to dominate the attention of a human brain, it must do so at the expense of 'rival' memes" (1989, p. 197). "Selection favours memes which exploit their cultural environment to their own advantage" (p. 199). He hopes that, once we learn about both genes and memes, "we have the power to defy the selfish genes of our birth and, if necessary, the selfish memes of our indoctrination" (p. 200).

But it is not at all clear, at the latter defiance, what we would defy these selfish doctrinal memes with. Would not their replacements, ideas that we come up with on our own perhaps, become other selfish memes trying to oust their rivals? There cannot be anything amiss just because some ideas are repeatedly transmitted (shared, we might say, or with which we are "indoctrinated"). That is the essence of any transmitted culture; we do not wish to overthrow all the ideas that have been most successful in getting themselves inherited over the centuries. We wish to replace only the ones for which there appear better replacements. Until we get some account of critical ideas, of nonmemes, or at least nonselfish memes, or of good selfish memes versus bad selfish memes, it is hard to know how to proceed with this overthrow.

Everyone knows that ideas, whether inherited or of our own origination, can be bad as well as good for us, and we ought to be circumspect about which ideas we let influence our behavior. So there can be good and bad memes, which is only to say good and bad ideas. But the idea that some or all of these memes are selfish is unhelpful. Already in genetics we found that the idea of an allegedly selfish gene was just as readily interpretable as the idea of a shared and sharing gene, and memes are even more nonrival than genes. Einstein published his theory of special relativity and found within a few years that it was accepted by thousands of physicists all over the world. You can say if you like that his "selfish" theory, launched in one mind, had gotten loose and reproduced itself in all these other minds (something like a parasitical virus). It had competitively eliminated the Newtonian "memes" previously in those minds (if reducing the Newtonian theory to a special case of relativity theory can be said to eliminate it).

But why not just say, as we usually do, that one person, with his "genius," had shared his great thoughts with many others, who critically evaluated his arguments and found them persuasive, superior to the accounts they previously had? The "sharing" paradigm is quite as plausible as the "selfish" paradigm, when ideas are spread around. If Jesus' preaching the Golden Rule (eliminating selfish ideas), Martin Luther King, Jr.'s, having a dream of civil rights for all (eliminating discrimination and segregation), Thomas Edison's inventing the light bulb (eliminating kerosene lamps), Pasteur's discovering inoculation (eliminating smallpox and ignorance about its causes), and so on, is the selfish propagation of ideas, then let us have more of such beneficial, moral, and humane selfishness.

Two things are happening here. One is that genetic Darwinism is being subsumed under a more general selection theory (that x's are generated and tested, the better ones selected for reproduction), so that genetic Darwinism is only one application, in biology, of a more comprehensive formal theory of heritable variation. The application in culture is radically different, as will become evident when we turn more specifically to science, ethics, and religion. The leash is broken; biology and culture are two dramatically different events – even though culture is superposed on biology, and even though they both can be subsumed under a formal selection theory. With these concessions, such biologists have in effect conceded all that social scientists, or philosophers, theologians, politicians, and other humanists, wish to defend by way of the independence of culture. For analysts of culture nowhere wish to despise their biology; rather, they admire it.

They also insist on an adequate account of culture as possessing a "genius" irreducible to biology.

If humans variously have different abilities, genetically inherited, and if in culture there arises a differentiation of roles such that each can do his or her own thing best, with the benefits of this specialized productivity widely distributed throughout the community (for instance through markets and schools), that is reason to think culture has overleaped genetics. Each develops his or her own abilities to the fullest and, where ability is lacking, benefits when the specialized abilities of others are culturally shared. There is nothing new or controversial in the claim that different people have different abilities and that we enjoy the community of talents. There is nothing suspect ("selfish") about a heritage of life-sustaining information being discovered, selected, and transmitted culturally (in crafts, politics, philosophy, art, religion, and so on), any more than there is anything immoral in a heritage of information being discovered, selected, and transmitted genetically. Rather, to anticipate religious interpretations, if anything is to count as being sacred, this could plausibly be either or both of these creative processes by which truth is discovered and carried forward.

6. HISTORICAL VERSUS UNIVERSAL EXPLANATIONS IN NATURE AND CULTURE

There is a long-standing problem with any science of human nature when it attempts to explain the novel kinds of genesis made possible by such human genius. Scientific explanations are paradigmatically lawlike, universal, causal, repeatable; they interpret particular cases as instances of general theories, under specific initial conditions. Human affairs, by contrast, are historical. There are recurrent themes: wars, recessions, famines, regimes overthrown, and so on; yet history does not simply repeat itself again and again over the centuries and across the continents. Can there be a lawlike science of human nature that explains all the variety, diversity, and decision making that have characterized cultures?

Perhaps this will be one universal theory reapplied to differing sets of initial conditions. But what if the initial conditions are cumulating to govern outcomes in ways with which the lawlike theory is unable to deal? Something is building up, information accumulating, decisions launching particular courses of travel, and the results are being transmitted and evaluated. Something is being "initiated" that "conditions" the theory, rather than a theory's just being reapplied to initial conditions. The result is the emergence, the genesis, of novelty, and something is escaping the analysis of the lawlike, universal theory, something narrative and distinctive to particular cultures. The one extreme is nomothetic; the other is idiographic.

We will be better prepared to take the historicity in culture seriously when we realize how we must first take history in biology seriously. There it is made possible by genes, and these genes generate novelties that are not fully explained by laws operating on initial conditions. Physics and chemistry are full of universal laws. Biology is an Earth-bound science, historically conditioned. Celestial mechanics predicts eclipses, centuries ahead, to within seconds. Einstein's theory of relativity is true all over the universe; the atomic table explains thousands of chemical reactions yesterday, today, tomorrow, here, there, everywhere. A mineralogist will want to take his copy of *Dana's Manual* to the Moon. But no botanist would take *Gray's Manual* there. Geneticists do not suppose that heritability will have to be Mendelian on other planets, with two sexual parents each contributing one of two genes at a locus, with independent assortment. The products that these genes make (ribosomes, Golgi apparatus, cytochrome-*c* molecules, acetylcholine molecules, hearts, livers, ladyslipper orchids, tigers, humans) are historically derived to function in the Earth environment.

In biology, the big-scale laws of chemistry and physics are inevitable and never violated, but they are also so loose that the vicissitudes of an individual organism's fate rattle around within the overarching laws. When the achene of a fruiting pasqueflower *(Pulsatilla)* articulates from the head, it falls subject to the laws of gravity, and its downward acceleration is governed by the equation $D = 1/2\ at^2$ (where D is distance downward, a is acceleration due to gravity, and t is time). But that does little to help us to predict the seed's fate as it is blown about by the wind, lands in a favorable location, or perhaps not; may be eaten by a passing rodent, or not; may germinate if it chances to rain, or get stepped on by an elk. Even if one averages out the fates of a thousand such seeds, so far as they are controlled by the law of gravity, the results of the equation are worthless in understanding the ecology of *Pulsatilla*.

In the DNA coding, physics and chemistry are never violated, but there is no information in any physics or chemistry book that will enable us to deduce what protein structures will evolve when an organism adjusts its immune system to a newly introduced disease organism. Nothing there helps a biologist predict that there will be heterozygotes, or predators, or males and females, or vertebrates and invertebrates, or primates or people. New factors arise in the higher science (biology) with which the lower science (physics) is unable to deal. Idealized assumptions and generalized laws are not untrue, but they are so simple that the more general science fails to discriminate what makes the critical difference in the phenomena examined in the more specific science. The lower-level science does not have the relevant information in the categories of explanation available to it to explain the emergent events.

In merely physicochemical materials there is no information discovery, transfer, testing, and accumulation over time. But this pervades biology; it is the essence of life. The later chapters of the story are different because the earlier discoveries get folded into the later inventions. Something "initiated" in one set of initial conditions instructs a later set of initial conditions; in turn, something is further "initiated" there that is transmitted thereafter. The evolution of photosynthesis could not be inferred by natural selection operating on prephotosynthetic life, even though one might have anticipated that solar energy would somehow be tapped. Yet, when photosynthesis appears, first as nonaerobic photosynthesis, the initial conditions on Earth are thereafter different from what they were before, because the energy available for life is radically increased, altering trophic pyramids. Later, aerobic photosynthesis appears, and aerobic respiration evolves, altering the atmosphere, and, again, the initial conditions are thereafter changed by the new discovery, although the new discovery is no implication of law plus previous initial conditions.

The origin of calcium-containing shells cannot be predicted from the previous use of silicon by diatoms to make protective shells. From such use of external shell calcium, there seem to have originated calcium-containing endoskeletons, making possible all higher animal life. But this is no implication of calcium shells. The biological use of calcium increases the production of limestones, calcium carbonate. This removes carbon from the atmosphere and additionally builds up the amount of free oxygen, favoring aerobic respiration and photosynthesis. Such breakthroughs change the initial conditions and reset the trends of life. Although, throughout the whole epic, the fittest survive, and the law of natural selection reigns unaltered, natural selection theory cannot predict or retrodict, in the sense of designating the critical difference between the roads that are taken in contrast to roads not taken in this cumulation of historically novel initiatives, generated steadily along the evolutionary course. Explaining genes and genesis, the novelty in the emergents is more significant than the recurrent law. Laws plus initial conditions are not good at explaining how more evolves out of less.

Biologists start out thinking that what they have to examine are structure, anatomy, morphology. Soon, they have to examine process, physiology, metabolism, ecology. Finally, they have to recount history. Proximate causes are embraced within global development patterns. To explain an event – the coming and going of the dinosaurs – is not so much a matter of putting it under a covering law, nor of analyzing premise and conclusion, much less of putting it into a mathematical equation; it is a matter of setting it in a story line. All the laws of physics and chemistry are unviolated, yet there are emergent biological phenomena. An organism has to "know how" to nurse young, hunt prey: to code and cope. Physics and chemistry, the two prior sciences, do not provide the critically important interpretive categories, those of information discovery and transfer, with which to understand the superposed phenomena.

Biological science seeks the maximal mix of generality, lawlike repeatability, predictable biochemistries, ecological successions, long-standing trends, all on the nomothetic side, mixed with particular historical discoveries and recompounding cybernetic achievements on Earth, on the idiographic side. That provides a science of life on Earth that is deeply infected with history. "The evolutionary process itself," concludes Robert Rosen after a study of the fabrication of life, "is devoid of entailment, the province of history and not of science at all" (Rosen 1991, p. 279). There is what we called (Chapter 1) a creative construction, conservation, and sharing of values. In this historicity, there is no "leashing" of these biological emergents by the more fundamental physicochemical sciences.

Is there "leashing" when, later on, culture emerges from biology? A human science (such as psychology,[14] anthropology, and sociology) is a science restricted to *Homo sapiens*, the one Earth species that

---

[14] Behavioral science treats all species whose behavior can be conditioned by learning. Some psychology is of animals; still the focus of psychology overall is humans.

seems to surpass biology by forming culture. We face the previous question moved, so to speak, one further order of magnitude away from universal (physical) or global (biological) generality, on the nomothetic side, toward historical specificity, on the idiographic side. Is there more adventuring development that makes *Homo sapiens* with our special genius no longer subject to Earth-wide biological "laws" (if such there are), binding on all other species?

Even Wilson is forced to conclude, "By every conceivable measure, humanity is ecologically abnormal" (1992, p. 272). Humans are a hundred times more numerous than any land animal of comparable size; they make clothes and build fires and thus inhabit more diverse landscapes; they appropriate 20–40 percent of the photosynthesis of the planet. Agriculture and industry, especially fueled by scientific knowledge, enable humans to exploit natural resources at escalating rates, and, unfortunately, humans have accelerated extinction rates up to one thousand times over normal.

Well – we must ask Wilson – why not conclude that this ecological abnormality exists because humans are genetically not normal either, not one more species whose fortunes are determined by natural selection operating on genetic information. One would expect abnormality in the phenotype to be linked with abnormality in the coding genotype. Perhaps "abnormal" as used here really means "transcending" of ecology as this constrains all the other species, because of the innovations of culture.

Sociobiologists hope to give a scientific account of the "human qualities . . . insofar as they appear to be general traits of the species," the human "biogram" (Wilson 1975a, p. 548). Likewise, the evolutionary psychologists, though distancing themselves from too simplistic a genetic determination of culture, are hoping for "universal mechanisms" in the plural behavioral routines of their "adapted mind." Explanations should be based on "the underlying level of universal evolved architecture. . . . There is every reason to think that every human being (of a given sex) comes equipped with the same basic evolved design. . . . One observes variable manifest psychologies or behaviors between individuals and across cultures and views them as the product of a common underlying evolved psychology, operating under different circumstances" (Tooby and Cosmides 1992, p. 45).

Irven DeVore says:

I am interested in what is universally human, more than what's culturally variable. I am intrigued by the human biogram, the enormous areas of human behavior that are the same the world over. . . . Different cultures turn out only minor variations on the theme of the species. . . . Almost everything that's importantly human – including behavioral flexibility – is universal, and developed in the context of our shared genetic background. (DeVore and Morris 1977, p. 88)

Many theories of human nature – psychological, anthropological, sociological, economic, political, philosophical, theological – have sought human universals, and if biologists can discover any such universals, well and good, up to a point. But if such evolutionary mechanisms and genetic dispositions are claimed as the overriding explanatory category for all events in culture, complains Kenneth Bock, "human culture histories here emerge as fortuitous meanderings of people within bounds set by a human nature produced by organic evolution" (Bock 1980, p. 118). What happens results from selection pressures, genetic coding, epigenetic rules, instinctive dispositions, behavioral mechanisms, or even accidental mutations, and if something is not explained in such ways, it cannot or need not be explained at all. There are no higher relevant explanatory categories. The "manifest psychology," which is admittedly quite diverse, provides "ill-suited frames of reference," mere "surface" behavior, only to "obscure the underlying level of universal evolved architecture." Regrettably misplaced focus on the manifest psychology "has nearly precluded the accumulation of genuine knowledge about our universal design" in which the ultimate explanation lies (Tooby and Cosmides 1992, p. 45). The cultural variations are not interesting determinants.

Such insistence can, however, even more regrettably, miss the novel genesis of ideas, knowledge, beliefs, breakthroughs that emerges across developing historical cultures. In earlier societies women were often virtual slaves, and in contemporary societies they can be almost the equals of men. Blacks were slaves in the southern United States, freed in 1863 during the Civil War. What has made the critical difference? Whatever it is, sociobiologists, behavioral human ecologists, or evolutionary psychologists seem to think that it is not of interest so much as the fact that in all societies there are genetic differences that make women less assertive and less physically aggressive than men, or men have preference for youth in females, and females have preference for greater status in men. In all societies, there is a tendency to subjugate and exploit other races. But the difference between slavery and freedom is what one wants explained (as well as what one values), and a generic theory common to all *Homo sapiens*, biologically based, cannot explain the struggle from slavery to freedom by applying a universal theory to variant initial cultural conditions. This is because the variant initial conditions launch movements that grow into dominant social forces, and these consequently determine the outcome. There was the historically novel movement to liberate the slaves in the last century; there has been a historic movement to liberate women in this century.

That amounts to saying that the biological theory, even if true as far as it goes, does not have the relevant categories within its scope to discriminate the ideological differences between cultures. Hence, from the viewpoint of such a vacant theory, these seem "meanderings" and "minor variations," "surface psychology." But such labels in fact report what the theory is incompetent to detect. The allegedly universal explanation is not robust enough to tell the particular critical stories of the exodus from slavery to freedom, of the liberation of women. The theory confesses implicitly to something that has escaped it. The claim of universal application is really a confession of ignorance. Those who are immersed in universals can never narrate history.

The critical difference lies in the historically emergent ethical conviction that slavery is wrong and freedom is right; that women and blacks are, in morally relevant respects, to be given equal opportunities and responsibilities with men and whites. These new-found convictions have little to do with selfish genes or instinctive adaptive mechanisms. Persons with essentially the same genetic makeup are being converted from one ethic to the other. The biological theory is not explaining this cultural development, any more than chemistry and physics explain instinct or conditioned learning.

Since biologists do not typically claim that in biology they have "escaped" physics and chemistry, they may dislike the idea that culture in turn "escapes" biology. Just as relativity theory, gravitation, the atomic table, and covalent bonding operate unexceptionably in biology, so natural selection theory – this claim runs – should operate unexceptionably in culture. Scientists do not like anomalies. But biology does escape physics and chemistry, in the sense that phenomena are generated there that are beyond their competence to recognize, explain, or govern. When biologists in turn find that human science, superposed on biology, claims to escape biology, *Déjà vu*! Just as life is a countercurrent to entropy, culture may be a countercurrent to natural selection.

History, the cultural software that is run on the biological hardware, does make a critical difference. Consider, in the Western heritage, the rise of Israel, the crucifixion of Jesus, the signing of the Magna Carta, the rise of science, Martin Luther and the Protestant Reformation, the defeat of the Spanish Armada, the American Revolution, James Watt's inventing the steam engine and revolutionizing transportation, Abraham Lincoln's setting the slaves free, the discovery of nuclear weapons, or the computer revolution. Without genes, none of these events would have taken place. But with our genes none is explained in contrast to other events that might have taken place but did not – or in contrast to the different events that took place in other cultures elsewhere on Earth. None is explained, when it does take place, pivotal in history, as the outcome of selfish genes.

When Andrew Young was inaugurated as mayor of Atlanta, he reminisced that as a black child on city streets in New Orleans, dreaming that he might grow up to be Atlanta's mayor was beyond his wildest fantasy. As a result of their biological anatomy and physiology, blacks and whites have different skin colors that, in former environments, had survival advantages. In common, blacks and whites may both see four primary colors and avoid incest. Such things may be blueprinted in their genes. But the civil rights revolution was not. Young and the Atlanta citizens who elected him to office all have the same genes now they had in their childhood four decades ago; meanwhile they have adopted a new ethical position. The new position is for a more just and free society, and therefore a more stable one; ideally it is henceforth coded into the ethics and politics of Atlanta for better dealing with such challenges. But the change is not genetically determined, nor genetically transmitted; nor is it a universal behavioral mechanism; nor does it enable anyone to leave more offspring in the next generation. The evidence of history indicates that two human societies can be genetically indistinguishable on average, and yet the outcomes of those societies can differ radically, because the outcomes turn on the historically cumulative traditions in which they stand.

In writing a human science, it is plausible to expect that there are psychological and sociological traits universal to *Homo sapiens*. One should welcome insightful inquiries about the extent of these (Brown 1991). There is nothing irrational or disagreeable about there being a common human nature, so that across the countries and cultures, across the centuries and millennia, people can sympathize with each other, follow rational arguments, translate each other's languages, have a conscience, enjoy the fellowship of other races, and so on. These universals might well be due to their historical genesis in our evolutionary history, but they might be due to other causes. People universally believe that three sticks plus two sticks equals five sticks, but not because they all evolved from hominid primates. Some human virtues, and vices, might be transcultural. The universal religions have claimed truths relevant to all ages, to all conditions and races of humans. Ethical systems ought to be universalizable.

To some extent this universal humanity will be hard-wired into genetics; to some extent it will rest on a common possibility for cultural softwares that can be run on the common biological endowment. The first part of the story is how the hardware evolves; the conclusion turns on what software is run. One can write many budgets on one Lotus program, one can write new novels indefinitely on WordPerfect. Nature could build a billion species with twenty amino acids, just as humans can write a million novels with twenty-six letters in an alphabet. Culture stays linked with the biological hardware, but it opens up unending possibilities for cultural history.

When one has written such a human science, is our explanation of human life finished? No. Even a generic account of human culture falls short of narrating history along any particular world line. We cannot take universal cultural categories, apply them to the initial conditions existing at particular moments of history, and predict the course of the future. The pace of historical development in human culture is repeatedly compounded; past learning is folded into the present, and present discoveries are folded into the future, with technological, political, philosophical, moral, and religious decisions determining narrative world lines.

Events move one order of magnitude when elevated from physics and chemistry to biology. Continuing that analogy, events move a second order of magnitude when elevated from biology to human science. Events move now a third order of magnitude when elevated from nomothetic human science to idiographic human history. The historically accumulating story – the visions peoples have of their destinies, their choices, resolutions, achievements, and their good and bad faith – is determining outcomes in behavior, making the critical difference, whereas the universal human laws, though true enough, have limited explanatory power.

José Ortega y Gasset exclaimed, *"Man, in a word, has no nature; what he has is . . . history"* (1961, p. 217).[15] That is wrong if it forgets that humans evolve out of a nature in which they remain, and that nature itself has a natural history over evolutionary time. But Ortega y Gasset is right to insist that humans have made an exodus from nature into culture, where they are no longer determined by natural history; rather they make their cultural history. "Expressed differently," continues Ortega y Gasset, "what nature is to things, history . . . is to man." Expressed more accurately still, what natural history is to biological life, cultural history is to human life.

If A issues in B, one may, in some circumstances, take A as foundational and explain how B results from A, interpreted in terms of the categories through which we understand A. Given the positions of Earth, the sun, the moon, and gravitational pull on the oceans, tides result. Nomothetic scientific explanation looks for general theories and initial conditions, then follows a causal chain to find the results. Deductive logic, inductive logic, and probabilistic statistical analyses are key tools of analysis. The end is understood in terms of the beginning.

In other circumstances, one may take A as a developmental stage of what matures in B. The acorn produces an oak. "Genetic" explanations, taking *genesis* seriously (the root idea in "genes"), are as likely to be developmental as foundational. Otherwise, we can commit the genetic fallacy, holding that what now is in culture cannot be anything more than what once was, and still is, in biology. Developmental explanations can be cyclic: acorns, oaks, acorns. But narrative explanation follows story lines; initial events may or may not develop to historical conclusions. There are surprises and contingencies, emergents, critical achievements and discoveries, appearance of novel information, and, in human affairs, resolutions, intentions, and decisions. There are tragedy and failure, and beginning again. Where there is development, the beginning has to be understood in terms of the end.

Thus there is a story to be told from the Magna Carta through the

---

[15] Emphasis and ellipsis in the original.

American Revolution through the Civil War to the civil rights movement, the election of a black mayor in Atlanta, and on (perhaps) to the election of the first woman president in the United States, a story that traces increasing freedom, although there is no theory plus initial conditions, no deductive or inductive argument, no statistical analysis, that will yield from the Magna Carta as premise women's liberation and Young's election as conclusion.

We certainly wish to know the antecedents of human mentality across evolutionary history, and how this "resulted" in culture. If we put that sequence into a narrative framework, we want to interpret the beginning in terms of the end, not the other way around. That has to be done with care; taking emergence seriously may mean that one cannot explain the latter in terms unavailable earlier. Thus I am not going to understand guilt and forgiveness, which appear in humans, by studying trees and flowers, where neither has emerged. Similarly, I am not going to understand trees and flowers by studying rocks and minerals, where photosynthetic life has not yet emerged. Each level has to be understood for what it is, for the actual achievements reached to that point, but each has also to be understood in terms of its potential in developing story lines.

Explanation from below gets the explanatory sequence backward. It assumes that biology is the higher-order premise, from which the lower-order phenomena (cultural behavior) can be shown to follow. This is like assuming that physics is the still higher-order premise, from which the lower-order conclusion (biology) can be shown to follow. But in fact the culture follows the biology not inferentially but dramatically. One does not know all about atoms and chemistry until one knows what they become in biomolecules. One does not know all about nature unless one advances to psychology and the capacity for felt experience. Nor will I understand culture, if I remain in natural science. Story explanation is always of this form; in narrative the later events interpret the former. In the chapters that follow, we will explore this historically creative "genetic" explanation, trying to explain nature as an antecedent to, rather than as a determinant of, culture.

Wilson claims, "We are biological and our souls cannot fly free" (1978, p. 1). Is this an aphorism that stimulates thought or a confusion that hides equivocation? The claim seems modern and scientific, rejecting a supposedly archaic medieval or Cartesian dualism of body and soul. But it also rejects any liberating breakthroughs in the development of culture out of biology. Elliot Sober replies, representing a developing consensus of both philosophers and geneticists, "Biological selection produced the brain, but the brain has set into motion a powerful process that can counteract the pressures of biological selection. . . . Natural selection has given birth to a selection process that has floated free" (Sober 1993a, p. 215). The biological journey, from matter to life, on from protozoans to primates, is startling enough, and then, among the primates, one species roams the world. Humans are free to make discoveries and to transmit information neurally, past the limitations of genetically based information discovery and transfer. Humans are only part of the world in biological and ecological senses, but they are the only part of the world free to orient itself with respect to a theory of it. A culture is, in essence, such a world orientation, cognitively and behaviorally.

Chapter 4

Science: Naturalized, Socialized, Evaluated

On the path we next travel, we explore three principal fields of achievement in cultural history, events on a storied Earth with its plural histories, which began with the achievements in evolutionary natural history. In any philosophical overview of both the natural and the cultural landscapes, one must take account of the genesis of science, ethics, and religion, for all three are undeniably now present on Earth in considerable force. One must explain the remarkable creativity with which humans find themselves surrounded, but this creativity also reaches within to include our own human activities. Up to this point, genetic creativity has been our focus, but this may be only one kind; cultural creativity may be another, including its scientific, ethical, and religious forms.

Evidence of such scientific creativity is immediately at hand in the genius (Darwin, Mendel, Watson and Crick) that has discovered genetic creativity. After all, the place we learn about genes and their genesis is science. Persons at all times and places can look around them and see that something has been somehow created; they can puzzle and speculate about this. But only with the genius of science do humans learn of the remarkable genesis that has taken place in evolutionary natural history. Two things now need to be fitted into a comprehensive worldview: such natural history generated, and the human mind, generated out of such natural history, with the capacity to generate science: to look over, understand, and evaluate such events. In this sense, science too, like ethics and religion, is among the humanities – a human enterprise needing appropriate evaluation for what it reveals about nature and about culture.

Science, we argue here, is a new chapter in the generation and distribution of value on Earth, with antecedents in natural history, with benefits to humans in their cultures, and yet with a logic transcending both nature and any specific culture.[1] In chapters to follow we continue with ethics and religion to see whether they too, differently than or similarly to science, transcend their biological frameworks, at the same time that they understand and evaluate such biological origins. This order may reverse expectations. Religion and ethics, it might be thought, have had evident survival value, enabling people to function in their ten thousand cultures across the millennia. But science is of recent Western origin; only a few cultures have been scientific.

Such scientific culture is quite transcultural (many say); here humans break through for the first time to objective knowledge. In Darwinian science, when people learn about the genetic leash, they gain the power to break it. Science should come at the end of the story. But science is never the end of the story, because science cannot teach humans what they most need to know: the meaning of life and how to value it. The sciences are as practical as theoretical; science has evident survival value, teaching us how to gain benefits that we desire. But what ought we to desire? Our enlightened self-interest? Our genetic self-interest? More children? More science? The conservation of biodiversity? Sustainable development? A sustainable biosphere? The love of neighbor? The love of God? Justice? Equity? Charity?

Those more comprehensive questions prove to be tough questions that do not have completely scientific, much less biological answers. Medically, we can abort fetuses, but should we? With biotechnology, we can produce much food, but is it equitably distributed? We can use science to save endangered species, or to transform tropical forests into farmlands for a growing human population, but which is better? Science, ethics, and religion all have to do with sharing what is valuable; science is itself valuable and enables us to generate more value.[2] But science alone does not teach us all we need to know about sharing values.

For all its recent brilliance, science has proved penultimate to ethics and religion. The theoretical sciences typically are supposed to be value-free, and the applied sciences are notoriously deficient in orienting values; science is know-how without know-whether. Sci-

---

[1] Science brings costs as well as benefits, but a cost-benefit evaluation will require ethics and perhaps also religion.

[2] Or perhaps to aggrandize ourselves by exploiting the values of others.

ence describes what is (or was, or will be), not what ought to be. Scientists, qua scientists alone, are not ipso facto wise. After science, we still need help deciding what to value; what is right and wrong, good and evil; how to behave as we cope. The end of life still lies in its meaning, the domain of religion and ethics.

Science does teach us a great deal, however, both theoretically and practically, and one needs a philosophical account. Science ought to give us an important example of what kind of knowledge the mind is capable of attaining, and there we will have to demand of evolutionary theory that it give account of how such knowledge-acquiring capacity arose or, failing such account, that it admit its incompleteness. Perhaps, although we do not know how such capacity once arose, we can still give a scientific account of how scientific knowing today operates. A frequent, plausible (and fashionable) hope is to naturalize epistemology.

Science has been increasingly impressive in its explanatory categories, a powerful probe into the nature and origins of things, illustrated in astrophysical cosmology, and in evolutionary biology, our particular interest. Science describes the chemical evolution of life, its biological elaboration with increasing diversity and complexity over the millennia, and, at length, the evolution of brains, with a result in human minds' creating their cultures. Out of those cultures, especially Western European cultures, science has been generated. Scientific explanations have become increasingly naturalistic – at least those in the natural sciences have – and many seek a Darwinized or naturalized account of the human events the social sciences study as well. That introduces a still further question, whether the discipline of scientific explanation, the logic of a naturalized science, can be turned back on itself, so to speak, and become self-referential.

Science is indeed our latest and (many say) our most brilliant cultural achievement, and we can expect it best to test any comprehensive Darwinized theory of everything. In science the human genius (*Geist*) is undeniably displayed (in Galileo, Newton, Darwin, Einstein). There are many sciences, with differing logics, but they share much in common. If we can find some logic for science with reference to its biological bases, perhaps we can gain clues for the logic of religion and ethics. If the sciences rise free from their biological roots, so too may ethics and religion. Perhaps science can critique or complement the other two. These three great cultural behaviors – science, ethics, and religion – might each and all stand or fall together.

The names of the sciences have, nicely, a double connotation and denotation: the discipline of study, carried on by humans, and the events in nature, carried on without humans. Humans live in a world in which there is first a "physics" in the stars, in planetary motions and Earth processes; and later a scientific discipline that describes such phenomena. "Biology" goes on during laboratory exercises and field trips; one finds it in textbooks and journals. "Biology" also goes on in the life metabolisms and ecological relations "out there" for millennia before humans evolved and that continue independently of humans. The challenge is to explain both the subjective science and the objective processes that science studies, what takes place in scientists' heads and its relation to what takes place in natural history. One needs to get biology in both senses related to biologists. One must ask whether the biologists with their biology, who have so competently discovered natural history, are competent to do this, without crossing over, sometimes unawares, into philosophy.

The analytical capacities of the human mind are quite startling, compared with the mental capacities of the other five million (or so) species generated by evolutionary natural history. Max Delbrück, the father of molecular genetics and a Nobel laureate, whom we earlier heard puzzled about the historically unique genesis in biology (Chapter 1, Section 8), finds even more deeply puzzling how human rationality has evolved out of natural history, selected for better surviving in the jungle, for producing more offspring, yet providing an exodus by which we transcend our origins to probe the depths of the universe:

> Evolutionary thinking ... suggests, in fact it demands, that our concrete mental operations are indeed adaptations to the mode of life in which we had to compete for survival a long, long time before science. As such we are saddled with them, just as we are with our organs of locomotion and our eyes and ears. But in science we can transcend them, as electronics transcends our sense organs.
>
> Why, then, do the formal operations of the mind carry us so much further? Were those abilities not also matters of biological evolution? If they, too, evolved to let us get along in the cave,

how can it be that they permit us to obtain deep insights into cosmology, elementary particles, molecular genetics, number theory? To this question I have no answer. (1978, p. 353; cf. 1986, p. 280)

Humans do seem to be an exceptional species.

Einstein agreed: "The eternal mystery of the world is its comprehensibility.... The fact that it is comprehensible is a miracle" (1954, p. 292). Indeed, the question how the mind comprehends such an intelligible world is among the most challenging that the intellectual community now faces, but others do suggest answers, which we must consider. Look first at what goes on in the generation of science.

1. SCIENCE: GENERATING AND SELECTING THEORIES

Whether there exists an overall scientific method is open to question, since the procedures of electronic engineers, astrophysicists, plant taxonomists, sociobiologists, and geneticists are so diverse. Still, in a general way, scientists share a common procedure: they generate hypotheses and test them. Elaborated hypotheses are theories. Depicting this in a schematic (Fig. 4.1), a scientist faces a problem to be solved, as a result of puzzling over facts, and proposes a theory. This is followed by deduction back down to the existing facts, and to further empirical-level expectations, which then are related back to observations to confirm or disconfirm the theory, more or less, and to generate revised theory, from which new conclusions are drawn, after which the facts are again consulted. Science operates in an if-then mode logically, which is a generate-and-test mode practically.

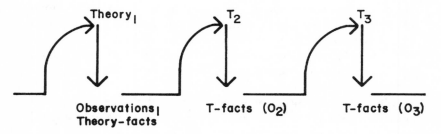

Figure 4.1. Generating and testing scientific theories.

This is sometimes called the hypothetico-deductive model (Hempel 1966), of which we are using an expanded version, and noticing already that a theory comes to have a developmental history.

How such theories are originated, as distinct from their subsequent verification, has proved troublesome to analyze, not surprisingly, since much of the creativity in science lies here. This context of discovery is just as important as, and, for those interested in the genesis of ideas, more interesting than, the later context of justification. Given a certain set of observations, what theory will fit them? In cataloging natural types or in formulating simple regularities one is tempted to say that science works by induction. Here the contribution of the scientist can seem minimal. But the generating of theories is more complex; the scientist "comes up with" models and abstractions, such as "lines of force in an electromagnetic field," or "covalent bonding," or "black holes," or "natural selection," or "dominant and recessive genes," or "kin selection," concepts that no doubt arise from mulling over the data, but in which the scientist also contributes creative hypotheses that require the stroke of genius.

Albert Einstein reported that he initiated his relativity theory, partly at least, "in vision" late one night, and he greatly emphasized the free play of the imagination, first and charismatic, only later to be put sternly to observational test (Marianoff 1944, p. 68). Hans Adolf Krebs, on the other hand, reported a long and steady step-by-step deciphering of the citric acid cycle (Stryer 1975, pp. 327–328). Both elements are present in Darwin and difficult to separate (Gould 1979). There is much inspiration whenever a fertile hypothesis is born. Neither induction nor deduction is sufficient. Although some tips can be given by others who have succeeded, there seems to be no recipe for cooking up theories, perhaps necessarily so proportionately as this is creative. Revolutionary science is more demanding here than is normal science.

The test of theory comes with its verification. Given a theory (T), what observations (O) follow? Here deduction is in order, at least in a broad sense; logic leads from premised general principles to particular conclusions. In the mathematical phases of science, where one has formal laws and initial conditions, this can be exact and necessary deduction, but elsewhere it is less so. Atomic theory is only partially metric, and what could be deduced from the atomic table about the properties of as yet unfound elements was suggestive and imprecise. Often a theory permits the deduction only of a range of possible alternatives, and scientists must sometimes deduce in a weak, nontight sense. Still, a fertile theory will accommodate existing data and suggest new observations that can be made to check it. Here the scientist often presumes that his or her logic is paralleling a causal chain; that a cause, in a lawlike way, produces an observed event, the narrower sense of the hypothetico-deductive or covering-law model. But the principle more broadly includes whatever particular events or observational structures follow from general theoretical models: If T, then O. Given: O. Therefore: T.

This procedure commits the logical fallacy of affirming the consequent, since some variant theory (T') might as well or better explain the observations in question, and the history of science is replete with examples of such theory replacement. As we will be emphasizing, most trial theories fail the test, later if not sooner. There is no logical route from fact "up to" theory; that requires the creativity just noted. The logic must be from the theory back "down to" the facts. So, although technically fallacious, science has no better way to proceed. On the other hand, if the observations fail (not-O), then the theory might seem refuted, by modus tollens, an elementary principle of valid argument. This asymmetry has led some scientists to concentrate on falsification, counting disconfirming instances as more weighty than confirming cases (Popper 1968).

In actual science positive observations do in some way tend to establish the theory, although it is difficult logically to specify just how. It is tempting to say that positive observations by induction render the theory probable, while conceding that this is never hard proof even in science and cautioning that future predictions from the theory involve a kind of backing into the future. Positive observations corroborate or strengthen the theory, although they cannot clinch it. Scientists get no proofs for their big theories; they get at best plausibility arguments. The negative observations that first appear to offer hard disproof also soften on closer inspection. Theories are not tested purely and simply but in conjunction with various presumed or unknown intermediate factors, called auxiliary hypotheses (A), those pertaining to instruments, to irrelevant or absent influences, and so forth, and one can typically adjust for upsetting circumstances so as to salvage the central theory. The scientist has to figure out whether the problem is in the auxiliary belt of surrounding hypotheses or in the body of the theory itself. Every theory is held in the face of certain anomalies, margins of error, and so on.

In more complex and partly established theory there are large amounts of confirming and some disconfirming observations, and one has to decide just how good the evidence is. That decision is rational, perhaps progressively corroborated as science settles into a theory, but often it is significantly discretionary. Every comprehensive theory has got to argue away some of the evidence it faces. Sometimes we do not believe the theory because it is not confirmed by the facts, but sometimes we do not believe the "facts" because there is no theory that confirms or predicts them and they go against a well-established theory that we have. Meanwhile, an anomaly makes a poor logical fit in what theory we do have. Experiments can be quite repeatable and quite wrong, where the conceptual framework repeatedly gives you the wrong result. (You can step on a bathroom scale and get 150 pounds every time, when your weight is really 160 pounds.) Theories cast light but may also put some things in shadow.

The evidence for the big theories, which make any philosophical or ethical difference, is never of the here-and-now, before-your-very-eyes sort. What counts for a good theory is its ability to draw together and make sense of the available experiential material, and in this the relationship between theory and observation is often indirect and interactional.

2. SCIENCE NATURALIZED?

After this survey of science and its method, we turn to the genesis of science, to press the issue of a naturalized science. Peter Skagestad remarks: "Man has generated a novel mode of evolution – the evolution of culture, including that of science. . . . The crucial question of evolutionary epistemology is the question of how evolution by natural selection was able to generate, in one biological species, a mode of evolution *not* operating through natural selection, and yet contributing to the survival of the species in question" (1978, p. 620). If not through natural selection is there some broader sense in which the development of science can be naturalized (Bradie 1986; Wuketits 1984)?

Darwin's theory is one of variation, reproduction, selection, and success, and if we find these elements in science, then perhaps science can be naturalized. But science involves such processes at levels other than the genetic or the natural; they operate at cultural and rational levels. Perhaps Darwin's theory needs to be incorporated into a fuller story? Here we will carry forward our argument that the better gestalt is that something of value is being created, defended, distributed, shared, and thereby surviving. If not genetically, how are the events that generate the sciences to be related to the genesis in natural history? Getting clearer about this can also help us understand, when we look back from this deepening perspective, whether the precursor natural processes were as "selfish" or "blind" as was earlier alleged.

(1) Darwinized Science?

Science evolves over historical time. Concepts come into being and may pass away; some "survive" and others do not; and there can be competition between ideas. Some win; others lose; still others get transformed (evolve) into new forms. Is this evolution of science illuminated by natural selection theory? T. H. Huxley observed, "The struggle for existence holds as much in the intellectual as in the physical world. A theory is a species of thinking, and its right to exist is coextensive with its power of resisting extinction by its rivals" (1880, p. 229).[3] Karl Popper claims: "The growth of our knowledge is the result of a process closely resembling what Darwin called 'natural selection'; that is, the *natural selection of hypotheses*: our knowledge consists, at every moment, of those hypotheses which have shown their (comparative) fitness by surviving so far in their struggle for existence; a competitive struggle which eliminates those hypotheses which are unfit" (1972, p. 261; D. T. Campbell 1974).

Stephen E. Toulmin finds that science typically follows an "evolutionary" model. "Science develops . . . as the outcome of a double process: at each stage, a pool of competing intellectual variants is in circulation, and in each generation a selection process is going on, by which certain of these variants are accepted and incorporated into the science concerned, to be passed on to the next generation of workers as an integral element of the tradition" (1967, p. 465; Toulmin 1972). Interestingly, the same mathematical models can be used to analyze the spread of epidemic diseases, when microbes repro-

---

[3] Several earlier scholars had occasionally noticed such an analogy between the development of science and biological evolution, among them C. S. Peirce and William Whewell.

duce themselves, and the transmission of ideas, when a new theory swiftly gains widespread acceptance (Goffman and Newill 1964; 1968).[4] Dawkins posits memes, analogues of genes, and these include scientific ideas (Chapter 3, Section 5). Computer scientists solve problems with genetic algorithms, patterned after natural selection, using random variation and retention of the best surviving solutions (Chapter 1, Section 5).

But we must be careful. Are the selection processes really similar? Do we have two different theories, or two versions of the same theory? Have we two species of coins or two sides of a single coin? What is competing with what, who with whom? What is being reproduced, selected, and surviving? Are the ideas reproduced as the offspring are reproduced? The question whether the sciences bear interesting analogies to natural selection with genetics in wild, nonhuman nature (addressed here in Section 2) is a different question from whether the sciences enable humans, including scientists, to have more offspring contending with nature or in competing cultures (addressed in Section 4). Science Darwinized in the selection-of-best-ideas sense need not imply that science had been, or could be, Darwinized in the selection-of-most-offspring sense.

There are indeed significant analogies between genetic learning and neural learning, and an analysis here can illuminate the processes of creativity in both nature and culture. The genetic mutation is a trial "idea" (*idea*, form, type), often neutral or detrimental, but sometimes beneficial in function because it enables the organism to handle itself better in corresponding with its environment. If so, it is selected and the result is better coding for better coping. The biological organism, in its genetic heritage from the past, is a programmed cybernetic center, under informational control. It is also, in its prospect for the future, a nonprogrammed learning center, "programmed" to the open search for something else, to an improved fitness, or sometimes to a little advance toward complexity, mobility,

---

[4] But little may follow about the similarity of these two processes. The same statistical bell curve can be used to analyze data scatter in nuclear physics and the average length of time that workers in factories stay on their jobs, as well as thousands of other statistical phenomena, but we do not suppose they are all closely related. The same fluid dynamics equations can be used to analyze river currents and traffic flow patterns on freeways in cities, but, in any fuller analysis of what is going on, the differences are more important than the similarities.

sentience, or intelligence. On its cutting edge, the speciating process is drifting, but drifting through an information search and edited for its discoveries of information. The editing is for survival, but it also scans and produces new arrivals, some (though not most) of which climb toward complexity and sentience and evolve brains.

In at least one species of primate, the brain becomes complex enough to sponsor mind, and culture; some of these minds become scientists. Throughout, the production of variants, mostly proving to be errors, produces knowledge. The whole system is a context of instruction. Deliberative thought, including that in the minds of these scientists, is also the launching of many trial ideas, and the selective testing of these in experience. ("Cognitive" is from a Latin root, *cog-ito*, to think; it is formed from words meaning "to shake together." "Intelligent" is from a root, *intelligo*, "to select among" [Hadamard 1949, p. 29].) Most such innovations are abandoned; very few prove worthy to be transmitted to posterity. In the invention and engineering of the internal combustion engine or the aircraft there lie abandoned a thousand dreams and attempts for every component that we now inherit. Similarly, there were eliminated a thousand mutations for each one now coded in our DNA. In that sense the entire scientific enterprise moves by throwing forward hypotheses on the forefront of experience, by testing these, and by preserving only those few that succeed.

(2) Genetic Heredity and Neural Heredity

Yet there are remarkable differences. An idea is not an organism; it is an idea in the mind of one specific kind of organism, *Homo sapiens*, who is embedded in transmissible cultures. An idea does not modify itself, but its owner may, and the idea may get deliberately (or, sometimes, accidentally) modified in the course of its transmission.

Consider the language that makes this possible. Language, involving the neural transmission of ideas, is requisite for all culture. It figures especially in scientific thinking, with its literature, logic, criticism, cumulative education. Human language is importantly different from the genetic "language": there is complex syntactic structure; there are subjects, predicates, objects. There is case: nominative, accusative, dative. There are imperatives, questions, performative statements. There are nouns, pronouns, adjectives, verbs, adverbs, prepositions. There are active and passive voice; present, past, future, and other tenses. There are subjunctives and conditionals. There is a deep grammar (Chomsky) that is not present in genetic transmission. All this makes argument possible.

Perhaps there exists something like a genetic "grammar"; certainly there is complex organization of information in the genes. Enzymes and catalysts serve as modifiers. Critical molecular conformations are receptors that define their biochemical targets, but the syntax and semantics in human language are novel in their achievement. There is abstractive, discursive, inferential, representative, performative power unprecedented in the genes. Genes do not talk to each other, though the genes within any one organism must coact with each other functionally. There are sexual genetic exchanges (crosstalk, if you like), but genetic makeup is fixed at birth. Humans do talk to each other and make up their minds accordingly, exchanging and changing ideas, sometimes doing this critically, not only in science, but also in ethics, religion, and other disciplines.

Genetic sets are preset at conception when they are transmitted biologically.[5] Despite any necessary grammar, in language there is an openness, the free assignment of meanings optional to diverse cultures, which is not preset by human biology. This is proved by the bewildering variety in thousands of human languages, historically derived, which differ even within particular cultural traditions. Yet all these language variants are superimposed on a common human genotype. Although infants in their cultures "naturally" learn to speak, an infant taken from one culture, innate genetic set notwithstanding, can easily be taught the language of another culture. Language is embedded in culture as much as in nature.

A form of language especially well developed in some sciences is mathematics. Humans have the neural power to add, subtract, divide, multiply; to do fractions and percentages, to measure angles and compute areas and volumes; to work with logarithms; to use coordinate systems and transformations; to integrate and differentiate; to solve equations. Such mathematics can be used to communicate with other humans, to argue, to demonstrate, to persuade; indeed quantified argument is especially persuasive in the sciences. There is control of quantity in genetic systems, when genetic information regulates biochemical functioning, and the mathematics that

---

[5] Though there are somatic mutations, especially in plants, where the gene line is not separated from the somatic line.

humans devise can be overlaid on natural systems to find that nature is mathematical. Genetic transmission has no analogue to the mathematizing of language that makes theoretical scientific analysis possible.

Consider other differences. In the evolution of science, there is no analogue to sexuality. In natural history, sexuality offers the capacity to interweave the transmission of otherwise branching lines, the capacity to exchange information. In culture information is transmitted and exchanged neurally, and hence the sexual component is left behind; so there is no fifty–fifty splitting of inherited information, as at meiosis.

At the neural level ideas are produced in one place (in human minds, in brains, ultimately at the microscopic level in neural circuits) and tested in another place (in macroscopic human behavior in the world, in society), but this is not an evident analogue in science to the genotype-phenotype distinction in biology, because of the two-way flow in the one case and not the other. Ideas from outside flow into the mind and remake it throughout any one human's lifetime, during which the genotype is unchanged. As these differences accumulate, we can begin to see how there have been breakthroughs that make the sciences difficult to Darwinize, either in the human-offspring-maximizing sense or in the ideas-like-offspring-surviving sense.

(3) Randomness and Intentionality

Perhaps the most critical difference between genetic and neural creativity, especially as the latter is manifested in the analytical language of the sciences in Western cultures, involves intentional action. Even though the genetic processes are under constraints on the space to be searched, and though "smart" genes use randomness to generate clever problem-solving algorithms (Chapter 1, Sections 3–4), the genetic process is nondeliberate, random, or "blind" (no conscious envisioning of the future). There's "nobody there" thinking about it. By contrast the brain-based scientific process is consciously rational and deliberate. Natural selection is radically transcended because scientists "know what they are doing," whereas the genes do not – not at least in this intentional sense.

This is true, and impressive, but the fuller account requires a closer look, finding both continuities as well as differences. By the lack of deliberation, we do not just mean that no agent is thought-fully overseeing genetic events; we mean (it first seems, at least) that there is no cybernetic program homing in on any goal. The genetic variations that bubble up, resulting from mutations, recombinations, and accidental reshufflings (crossovers) of the genetic material during mating, are independent of all needs of the organism, both skin in, that is, somatic and metabolic needs, and skin out, that is, ecological requirements and survival skills in the niche that the organism inhabits. The variations are not induced by either the organism as a whole or the environment favorably to the organism. They just happen at the microscopic level and have their "blind" effects at the macroscopic level, where they are mostly worthless or detrimental, rarely beneficial.

A later variation (so we might think) is not linked to an earlier one by any trend that is intrinsic to the microscopic biochemistries; the variations are not orthogenetic but a random walk. Nor is a later variation linked to an earlier one by any feedback loop, so that a later variation is corrected by the direction of error of the earlier one. The result (it seems) is that there is just aimless scanning, and, rarely, slowly, the occasional organism lucks into a beneficial result, while the myriad most survive with neutral mutations or limp along with detrimental mutations. When such a beneficial mutation does rarely happen, natural selection locks on, and this mutation is preserved in the genetic coding.

Also, there is no means through which any beneficial acquired characteristics, stored neurally in the memory and on recall during the organism's somatic life, can instruct the genes with the result that this information is passed on to the next generation.[6] Unless the organism can pass this information behaviorally (as is rarely the case in the animal world), acquired information simply collapses with the death of the organism. The lack of feedback from whole organism needs, from environmental needs, or from acquired traits – all this seems in stark contrast to scientists doing their science. Scientific creativity is radically different.

Scientists routinely state the nature of the problem that needs to be attacked. They start by reviewing what others have learned, they

---

[6] There does seem to be some genetic inheritance of acquired somatic characteristics involving immune systems (Gorczynski and Steele 1980), but this does not involve neurally acquired information.

do their research, and a standard conclusion suggests what research needs to be done next. Models and paradigms focus our attention on likely revisions (mutations) of the theory. Conceptual hypotheses can be quite diverse, but scientists do not seriously examine all of these (for example, that an offended witch caused the anomalous meter reading); scientists are guided by heuristic rules that track in likely directions. They are directed by their various research programs. Without such intentionality, can we still profit from considering the analogy with evolutionary natural history?

At the level of the developing organism, there is much feedback and directional development. During embryogenesis, the developing fetus checks performance against heading, and switches control enzymes on or off accordingly. The organism, throughout its lifetime, maintains its metabolisms constantly. Though intentionality is present in only a few species of organisms, all of them have a "life program" maintained genetically – teleonomic, though not deliberatively teleological.

In evolutionary history, as soon as enough mobility arises, the organism begins to select its environment as much as the environment selects the organism. The animal moves to another place. Since some sort of search program is possible in all motile organisms (any with flagella or cilia, or fins, legs, wings), this is quite pervasive through the faunal world. In simple organisms, random directional trials may find, first by accident, but then lock onto, a gradient into a new environment, where there is more food. In neural organisms, where there arises enough psychic life to drive a search for a home environment, the focus is intensified by a subjective life. The warbler "likes" (has satisfactory experiences in) the treetops and "dislikes" (is uneasy and frustrated if it finds itself in) the grasslands. The coyote hunts prey or returns to its den, mixing genetically based instincts with perception of cues in the world, and corrects its trials and errors with feedback information. Though there is a sense in which the operants ventured are blind – and regardless of whether coyotes, warblers, or amoebas think much about it – these feedback loops put a cognitive focus on the search. The organism takes on, as it were, a research program; it concentrates on learning about and achieving an adapted fit in a particular niche.

Ecosystems, although not under any centralized neural or genetic control, are sometimes equilibrating communities in which there are feedback loops that stabilize the system. When ungulates over-

174

populate, their predators or parasites increase, or unpalatable forbs and grasses replace palatable ones over their range. Then undernourished yearlings do not survive the winter, and this trims the population back to carrying capacity. So one cannot say that there is no directional feedback in biology, or that this is only found in deliberated science. Though they are "merely" metabolic and physiological, that is, not self-conscious, these processes can be cybernetic.

But what about evolutionary development? Here the innovative part of the process seems especially random and blind, in contrast with scientific exploration. Incremental blind trials (making a mutation at random in the cogs and wheels of a watch) fail with high probability, but deliberated trials (replacing a gear that is failing frequently with one made of stronger alloy) often succeed because they are made with an overview of the entire mechanism and an analysis of where the problem area is located. Blind trial and error is devoid of any gestalt that controls educated guesses about what improvements in theory or practices might work and why. Incremental, deliberated experiments are controlled from the top down, holistically, by an overall pattern that is partially already in place or envisioned. By contrast, an incremental genetic mutation bubbles up from below at random with regard to the whole. Scientists do grope, but they can and must grope for an overall pattern in terms of which they can structure a theoretical understanding and form a set of laws or an integrated theory.

In natural selection, if it can't be done incrementally, it can't be done. Development can only be by piecemeal modification. The organism has to work while it is being made. The evolving organism has to be able to get there from here, step by step, with the organism alive every step of the way. Not only does the species have to live during its evolution, it has constantly to compete and win. This demand for immediate advantage can prove a long-term disadvantage. In natural selection, it is difficult to get from local optima over to much better optima some distance away, because these cannot be reached incrementally without a downhill traverse over nonoptimal paths. So we get not the fittest possible, only the fittest that could be reached incrementally and historically from where ancestors had managed to get. There is "optimization subject to constraints, and in particular the constraints of historical contingency" (Lewontin 1987, p. 158).

If science were like this, the only boats that could be built would have to float while under construction; the only airplanes that could fly would have to be assembled up in the air. Even more, it is as if these boats and planes had to race while being made. But engineers can envision an entire boat or plane, plan it, assemble all the parts, and then turn on the switch and start it up. If it fails to perform as hoped, if it needs repairs, one can shut it down, bring it into the shop, and overhaul it completely. That does make possible radically new capacities for engineering.

Nevertheless, the engineering form of creativity, appropriate for artifacts and machines, may not always be the better form, and it may not entirely characterize live scientists in their research programs, since these go on "in their minds," which are in bodies that do have to stay alive while the science is being developed. Scientists do not engineer their artifacts this way, but scientists (and philosophers or theologians) can themselves only be made out of newborn infants, reared, nurtured as children, educated in college, learning the "art" of science in the culture they happen to be in, applying for grants and fellowships, taking some wrong turns as well as some right ones, groping for success, having to get there from here, struggling to become successful scientists, as well as to stay alive over decades. Some of the artifacts they make with their sciences can be engineered overall and then switched on, but the science that is formed in a living scientist's mind does have to be integrated with the personal survival process.

There is required historical development both in natural history and also in human science. There are many vicissitudes in both, wayward turns as well as achievements gained. The big picture is development along a story line. Lives have to be narrated, not engineered. Scientists may engineer their artifacts, but the lives of scientists (and all human persons) have to be biographies. Life has its revolutions and conversions, its dramatic crises; still it has to be lived incrementally and vitally day by day. Robots can be assembled and switched on; persons have to be assembled while they are living. The continuity with natural history is that such is the nature of all self-generation, of plants and animals as well as of people, including scientists.

We concluded earlier that the genes, if in some sense blind, are in still more interesting senses "smart" (Chapter 1, Section 4). The generated novelties are not formed in blind chaos: there are many genetic and enzymatic controls on the mutation rates, on the genetic reshuffling in successful blocks, and there are repair mechanisms. Only the nearby search space is searched. The genetic set is selected to permit versatility coupled with stability, saving what has worked in the past (coded in the DNA), but trying variations on it to see what might work better. Genetic vitality is in fact a rather sophisticated problem-solving process; many achievements there are not yet possible for scientists to duplicate. Genetic creativity is quite startling in what it has produced: many millions of species all the way from microbes to persons, coded for living in all kinds of environments. So we are too swift if we think that there is no research program in the genes.

Likewise, we are too swift if we think that there is no trial and error in scientific problem solving, no groping about in the dark. When R. E. Monro, a molecular biologist, reflected over the development of biology, he concluded: "An essential characteristic of scientific research, in its more revolutionary aspect is that the scientist is searching for the unknown or, in other words, he does not know what he is searching for" (1974, p. 119). Baruj Benacerraf, reflecting over his career in immunology, for which he was awarded a Nobel prize, agreed: "After more than 40 years in research and over 600 publications, I have learned that discoveries are determined primarily by chance observations and are conditioned by past experience and advances in technology" (1991, p. 6). The research papers describe an orderly, deliberated, simplified logic of discovery, proceeding from problem analysis to experiment, to data collection, analysis, and conclusions, but this is often a story that never happened. What did happen is far more complex, wandering, provisional, tentative, exploratory, lucky (Latour 1987).

The *justification* of variants, the testing of them, can sometimes be highly selective, but the *discovery* of variants, the generation of them, cannot be very selective and is perhaps not selective at all when one is really stymied about where to go next. In the midst of a search, novel ideas are often just stumbled upon by accident. Luigi Galvani happened to cause a spark near a frog specimen, which happened to cause the leg to jump, and electricity was discovered. Alexander Fleming happened to notice a Petri dish of staphylococci, which happened to be contaminated by a mold, and penicillin was discovered. Henri Becquerel was experimenting with fluorescence, incited by sunlight, wondering whether the sunlight could also induce X-rays. Bored during a series of cloudy days, he happened to put wrapped photographic plates into a drawer with potassium uranyl sulfate (containing uranium), to discover that they became fogged. Thereby he discovered natural radioactivity, destroyed the nineteenth-century conception of atomic structure, and launched twentieth-century nuclear physics.

Sometimes scientists stumble over things; other times they look hard for something their theory says is there. At other times, they abandon their theories, after too many phenomena turn up that the theory does not predict. The story of science, like so many good narratives, is the story of searching, but often too it is the story of lucky turns of events in puzzling situations, of surprising directions of resolution when conflict deepens (Beveridge 1957; Taton 1957; Shrader 1980; Simonton 1988).

The story is, we must also notice, repeatedly the story of failures, sometimes tragic ones (though "tragic" is more an evaluative than a scientific word). In 1842, Dr. John Croghan set up a hospital sanitorium inside Mammoth Cave in Kentucky hoping to cure tuberculosis patients in the near constant temperature and humidity of the cave. The experiment was a reasonable one in the light of what was then known, but it failed. Smoke in the cave made his patients' condition worse. His patients died, and he himself contracted tuberculosis and later died, a victim of the disease he sought desperately to cure (Sides and Meloy 1971).

"Generate and test" is standard scientific procedure, not only when computer scientists set up genetic algorithms, but regularly when they undertake research programs. Normally a scientist does want to search the nearby space for possibilities of development. On the other hand, in more radical research, a systematic search is a waste of time if you are nowhere near the zone of good answers, in which case a little random probing around in supposedly wild places may be a useful heuristic. In creative desperation, one tries anything novel; one cuts loose from the degenerating research program and casts around in what were hitherto regarded as unlikely directions, or directions in which one had before never looked at all. New ideas may be recombinations of old ideas, but they may come from places entirely out of the range of the old theory. When one goes beyond the range of what is already known or suspected, one proceeds blindly. There is now a kind of random trial and error, with most of the ideas worthless or irrational, but the occasional bubbling up of one that has promise. That rare, lucky idea is locked onto by rational selection, and then that science turns in hitherto unanticipated directions.

Nor can we assume that, though the context of discovery and generation has random elements, the context of justification, the testing, is always admirably rational. Rare and right but unlucky ideas get launched only to be ignored by a scientific community unprepared to hear them. Much depends on the circumstances of launching; the sensitivities of initial referees and critics; the academic posts and laboratories that happen to be involved; editorial and funding decisions; the wealth, health, and persuasiveness of the scientist–discoverer; the perceived relevance of the discovery in applied science; ideological implications; contingencies of timing; and so on.

Theories get misjudged because scientists are flattered or jealous, because they are in too much hurry patiently to digest the evidence, or because they are distracted by peripheral interests and convictions. Mendel's work in genetics was ignored (Glass 1974); an early molecular theory of gases by J. J. Waterston was said to be "nothing but nonsense" by the referee of the Royal Society, though it anticipated the work, years later, of the eminent physicists Joule, Clausius, and Clerk Maxwell (Strutt 1968, pp. 169–171). Alfred Wegener published a theory in 1915 that anticipated plate tectonics and supported it with much geological research, but he was ridiculed by his colleagues and died in 1930 an intellectual outcast. Half a century later, his idea became the paradigm that made geology a unified science. Many discoveries have been stillborn or smothered; we know only those that survived by mix of plausibility, push, and luck. None of these human foibles operates at the genetic level, but they serve to diminish the contrast between the rational in science and the contingent in nature.

Nevertheless, one way or another, on occasions there are profound redirections in science, and the really creative turns, hoped for and sought, are also unexpected. Deep revolutions have come in the sciences (electricity, radioactivity) and are still coming, and when they come they will entail unforeseen changes in the way we think. Darwin's creative discovery of the theory of natural selection and the incremental evolution of species, replacing the previous paradigm of fixed species, is, in this book you are now reading, stretching into our whole worldview. Darwin probably did not seriously entertain some sorts of explanation (that demons or astrological forces caused the speciation); he was looking for a naturalistic explanation. Nevertheless, within that focus, he was groping. What he found changed history. When, in the future, evolutionary theory is transcended, it will be by ideas that initially seem to be in the twilight zone. Such ideas, however, will be evaluated by rational selection, not by natural selection.

We began by making the point that there are forms of creativity available in the sciences that are impossible genetically. Science is a conscious process, being neural, whereas natural selection is non-conscious, being genetic. There are feedback and coupling in science that transcend any in biology. Both processes are cybernetic with elements of trial and error, such that the trials and errors are requisite to epistemic growth. When natural selection is elevated into rational selection there is a new chapter in the story of knowledge, but some themes are pervasive throughout both epistemic adventures. There is a narrative continuity as well as an emergent novelty. Science does transcend what goes on genetically; scientific epistemology is not, in any sense we have been able to find at least, yet fully naturalized, much less Darwinized.

The result in the sciences is a vastly intensified and even radically novel capacity for creativity, for making discoveries of value. A vital component in this is the way in which ideas can be communicated socially, flow back and forth in feedback loops, with their value being evaluated by self-consciously critical communities, any acquired and worthwhile ideas transmitted to others, kindred or not, and to future generations, kindred or not. In terms of our larger paradigm, there is "sharing" ("distribution," "dissemination," "multiplication," "reproduction") of such ideas and benefits that result, certainly something more than "selfish genes" programming an organism to maximize its offspring.

(4) Biological Diversity and Unified Science

Comparing science with natural history, there seems another relevant difference in the creativity in each. Over the millennia evolutionary history has gone from a first (or a few), simple form(s) of life to the many millions of species now on Earth. Natural selection diversifies and results in more complex organisms. By contrast science seemingly works the other way; over the history of a science the many speculations that exist at first, or that are proposed along the way, get eliminated in favor of fewer and fewer theories, until at length (if the science is successful) there is only one theory, and under this theory many complex phenomena are simplified and understood as outcomes of a small number of principles and laws. Thus millions of chemical reactions are all understood in the light of the one atomic table paradigmatic to the science. Millions of kinds of proteins are coded by one system, DNA and the triplet codons that sequence the twenty amino acids, used to assemble the myriad proteins. Scientific rationality unifies; natural selection diversifies.

Is this really a disanalogy? Often, such science is only backtracking what is happening in nature. In nature a few simple units regularly construct more complex things, level after level. Protons, electrons, and neutrons form into ninety-two atoms; these atoms generate myriad chemical compounds, ringing the changes permitted by the atomic table. They further generate the precursor biochemical molecules, amino acids; these further generate (now by cybernetic coding, exceeding anything explained by the atomic table) all the millions of kinds of proteins, which form all the millions of species of life, governed by the forces of natural selection. Humans also emerge out of this biological history and build their myriad cultures and live out all the stories of world history.

The sciences track out the simplicity that underlies all this, but science in so doing is only running backward the film of a story that ran forward in nature, and such science is quite as interested in describing the outcomes as the inceptions, the superstructures as the foundations, the diversity as the simplicity. The sciences sometimes learn last, epistemologically, what happened in nature first, ontologically, finding out what is at the bottom of the production of all these phenomena. Describing how the complex and diverse originates out of the simple is fully a part of the story of science. So there are both plurality and unity in both science and natural history.

Sometimes science learns late of the natural diversity, as when it rather recently discovered microscopes and the varieties of microbial life, which were long unknown, or when science today explores tropical forests and the depths of the sea to discover previously undescribed species of insects and diatoms. The interactions that govern a forest ecosystem have proved complex and elusive, and scientists have hardly yet figured them out. Science builds telescopes and sees galaxies once unknown or builds radio telescopes and detects radiation previously unsuspected. One theme in the story of science leaves the world simpler and simpler because it is explained by unified theories; another part steadily finds phenomena of which we were earlier ignorant, a nature that is bigger, older, more diverse, more multileveled and complex than we had suspected.

The sciences discover for nonscientists as well as scientists the myriad species and processes that natural history once produced on its own. In terms of what is to be valued, science is valuable, and not just because it helps humans survive. The sciences are valuable because they discover the powers in nature that have generated all the natural history out of which we humans have come. We stand in considerable amazement at this genesis, a natural history continuing in an ecology with which we continue to find our own destinies entwined. The scientist may be unable, qua unaided scientist, to reach that value judgment either about the cultural activity of science or about the natural history that science studies. The epistemology of science, naturalized or not, leaves that problem for ethics and religion.

3. SCIENCE SOCIALIZED

(1) Paradigm and Theory in Scientific Communities

Scientific facts quickly become theory-laden. Scientists first generate theories over facts but soon find that these theories generate new facts. When the engineer reports that the current through the meter is ten amperes, or the zoologist discovers that the vertebrates are related to the tunicates, the larval notochord of the latter and the spinal cord of the former having evolved from a long-extinct hypothetical ancestor, their facts come within and are partially products of their theoretical frameworks. Fabricated concepts and laws are used to trace and to classify natural events, and the facts so obtained do not come nakedly but rather filtered through these constructs. The geneticist maps a gene by back inference from statistical phenotypic expressions; the behavioral ecologist takes certain primate behavior as evidence of genes for infanticide. The biochemist decodes the amino acid sequence in a protein by observing certain colored stains or layers of material in an ultracentrifuge, or, more recently, by reading the output of a sequenator.

The theories are all conceptual overlays on nature, constructs, or, as we will term them more provocatively in the following section, "social constructs" within a scientific culture. The "center of gravity" in a rock is as much assigned as discovered. An atomic "particle" such as an electron is said to have a "spin," though also said to be like a "wave" and to have certain "quantum numbers." An "adaptive landscape" is a graphical way of representing the "living space" through which an evolving species moves, with "fitness peaks" to climb as the species increases its fitness. A species occupies a "niche," both a location and an occupation in an ecosystem. Still, many of these constructs seem to map some evident natural kinds; there are tunicates and genes, there were trilobites in the Cambrian period, and Yosemite's Half Dome is made of quartz monzonite.

Much of the creativity in physical science lies in its mathematical capacities. The whole numbers may seem natural enough until we add, divide, and multiply by zero and infinity, and with some artificial innovation must define what these operations will mean. Biological science is less mathematical, but it greatly depends on measurement and statistical analysis of phenomena such as gene frequency or birth and death rates. Much of its instrumentation is highly metric. Scientists invent differential calculus for their physics or adopt regression analysis for their biology. These mathematical abilities become part of the theoretical structure transmitted from generation to generation, shared by the community of scientists.

The point in science is to generate hypotheses and theories that mix with facts and processes appropriately, and not to pretend that they can be insulated from each other. The "facts" are always to some extent "artifacts" of the theory. The "facts" are preceded by creative "acts" that set up the facts. "Seeing" is universally "seeing as." Scientists interpret what they see in order to see it. To tell what is going on, to see what is taking place, observations are formed within gestalts. In science an event makes sense not merely as the senses register it but as it is found to be intelligible within certain established patterns of expectation, which, though partially sensory and biological, are significantly culturally constructed. The understanding cannot see and the senses cannot think; cognizing and perceiving are wired up together. This interpretive seeing is sometimes thought to contrast with hypothetico-deductive science but really complements it. Scientists come to see processes and objects as instances of types, universals, forces, natural kinds.

As these models become increasingly dominant, they become paradigms. Paradigms are governing models that, in some fairly broad range of experience, set the context of explanation and intelligibility.

Their holders have inherited them in the cumulative scientific culture in which they have been educated; they wish to conserve these basic referent theories so far as they can by using them to interpret new experience or, in the event of counterexperiences, by introducing subsidiary hypotheses that allow the theory's conservation by peripheral adjustments. Paradigms are abandoned reluctantly, especially if they have hitherto been quite successful. In some cases, it proves difficult to specify just what qualifies as a paradigm; paradigms have sometimes broader, sometimes narrower scope, and there may be a hierarchical interweaving of major and minor paradigms. But the basic idea of a controlling patterned seeing does seem to characterize the history of science.

Much of the creativity in science, as well as the controversy and upheaval, arises at periods of major paradigm shifts (Kuhn 1970). Notice too that the scientist is not entirely oriented here by cognitive knowing; by following the techniques and methods of his predecessors and peers a scientist also gets a "know how" to do, as well as a "know that" something is so. Also, there are tacit as well as explicit elements in how a scientist is controlled by a paradigm. A paradigm is a "disciplinary matrix" as well as a theoretical viewpoint.

This element in science is without serious analogy in natural history, although, of course, genetic dispositions set certain patterns in animal behavior. Earlier we noted that primates, though they are highly socialized, do not seem to be capable of higher-order intentionality but live in private mental worlds. They do not attempt to change the mind, as distinguished from the behavior, of other primates. They do not, in this sense, either teach or seek to be taught, with awareness of acquired ideas passing from mind to mind, that is, of an inherited paradigm framing what is understood, and of the possibility of creatively critiquing such ideology. Although they signal, primates do not, in this sense, communicate (Chapter 3, Section 1); or, we can now say, they are incapable of "sharing" thoughts from mind to mind, which is the genius that makes culture possible, and, most especially, a scientific community possible, as when one learns differential calculus or natural selection theory. Such a community will be of genetically unrelated persons, bound together ideationally. For all the similarities that science bears to natural selection, this disanalogy is quite telling. Newly emergent capacities have been generated, exceeding any precedent in natural history. This generation, distribution, and sharing of such paradigmatic knowledge within a community lie at the center of value in science.

(2) The Social Construction of Science

Such construction at once makes science possible, with genius and insight, and also makes possible a social filtering or coloring of nature. On the one hand, science is constrained by the objects and processes under study; the trial ideas ventured are tested against a natural world that is there independently of humans. On the other hand, science is constructed in the human mind, subject to human perceptual and conceptual abilities, as these have been elaborated, frequently instrumented, sometimes mathematized, in the cumulative scientific community, which is part of a larger cultural community. In the critical analysis of science, one has to ask not only whether science escapes genetic determinants; one has quite as much to ask whether science escapes social determinants.[7]

Science is a series of changing historically produced representations of nature. In such a socialized science, one has to be alert for the bias introduced by the social construction of interpreted "natural" events. Everyone from any culture "sees" lions; a zoologist will "see that" lion behavior is "territorial defense," and some particular lioness is the "dominant matriarch" in the pride. That seems true enough, even if such terms have been constructed. But sometimes scientists can read into nature as much as they read out of it. Dawkins sees that lions are full of "selfish genes," and we earlier noted how he takes this as an "admirable" symbol of biological nature as a whole, which is "red in tooth and claw" (Chapter 2, Section 2[2]; 1989, p. 2).

Three forces are thought to be principal determinants of what goes

---

[7] Some hold that science is a social construct of the West, with only presumed universal intent or objective knowledge; it is actually a technique that Europeans and those they have converted to their view have used to dominate and exploit nature and non-Western peoples (an especially potent survival tool for leaving more offspring in the next generation). Whatever the merits of such a view explaining some of the sciences, this seems an incomplete account of the truth claims in evolutionary and ecological biology, often describing events from which humans were and still may be absent. Genetics has been applied to conquer nature, but genes and DNA are also descriptive maps of objective nature.

on in biological nature: competition, predation, and symbiosis, or mutualism.[8] In a series on the three, Paul Keddy, author of *Competition* (1989), notices, after he finishes, that mutualism is rarely mentioned in ecology textbooks, whereas competition and predation are everywhere featured. He puzzles over this, since mutualism and cooperation are everywhere in nature (Sapp 1994; Dugatkin 1997a). The explanation, he finds, is "that scientists are heavily influenced by their culture (consciously and subconsciously) when they . . . select models to describe nature. . . . With respect to research in ecology, we may be projecting our own cultural biases upon nature rather than studying forces in relative proportion to their importance in nature itself" (Keddy 1990). Western science, for instance, particularly in the formative era of Darwinism, came out of a culture of conflict, colonialism, capitalism, conquest of nature, British empire; Darwin drew his main metaphor from the economist Thomas Malthus's ideas about human resources, overpopulation, and resulting social struggles.

In some future biology, a decade hence, the theories could be different, emphasizing perhaps the pride's cooperation, the harmonious balances between predator and prey, or the comparative unimportance of predators, or population control by parasites, or how the fate of the lions, at the top of the food chain, is more an accident of rainfall and thus grass for wildebeest to eat, than of successful competition, dominant matriarchs, selfish genes, or red teeth and claws. A scientific account today can be as culturally constructed – especially when metaphorically seeing organisms as "like Chicago gangsters" (Dawkins 1989, p. 2, summarizing his main argument) – as was seeing lions as "the king of beasts" yesterday, taken as the lordly power symbol of the British empire, or as the totem of an African tribe.

Frans de Waal complains that biologists regularly ascribe negative descriptions to primates; they are aggressive, or combative, or have enemies, or are selfish; they even cheat and deceive and are greedy. Such terms appear in scientific papers as acceptable terminology. But there is a simultaneous refusal to ascribe various positive traits to

---

[8] Technically, symbiosis is of three kinds: (1) mutualism, in which both organisms benefit; (2) commensalism, in which one benefits and the other is neither helped nor harmed; and (3) parasitism, in which the parasite benefits at the expense of the host.

them, to say that they cooperate, or have friends, or companions, much less that they share, or care for each other, or show sympathy, or are honest; any such language is unscientific and will be edited out of the journals. Biologists go to great lengths to argue away all evidence of animal altruism, interpreting it as disguised selfishness, or kin selection, or nepotism, and so on. But de Waal, in his studies of primate behavior, finds that they display an enormous spectrum of emotions and different kinds of relationships. Scientists ought to reflect this fact in a broad array of terms. If, for example, animals can have enemies, they can have friends. If they can hurt, they can help. The problem, he suspects, is that the scientists' socially constructed filter, ultra-Darwinism, prevents their seeing and properly interpreting behavior that is counterevidence to their theory (de Waal 1996).

Science does often work with analogies, or partial models, or makeshift sketches that scientists will replace, after more explorations, with better theories, and even the improved ones will be only approximate. Analogies can inform, and they can also misinform. Science can sometimes be framed by passing scientific fashions. So the problem, in gaining a plausible account of natural history and, simultaneously, a plausible account of the science with which humans have access to natural history, is to recognize that there is no unmediated nature, at the same time that one recognizes that the scientific media can sometimes reliably and descriptively transmit truths about what is there. Biologists do abstract, and this can result in failing to see what is left out of the abstractions. They invent the theories with which they see, and these may blind them to other things. But inventions can also help us see, as when "territorial defense" seems better to explain lion behavior than does seeing lions as a tribal totem. Scientists can regularly check their constructs against causal sequences in nature. All study of nature takes place from within some culture or other, but it does not follow that scientific study is not conformed to the objects it studies external to culture.

Does not Keddy move to correct the prevailing bias toward competition, because he is constrained by what is encountered, rather than just introduce a new fashion? So does de Waal. Better theories will result. Many well-established theories have already been so constructed; they have been tested against the constraints of nature; these do map what is there. Sketches they may be, because they focus on particular phenomena and model these only partially. Still, there is no feasible theory by which life on Earth is not carbon-based and energized by photosynthesis, nor by which water is not composed of hydrogen and oxygen, whose properties depend on its being a polar molecule. Glycolysis and the citric acid cycle, adenosine triphosphate (ATP) and adenosine diphosphate (ADP), will be taught in biology textbooks centuries hence, as well as lipid bilayers and immunoglobulin molecules. Oxygen will be carried by hemoglobin. Although biologists constructed these ideas, using many theories and instruments, they are right that CO_2 is released in oxidative phosphorylation and that this cycles through photosynthesis II and photosynthesis I, so that in the world there is a symbiotic relationship between plants and animals and that this is a vital ecosystemic interdependence (the mutualism that Keddy worries ecologists have ignored).

Even if some of these claims should be revised, as they will be, the general cluster of discoveries is not going to fail. True, the picturing of nature is only partial; we see it through a glass darkly. One doesn't have to know it all to know something. These claims are modest, specific, Earth-bound, even, if one insists, fragmented. They only catch up a part of a scene in which much else is going on, of which we are as yet unaware. They are mixed with error. They are not arrogant, universal (true in all worlds), total, grand, absolute. But they are still significantly true in that they describe what is going on here on Earth, objectively and specifically in Earth-bound organisms and ecosystems.

As such phenomena are more and more emplaced in a comprehensive picture of nature, the worldview is indeed always a construct, but, again, it does not follow that a philosophical overview cannot be better or worse in its account of events in Earth history. Science is socially constructed, and one of its insights is bringing a sense of solidarity with a larger biotic community with whom we humans share this Earth place. Nature may not be as given as naive realists suppose, but, upon finding this out, we make an equally naive mistake to think that science is so socially constructed that nature is not given at all. We are here seeking the best informed metaphors with which to interpret the genesis and the genes of evolutionary history, antecedent as these have been to the human cultural story. Our claim is that metaphors of the genesis and sharing of valued achievements are as fruitful as any other and serve well as a complement to currently fashionable metaphors. This is the hypothesis we are testing to explain both the facts and the values found both within ourselves and in the external world.

4. THE SELECTIVE ADVANTAGE OF SCIENCE

(1) Science and Survival

Those who think we should take Darwin seriously in every domain of life do not hesitate to claim that science is a biological product. Michael Ruse insists:

> The principles of scientific reasoning or methodology ... have their being and only justification in their Darwinian value, that is in their adaptive worth to us humans – or, at least, to our proto-human ancestors. In short, I argue that the principles which guide and mould science are rooted in our biology. ... Darwinian advantage reaches through science like bones through a vertebrate. (1986, pp. 155 and 161)[9]

Franz Wuketits concludes, "Since the human mind is a product of evolution ... the evolutionary approach can be extended to the *products of mind*, that is to say to epistemic activities such as *science*" (1984, p. 8, emphasis in original). Wilson opens his *Sociobiology* arguing that, since natural selection made the brain, an evolutionary explanation involving "winning genes" must be pursued "to explain ... epistemology and epistemologists, at all depths" (1975a, p. 3).

Though Ruse can sometimes be quite insistent about Darwinizing science, he can also be more ambivalent. On one page we read that "the humans-as-beyond-biology thesis was never that plausible" (1986, p. 147) and on another that "science as adaptive and science as beyond biology" is "a paradox" (p. 149). The paradox resolved is that "we get the tools through organic evolution. What we produce has a meaning of its own, transcending biology" (p. 206). With that he really concedes all for which we will be arguing about transcendent science (in Section 6 later). That seems to make the science-as-

[9] Ruse can hardly mean that the principles of scientific reasoning had adaptive value to protohuman primates who were not yet scientists. His claim must be that mental capacities that were of adaptive value to protohumans also prove useful today in science.

beyond-biology thesis rather plausible. There is an exodus from natural selection in science – at least in the meaning if not in the tools. And if there, why not elsewhere in culture?, as we will be asking in the chapters on ethics and religion.

There is something to be said for the adaptive significance of science. The argument is that general intelligence must be some kind of adaptation and that science is an especially impressive use of such intelligence and furthers such adaptation. Consider our protohuman ancestors. If one can "figure out" how to catch prey, knowing that tracks are a sign of an animal, and fresh tracks a sign of one nearby; if one can build a bow and shoot arrows, cook food, cover oneself with animal skins, one has an advantage over those who cannot. Selfish genes need to be smart. The conclusion seems to be that the smartest will have the greatest survival rate.

On the other hand, gaining intelligence is not the only survival strategy, since by far the most of the millions of species – bacteria, protozoans, trees, grasses, insects, crustaceans – survive without evolving much or any intelligence at the neural level. Most do not even evolve the capacity for conditioned learning. When *Stentor*, a protozoan, is irritated, it ducks and dodges on its stalk, randomly, and if the irritation continues, it dislodges and tumbles away until by chance a nonnoxious place is found, whereupon it stops moving as the random locomotor variation ceases. If irritated again, it will duck, dodge, tumble away once more. *Stentor* knows how to survive this way on the basis of genetic coding.

But *Stentor* has no memory; it cannot store previous solutions and invoke them at the next irritation. It cannot, for instance, "know" the next time to continue in the same direction as before for likely escape from irritation. In contrast, a coyote has a memory and conditioned learning; it can remember in which directions to run for cover. Smart coyotes, using their memory, have flourished despite the best efforts of humans to eradicate them. Meanwhile, there are more *Stentor*s in the world than coyotes.

Humans evolve still further. Conditioned learning must take place in actual environmental encounter, but humans can imagine encounters, project hypothetically, and learn from their imaginings. They have an idea space, in their minds, that they can use as a trial-and-error simulator and test behaviors in thought experience. Such an idea evaluator is faster and safer. The mental simulator can project outcomes were such trials conducted in the real world and choose the best ones to test. Even the higher animals can do some of this, but human rationality enables humans to anticipate quite novel futures; to choose potential options; to plan for decades according to chosen simulations, or policies; and to rebuild their environments accordingly. They transmit and evaluate their cultures, and there are more humans in the world than coyotes.

Perhaps those primates who never evolved anthropoid brains flourished well enough in the African jungles and savannas. Still, when rationality does come, it proves useful, and humans have often replaced primates when in competition with them. Humans have spread widely around the globe, displacing and replacing, cultivating and using, plants and animals wherever humans have gone with their rebuilt environments. This idea-simulating and evaluating capacity, the genius of mind, has produced technology across many centuries, including scientific technology over the last two centuries, and this has made this conquest of nature possible. So, taking rationality at its apex in scientific achievement, it seems plausible that science has survival value.

A scientifically based culture, in competition with a prescientific one, will on average have improved nutrition, better medical care, more industrial capacity, rapid transportation and communication, all kinds of know-how that give it advantages over less knowledgeable cultures; that will mean that the children survive. Within a culture, those who take advantage of such science-based benefits will win over those who do not. Darwin claimed, "The more intelligent members within the same community will succeed better in the long run than the inferior, and leave a more numerous progeny, and this is a form of natural selection" (Darwin 1874, p. 143). Discovered and applied by intelligence, science benefits people and increases their fertility.

On the other hand, at least until recent decades, the "pure science" of any era has never been central to the survival of that culture; to the contrary it has usually been rather much an epiphenomenon, on the periphery of agriculture, industry, politics, and commerce. Darwin tagging along on the *Beagle* or Einstein publishing his papers on relativity, while working at the Swiss patent office, had little to do with the welfare of the British empire or the gross national product of the Swiss nation.

(2) Science and Fertility

The sciences have helped humans survive, and yet the ties of science to survival are rather loosely genetic. Because the sciences are transmitted neurally rather than genetically, scientific rationality may increase survival rates in a culture as a whole, but it is quite dissociated from the immediate survival of offspring of those who are scientists. We cannot just expect scientists qua scientists to have more children than nonscientists. Desiring to have many children is not one of the criteria for being a good scientist; to the contrary, Nobel prizewinners on average probably have fewer children than Italian immigrant steelworkers. There is evidence that, in certain societies, wealthy persons produce more children than less wealthy ones (Dickemann 1979), but not scientists.

When scientists do have children, no matter how smart the parents are as scientists, both parents and children also have to fit in with the rest of their culture. Scientists must marry, find a job with income, buy a home, manage their money, get along with their mate, rear their children, persuade them to share their values. They have to vote and maintain government and promote science in society. They have to avoid war and deal with inflation. They have to worry about finding meaning in life, keeping promises, being honest, living with failure and guilt, forgiving others, and so on. So their sciences will be cross-coupled with all the other skills of cultural success, including ethical, philosophical, and religious ones. The intelligence that succeeds is not just theoretical scientific intelligence; it must apply science. But should this be with assertive self-interest, or with savvy, wisdom, resolution, courage, patience, and love? Presumably they too will be looked over by potential mates for their kindness and understanding, and they will look for these qualities in their mates – as has been found in transcultural studies of mate preferences (Chapter 3, Section 3[3]).

It might be, for instance, that creationists with bad science will outreproduce evolutionary theorists with correct beliefs, because the creationists nevertheless have a lot of love and kindness for their children, who they believe are made in the image of God. Evolutionists meanwhile have only a belief in kin selection theory and this might dismay them because they do not know what they ought to do, given their innate genetic selfishness. Good science coupled with moral failure, or even moral indecision, could bring disaster. The sciences can contribute to cultural success, but they cannot be the sole determinant. Nevertheless parents who do care for their children (whether they are oriented by theological or kin selection beliefs) can better care for them if they have sodium sulfacetamide to use when their children have pinkeye, if they have fertilizer for high-yield hybrid wheat, if they have high-technology weapons to protect national boundaries.

Every generation has to reproduce itself, scientific generations included. But being smart no longer means maximizing the number of one's children. When the lesser-developed nations go modern, there is a population explosion. This does not always result in flourishing and prosperity; to the contrary, most modern nations curb population growth through birth control, and when European women had more birth control possible, thanks to science, they used it (Chapter 3, Section 3[2]).

In natural selection theory, that the fittest survive is a claim about genetics and differential reproduction of organisms, but if we try to think of science naturalized in this way, there is no genetic dimension. Scientists presumably often have children who make good scientists. Yet babies who are taken from nonscientific societies and educated in the West can make equally good scientists. Nor is there any differential reproduction between scientists and nonscientists within a culture. There is even a negative differential reproduction between the most highly scientific societies and the semiscientific ones. If there is any adaptive advantage to having science or being a scientist, it must be more indirect, linked into the rest of culture.

A critic might reply that the question is not of genetic fertility, but of epistemic fertility, as we have already established. Fitness for a scientist in culture means that his or her theory leaves more "offspring" in the next generation in the mind of scientists who adopt that theory. But epistemic fertility is quite different from genetic fertility, even if both procedures do involve a generate-and-test model, with a differential survival of variants. This radical difference is not just a matter of counting converts to the theory rather than children. One has to ask whether the theory comes to be believed because it is logically justified. An idea does not compete as it seeks prey or avoids predators; rather, it fits observations better or integrates more plausibly with theories in the other sciences. The fittest theory is not the one that helps its holders to reproduce, or even benefits its holders; the fittest theory is the one that reproduces itself because it best withstands critical attack when its holders argue for it in the light of its agreement with observations, logical coherence, predictive success, simplicity, non–adhocness, satisfactory sense of explanation, comprehensiveness, deployability, even beauty.

The theory may be adopted because it is fruitful, but not in the reproductive sense; rather it generates new hypotheses, further tests, applications, and so forth. That means not simply that the fittest theory survives because its holders survive, but also that it survives in part at least because this fittest theory better fits the truth about natural history. The explanation why the scientist chooses this particular theory comes to an end right there. She thinks it is the most plausible one, and one does not need to inquire how many children she wishes, or whether her nation has a rapidly increasing population.

Scientists think, do they not, that evolutionary theory is true in a correspondence sense, that it describes the course of events in the world, and that this is the ultimate reason for holding it, whether or not the theory holders leave more offspring (children), whether or not the theory holders have more disciples (students who adopt the theory) in the next generation, whether or not the nation with the most Darwinians flourishes? None of the latter is a relevant criterion for testing the theory. An evolutionist must to that degree be a realist in science, believing, for example, that incremental or punctuated development has occurred over the course of natural history, with variations supplied by genetic recombinations and mutations, and that survival of the better adapted has been a major determinant force in this history, perhaps also with some genetic drift.

It is difficult to imagine what a fully instrumentalist or pragmatic account of evolutionary theory would be like in genetic terms – holding that the theory is neither true nor false but only has usefulness for maximizing reproduction or for helping the society that believes in evolution to prosper over other societies that do not. It is easy to think that natural selection might select for theories in humans that have survival advantage, and that these theories might often be true but sometimes be false. A people might, for instance, become convinced that the gods favored them; believing this resolutely, they might (in self-fulfilling belief) prosper where others have less nerve. But what survival advantage does holding the theory of evolution itself convey? Perhaps some evolutionists, social Darwinians, once came to believe that evolution justified European capitalists in their colonial exploitations, and this urged them on. But nothing like that is being claimed here. Perhaps the theory only hitchhikes on other, more useful theories in biology, such as genetics that can be put to pragmatic use. Perhaps it will give its holders insight into their own biology and behavior and they will better be able to understand, discipline, and reform themselves, in competition with others.

But that is only a promissory note; there is no evidence that holders of the theory of evolution are now flourishing for these reasons while others languish. Even if they did so flourish, holders of the theory do not advocate it for its pragmatic utility only; they advocate it because they believe that it is a realist description of natural history. Adopting evolutionary theory is not just a way of coping; it is a way of copying (re-presenting) what has taken place in natural history. Those biologists who are urging us to accept these theories are not doing so because they thereby increase their offspring in the next generation. If they were, that is, if they themselves as scientists come within the scope of their theory about science, then there is no reason to think what they propose is true at all.

Really, it is rather surprising to find these academics, who spend their careers evaluating ideas for their intellectual merit and social significance, who live amid colleagues of the same mind, and who teach generations of students these skills, urging them to do likewise, here advocating theories of human behavior and the human mind that do not permit this life-style. They do not, in their own behavior or intellectual processes, exemplify their own theories of behavior and mind.

Rather, they have an evident trust in the objectivity of their own capacity for scientific reasoning, and they urge us to adopt their theories on such rational grounds. They do not invite us to adopt their opinions in order that we may outcompete them, or they us, or join with them against others in the competition for survival by producing more offspring. They invite us to join them if and because we believe that their convictions are supported with good reasons and correspond to the way natural history is. The historical fact that our conceptual and perceptual faculties have evolved and serve us for survival does not mean that nothing true appears when they are used.

If the issue is epistemic fertility, we are disoriented to ask whether scientists have more children biologically. The question to ask is whether scientists beget more scientists. Than whom? Than novelists beget novelists, businessmen beget businessmen, philosophers beget philosophers, theologians beget theologians? Or astrologers beget astrologers? If one is asking about proportionate sectors of the population, that is hard to say, because the skills of scientists, so far as they convert others and help rear the next generation, are widely dispersed; novelists, businessmen, and philosophers, as much as scientists, use agriculture, industry, and medicine to stay healthy, comfortable, and productive. Even astrologers can enjoy these benefits. The benefits that scientists produce, in the alternative paradigm we have preferred, are widely "distributed," "dispersed," "allocated," "proliferated," "divided," "multiplied," "transmitted," "recycled," or, in short, "shared."

Given a general welfare and a healthy incoming generation, the begetting process is intellectual, not genetic. This once again is a question of shared values, not of some selfish survival. Molecular biologists increase in number; vitalist biologists die out. Some scientists persuade their students and colleagues; others do not. Meanwhile, some astrologers remain around. Consider, for instance, the specialized science with which we are here most concerned, sociobiology. The proponents of sociobiological theory are promoting it because they think it is a valuable theory, giving insight into what has been and is going on, and that people widely ought to know this. They want their theory widely distributed, or shared.

To accomplish this they are asking us responsibly to evaluate whether their theory is credible on the evidence supporting it. But theories of this scope are self-referential. They apply to their proponents or opponents just as much as to everybody else. So, before we can intelligently share their theory, we will have to find some way to escape from the theory long enough to evaluate the theory, or else we cannot tell what is determining our convictions: the reasons we give, or the genes deep down inside us. But just this escape demanded to verify the theory falsifies the theory; it is a telling bit of counterevidence. (A similar escape, we later see, must be made in ethics and in religion.)

Otherwise, the sociobiologists themselves, as scientists, are asking us to join them in engaging in an activity that the theory they propose does not permit either us or them to engage in. If the theory is true, both advocates and critics are making whatever noises they make, claiming whatever they claim, protesting whatever they protest, as a fertility-optimizing stratagem. All this seeming "argument" is just so much fluff over proponents' genes battling opponents' genes. In that case we do not wish to, and ought not to, share their theory.

The scientists might claim a special exception for themselves; they, and they alone, enlightened, have escaped from the bondage of their own theory. That is real genius! But they seem also to be inviting their various readers to evaluate whether what they say is true, readers who are not yet members of this exclusive group that has become excluded from the bondage of the theory. The socio-biologists, well intended, argue that we ought to revise ethics throughout society, in the light of what they have discovered. And should anyone object to their theory, how would they know whether to take that objection seriously, or to put it down to a fer-tility-optimizing stratagem? They could say that it is their scientific truth versus our selfish genes. We could reply that it is our scientific truth against their selfish genes.

Moreover, when we hear these scientists frequently claim that persons are regularly deceived about their own intentions, motiva-tions, selfishness, altruism; when they hold that verbal reports are quite unreliable (as they will maintain in Chapter 5), we are going to need more than a little assurance that the scientists themselves are not self-deceived. We must have some account of their own ca-tharsis by which they alone can reach this purity of evaluation. If they supply such an account, we will have to be especially cautious. It will be difficult to take such an account seriously, lest it really just be their genes' putting up a good front, deceiving us about their real motives.

All this seems contorted. Surely it is simpler to think that when scientists and critics argue the merits of the theory, including socio-biology, both are evaluating a theory on its evidence, regardless of the reproductive consequences for either side. But just that one coun-terexample is a quite considerable leak in the whole theory. If such rational evaluation, separated from results in biological fertility, can be done for sociobiological theory, why not elsewhere in biology – in chemistry, in physics, in psychology, in sociology, in anthropology? Why not perhaps also in ethics and in religion? Why not recognize that there is a human genius, exemplified in science, that does tran-scend biology?

(3) Science and Selfishness

So far as the sciences serve our genetic or other human interests, can we think of this as being selfish? We have readily conceded that science brings many advantages; so, when humans use science to build automobiles, or to make fertilizer to grow high-yield wheat, or to cure pinkeye in their children, no one will deny that humans are doing this in their self-interest. In scientific societies infant mortality rates drop and there is often an increase in population. So we might say that the human "selfish" genes are using these sciences to keep the genetic material intact. The leash is long, but still there.

But acting in one's self-interest cannot always or even usually be labeled "selfish," or else it is selfish just to eat or to take shelter out of a cold rain. Using science for human benefit is not ipso facto selfish, nor is using science to help raise a next generation of offspring. Censurable selfishness involves some "self" acting or behaving in its own interests in an arena where other "selves" have interests that can be acted for or against, in such a way that the acting self, choosing from options, gains at a cost to other selves. Even then, we do not think that the person who shops wisely for the best bargains at the grocery store is being selfish, though he is acting in his self-interest, trimming the profit margins of the grocer, and consuming food and reducing the amount available for others. Censurable selfishness involves injustice, greed, inordinate attention to self-interest.

Certainly the sciences have been used for selfish purposes, for instance, when science is applied to carry out military aggression. But science has just as certainly been used for altruistic purposes, as when missionaries teach others about high-yield fertilizer and pinkeye drops. So there is nothing in principle about any science that is selfish, just because it is applied to serve human interests. In pure science, beyond applied science, selfishness is even less plausible. What one means when one says that a particular historical tradition is part of the "scientific" community of an era is that those who hold these ideas do so critically, motivated by a desire to be objective, disinterested about their tradition, having a desire to improve it, and, as a result of their careers, leave a better science than was in place at their arrival. We should hope that it is irrelevant to ask whether such scientists were selfishly keeping their genetic material intact.

If we did come to suspect that this was what was going on, that would be cause to suspect the theories they proposed. Scientists cannot hold the theories they hold merely because they are thereby increasing their fertility, and that applies even more if they are *selfishly* increasing their fertility. Scientists are going to have to transcend the category of selfish behavior, acting in their survival interests, before they can give us any reason to accept their theories. Otherwise, we have no cause to trust them any more than we trust anyone else, whether in science, ethics, or religion. Of course opinions may be (and often are) driven by self-interest in science, ethics, and religion (and theology has been quick to notice this). But that is a source of distrust and failure. Until the self-interest is set aside and the theory examined for its independent rational plausibility, there is no cause to share it.

Perhaps these scientists are using their sciences to succeed in their competitive cultures. Scientists care about who gets credited with their discoveries, and there is competition, a rush to be first in print, fights over who was first, and so forth. Sometimes this can be beneficial, though often it is disruptive. But scientists ought to separate the logic of the ideas (the science, whether it has good inference, adequate mathematical models, statistically reliable sampling procedures) from the psychology of the persons who hold these ideas (the scientists, driven to reproduce, to get promoted and receive a higher salary, to win a Nobel prize). Their ideas stand or fall independently of the authors' names, of the success of any "selfishness" in scientists who desire elevated status or seek the power that piggybacks on the rationality of their arguments. No scientist wants his or her theory accepted even if it is wrong, and even if one did, the critical processes of science preclude this. Any real and lasting fame still depends on rationality.

Uncommunicated research and scholarship are unimportant, because they can play no role in the development of a discipline. Uncommunicated knowledge is, shortly enough, no knowledge at all. Scholars without disciples are soon extinct. One publishes or perishes! But that is neural and social, not genetic; that is rational selection, not natural selection; that is information sharing, not selfishness. We are beginning to wonder whether it makes much sense to say that the selfish genes have scientific culture on a leash, whether long or short. It certainly makes better sense, evaluating science, to claim that the "truth value" is cognitive and epistemic, not biologically reproductive.

Social forces sometimes drive so-called pure science. The last three

decades of impressive advance in molecular biology, including genetics, have been funded in large part by medical interests. Nuclear
physics has been funded by military interests; computer science has
been driven by industrial data-processing needs. Skillful medicine, a
powerful military, and efficient industry help humans survive. At
the same time, the motivation behind these social forces is quite external to the inner logic of molecular biology, genetics, nuclear physics, or computer science. The rationality required for scientific genius
is not pragmatic and survival oriented.

5. PROGRESS IN SCIENTIFIC AND IN NATURAL HISTORY

Natural selection is for fitness for survival; the value conserved is for
producing more offspring. Fitness, which requires coping in the
world, is provided by information, coded in the genes (sometimes in
higher animals also coded in acquired, neural memory); so we might
suppose that there is a correlative relation: the more information, the
more fitness. But this is clearly not always so. The survivors over
evolutionary time are sometimes more rational (lemurs or chimpanzees); more often they are not (grasses and sedges). Natural selection
need not require the survivors to be any smarter.

But rational selection in science, though it contributes to survival,
involves getting smarter, and this is a relevant disanalogy between
scientists doing science and creatures evolving. Does this contrast set
natural and scientific history entirely apart, or can the two again be
related by factors common to the one event in nature, the other in
culture?

Is there historical progress in science? Scientists certainly think so
in theoretical science; that is essential both to its creative genius and
to its truth value. The theory of combustion as oxidation is an improvement over the theory of phlogiston, now abandoned. Relativity
theory, which reduces Newtonian physics to a special case, is a more
comprehensive description of nature than was Newtonian physics
alone. Scientists know about galaxies and the expanding universe,
unknown at the start of this century. They know about atoms, va-
lences, electrons, protons, whereas a century ago humans were entirely ignorant of nature at this structural level. We now know about
electricity and radioactivity. Species have evolved over evolutionary
time; people formerly and erroneously thought them to be fixed.

Scientists have discovered the depths of historical change over billions of years; people formerly supposed Earth was only a few thousand years old. There is progress here because we are closer to the truth – not to all of it but to some of it. So the sciences become more valuable over time.

In applied science, there is progress of a different kind. Humans have built automobiles replacing horses, diesel ships replacing sailing ships, jet planes for faster travel, telephones for communication. Scientific technology has resulted in higher yields in agriculture and automated production in industry. Physicians can cure diseases and perform surgery that they could not before. Electronic engineers can make compact disc (CD) players, surpassing the performance of phonographs. They manufacture computers with bigger memories and more graphics capacity. All this we commonly regard as progress, even though these applied sciences also result in some degeneration (lost topsoils, unreclaimed strip mines, manufactured junk, inane computer games, polluted air and water, global warming). Again, there is an increase of value, resulting from science.

Beyond the survival value in natural history, is there nothing more, no precursor to this progressive increase of value? Natural history is also a sphere in which there is some "progress," even though this is not a deliberated process (Chapter 1). The genes are "smart genes" with their search programs, which increase not only fitness but also diversity and complexity over the millennia. Although the simpler forms remain in the ecological trophic pyramids, required for their roles in the food networks and energy cycles, there is a discovery of new skills and achievements over evolutionary time. So one has to be discriminating about the contrast between progress in the sciences and the lack of it in natural history. There is creativity, increase of valuable achievement in both, though forms of progress do become possible in science that transcend the forms of progress of which natural selection is capable.

6. TRANSCENDENT SCIENCE

No one will deny that there can be a selective advantage to rationality in some sense of "advantage," not always a genetic sense. The "advantage," that is, increased value, lies in increased understanding. Knowledge is an instrument of biological and cultural adaptation, and yet the meaning of truth to a scientist is not just that which is biologically or culturally advantageous. Scientists still do their science even when they have plenty to eat for themselves and their children, regardless of whether there is relevant survival advantage. Scientific intelligence functions usefully inside the pragmatic natural and cultural worlds, but it also sees outside these world sectors that it locally inhabits; it studies other times and places, alien forms of life and phenomena. The sciences try to see bigger pictures, even to see the whole in overview. Those who emphasize the social construction of knowledge will insist that the sciences too belong inside some culture, that the scientists' views of nature are "constituted" by the sociology of science, a product of Western European culture, so perhaps there is no transcendence in any absolute sense.

But there is at least a transcendence of natural selection. Humans are able to discover what is going on in genetic, evolutionary natural history and to take an overview of it – a genius unique to *Homo sapiens*. In the sciences humans have the capacity, to an impressive extent, to transcend the niches in which they evolved by natural selection, even to transcend the cultures in which their sciences evolved. Humans do not live all over the universe. Vast parts are beyond their native range experience, parts with which they do not have to cope. But they know something about asteroids and Mars, about distant galaxies and supernovae. They know something about the expanding universe and the primeval big bang. They know that $E = mc^2$.

Humans do not live over all the Earth, though they travel widely over its surface. They do not live at the level of diatoms deep in the sea; they did not live among the now-extinct dinosaurs; they did not evolve any sense organs to detect neutrinos. They evolved at middle-range levels with appropriate perceptual organs – eyes, ears, noses. But with these faculties they have learned about diatoms, dinosaurs, and neutrinos. Cognitively through science, they have escaped their native levels and ranges. Since diatoms and dinosaurs, neutrinos and relativistic equations, are not things that humans were ever naturally selected to deal with, nor does culture have to reckon with them as part of any ordinary social needs, there is a sense in which scientists escape not only nature but even their cultures. This is true even though the instruments and theories they use are products of their culture.

The skills that natural selection produces are those of adaptive fit into a niche, but rationality enables humans to break out of their sector. Their knowledge is no longer simply niche-relative. Humans gain oversight of and insight into the whole – with limited success of course – symbolized well by the views of Earth from space, photographed by astronauts viewing the planet from space. The sciences they use were discovered in Western cultures, but with these they discover laws, mathematicize what is happening in the physical world; they discover the foundational unity of the phenomena. They decipher fossils and strata and form a view of evolutionary history. They build electron microscopes and decode cellular ultrastructures common to all life. Such an overview can drive a conquest of nature, nor should we overlook its cultural roots, but this is not merely and simply a genetically based, reproduction-maximizing conquest. It is a cognitively based, quality-of-life-optimizing conquest, in ideal and to some extent in real terms. It is functional, of course, in an expanded sense. But it is no longer just biological. Science has not so much been naturalized as it has transcended biology.

Michael Ruse, though he begins by claiming that "the principles of scientific reasoning or methodology . . . have their being and only justification in their Darwinian value,"[10] ends by going much further: "Ultimately, in its highest reaches, science pushes to the limits of culture where direct adaptive advantage sits lightly." "The methods of science are rooted in selective necessity, but . . . the product soars up gloriously into the highest reaches of culture, quite transcending its organic origins" (Ruse 1986, pp. 155, 161, and 149). So it seems that Ruse's "only justification in Darwinian value" is not the only justification after all, but that there is a glorious soaring that "quite transcends" biological origins. Score creativity high with this new genesis, but the higher the score the less there is genetic leashing of the genius of science. In a metaphor we prefer, there is an emergent taking off, soaring up to a new chapter in the epic story.

Although we earlier heard Karl Popper defend the "*natural selection* of hypotheses" in science, he is later clear that science requires the *rational selection* of hypotheses. "From the amoeba to Einstein there is only one step. Both work with the method of trial and error. . . . The step that the amoeba cannot take, but Einstein can, is to achieve a critical, a self-critical attitude, a critical approach. It is the

---

[10] Cited earlier, Section 4(1).

greatest of the virtues that the invention of the human language puts within our grasp" (1990, p. 51). Though T. H. Huxley does say, as we earlier heard,[11] that "the struggle for existence holds as much in the intellectual as in the physical world," he precedes this by insisting on the "scientific spirit" as being of supreme value. "Now the essence of the scientific spirit is criticism. It tells us that whenever a doctrine claims our attention we should reply, Take it if you can compel it" (1880, p. 229). A theory's right to exist is its compelling logical and empirical plausibility, and that is why it "reproduces" itself by claiming other minds and resists extinction by less persuasive theories. But this is not natural selection for reproductive fertility, leaving more genes in the next generation, at all. This is rational superiority in keenly critical minds.

Natural selection operates from outside the organism. After the variants are ventured from within the organism, by genetic recombinations, the environment selects the better enabled organism and thereby selects the inner coding, the genotype, that specifies such a phenotype. But in rational selection the scientist adds an inside to the outside. From among variants that may be supplied from within or without, a scientist selects the better theory, which means that the theory is more true to the outside world, a better empirical fit, but also this now means that the theory is an improved rational fit, more plausible because more coherent and with fewer internal inconsistencies and contradictions. This requires a selective process going on inside the head of the scientist, a conscious evaluating whether there are correspondence, coherence, agreement with evidence. Subjectively, there is evaluation of the objective world.

So there seems to be a rational dimension in the sciences that has quite escaped the genetic demands of survival by natural selection, one that even transcends use of the neural faculties for pragmatic advantage. Meanwhile we do not forget that the history of these scientific ideas, in any comprehensive account, is going to be a history of these persons who forged and carried these ideas in various social contexts. These scientific ideas, true enough in their own domain, are, in the full story, going to be linked with ethical and religious ideas, where there may be yet more truth. A science can "survive" only as it is fitted into a worldview held by scientists who have to survive. Such scientists have to make value judgments to operate

---

[11] Section 2(1).

in the world, as everybody else does. When the sciences are applied in life, the scientist's choosing which of various theoretical areas are the likeliest ones in which he or she should concentrate next may involve what the scientist has chosen to value pragmatically, politically, ethically, or religiously. Whether and how far the logic of science is separable from the psychology and sociology of science are issues to which we will return in our analysis of ethics (Chapter 5) and of religion (Chapter 6).

7. AN UNFOLDING STORY

The root idea in the word "science" is "knowledge" (*scientia*). Francisco Ayala observes, "The ability to perceive the environment, and to integrate, coordinate, and react flexibly to what is perceived, has attained its highest degree of development in man. Man is by this measure of biological progress the most progressive organism on the planet" (1974, p. 352). Perhaps also – by some measures of cultural progress – scientists have attained, among humans so far, the most ability to perceive the environment and to react flexibly to the natural world that surrounds us.

That privileges science, and it can sound arrogant. Cultural relativists will object; knowledge in science perhaps comes with the forgetting of indigenous wisdom, ancient truths. Philosophers may observe that the sciences bring only knowledge, not wisdom; ethicists and theologians will, in chapters to come, argue that ethics and religion also have their roles to play in measuring progress.

Meanwhile, the successes of science are impressive. They are quite valuable, and today no one is able to evaluate the world, to form a worldview, adequately without scientific knowledge. A significant part of what must be evaluated is the human mind, capable of such science. The operations of the mind, indeed useful in the jungle, on the savanna, and in the pragmatic world of culture, carry us much further. Rationality works not simply for middle-world, native-range living, in country and town; it works for building microscopes and studying *Stentor*, for decoding atoms and quarks, for doing calculus and statistical regression analysis, for solving equations that run time backward to the big bang and then philosophizing about cosmology, for postulating and trying to simulate the chemical origin of life in ancient seas. These activities were no part of the survival routines in the hunter–gatherer cultures in which the mind was formed; skills here are not complex mechanisms of an adapted mind, and so how did humans obtain these capacities that transcend any relevance to the environments in which they evolved?

Darwin concluded that the creative result of evolutionary history could not be accidental, only to wonder about the capacity of his own mind to make such judgments, since, if evolutionary history is true, we have minds descended from monkeys. He writes of "my inward conviction . . . that the Universe is not the result of chance. But then with me the horrid doubt always arises whether the convictions of man's mind, which has developed from the mind of the lower animals, are of any value or at all trustworthy. Would any one trust in the convictions of a monkey's mind, if there are any convictions in such a mind?" (1881, p. 285).

But Darwin here misses the historical development between monkeys and men; the query fails because it reduces man's mind to a monkey's mind and then distrusts it. What we are asked to trust is not the convictions of a monkey's mind or those of a man who is nothing but a monkey. We are asked to trust the assessments of critical minds, of genius–scientists (like Darwin), none of which monkeys can be. Darwin stumbles into the genetic fallacy. Much humility is still required, but nothing is gained by supposing that humans cannot now make such judgments because monkeys before them could not.

One might expect that the organs of knowing will have a biased focus on the native-range habitat, and perceptually that is true: eyes see, ears hear, noses smell at appropriate ranges for moving around in the local world. But conceptually the organ of reasoning has vaster powers. One might say that the explanation lies in instrumented intelligence; our natural intelligence, unaided, cannot reach any knowledge of elementary particles, molecular genetics, or cosmology, but intelligence augmented by the detection capacities of scientific instruments can discover these levels in nature. But how then does an intelligence that has evolved to function at native range push past that range to build instruments of this kind? These instruments are coupled with theoretical explorations; they are thought up and built by this natural intelligence evolved only to contend with the jungle. We might put it all down to serendipity. But a mind that evolves in the jungle and, as a result of the skills needed to find food there, happens by luck to be able to construct Gödel's proof of the formal undecidability of mathematical systems has certainly lucked into a lot of serendipity! Perhaps this "luck" is really a confession that one has no plausible causal, lawlike explanation of the mind's origin.

If, rather, we narrate these events as a dramatic story of the genesis of increasingly valuable achievements, we gain an evolutionary epic. That is the explanatory frame required. The justification is going to be by way of history, not law, much less luck. The justification will be incomplete at the level of causes; it will require rising to the level of meanings. Indeed, natural selection theory is troubled even to give a causal, much less a meaningful, pattern to the path from matter to life to mind. Those forms of life in which no mind materializes (grass, earthworms, diatoms) fit the theory quite as well as do those forms of life in which, luckily, this does happen. In the only form of life where there is this appearance, the result, mind, fits the theory, since mind conveys survival benefits; yet it explodes the theory, since the scientific mind so evidently transcends the necessity to produce offspring and to live in the native ranges that humans inhabit.

These striking evolutions can only be related as a story. Stories do progress. Theories and laws per se have no dramatic development, but natural history does. Both the natural selection of better organisms in nature and the rational selection of better theories in the sciences drive story lines. The story that begins in physics (the events studied by nuclear physics, astrophysics, geophysics) continues in chemistry (the events studied by molecular chemistry, organic chemistry) and rises to biology (the events of molecular biology, evolutionary history, and speciation). At length culture arises, and in culture there are multiple stories, including in science the discovery of this natural history from which we emerged. There is an ongoing effort to interpret this history and to understand ourselves in relation to it.

Emphasizing the contrasts, one can conclude that natural selection has a causal structure, whereas science has a rational structure. But natural selection is not just causal, although physics and chemistry are causal, and that is why biology is a historical science in ways that physics and chemistry are not. In the genes, information gets inherited, superimposed on merely physicochemical causes, and there is a cumulating genesis of discoveries and achievements over the millennia. The DNA codes designate something; they are symbols with biological meanings. This makes the logic of biological sciences differ from that of physical sciences.

A handful of forest humus contains thousands of insects, mites, nematode worms; millions of fungi, billions of bacteria, and trillions of protein molecules; ribosomes; Golgi apparatus. The "know-how" for all this is coded in genes. Each species living in the dirt is the product of millions of years of Earth history, having evolved as an adapted fit in its community of life. In a handful of dirt carried back by astronauts from the moon, there is none of this at all. Events that form and erode lunar rocks are lawlike; also, in their limited way, they are historical. But they are merely causal, whereas events that take place in forest humus are cybernetic. There is a story at levels not present on the moon. This is what the alleged "selfishness" of genetic conservation is really describing: the conservation of this storied information about defending somatic selves, embedded in a genetic process by means of which there emerges also the discovery of novel information, through adaptation and speciation, building more diverse and complex selves. This conservation requires, as we have been claiming, a sharing of values, values both conserved and distributed over the millennia.

Scientific explanation is typically causal and lawlike in the natural sciences – generalizing, finding repeated orders in the empirical world. The future will be like the past. Learning such causal regularity will help humans cope. This is true not only in physics and chemistry but also in biology, where, with transmitted information, superimposed on causes, acorns regularly produce oak trees, and beech trees regularly replace red maples, which replace gray birch in forest succession, some species having learned how to grow in the shade, others in the sun.

Though nature is causal in this way, this is only a half-truth. Nature is also historical; it is just as true that the future will not be like the past. Photosynthesis was discovered; calcium skeletons emerged; endoskeletons appeared; trilobites arrived where there were none before; dinosaurs came and went; there are turnover and increase (sometimes decrease) of species. Life moved from the sea onto the land, and into the sky. Warm-blooded animals appeared. Predators increased their speed and learned to switch prey. There is an arrow on evolutionary time, made possible because of genetics. At this point, the openness enhances the story. Life is destined to come as part of the narrative story, yet the exact routes it takes are open and subject to historical vicissitudes.

Likewise in culture, at the neural past the genetic level, there is history. The past frames the future but the future is more than the past. There is the scientific revolution, replacing medieval thought. There is evolutionary theory, replacing the fixity of species. There appear nuclear weapons to deal with, reshaping social power balances and reshaping the financial resources available with which to do physics. This affects the directions physical theory takes. Computers are developed on which to model climates, kin selection theory, and forest ecosystems. The women's rights movement arises, opening up new possibilities for women to become scientists or to give the sciences a feminist interpretation. Especially as a result of science with its cultural information transfer, technology makes the future different from the past. None of this is going to find an adequate explanation in terms of any human "biogram," common to us all, or of complex adaptive mechanisms, by which the fittest leave more offspring, and under which science can be subsumed.

When we interpret A as an analogue of B, we can also interpret B as an analogue of A. We have to ask again what is illuminating what? It may be that the biological phenomena in the world (A) are illuminating the character of the biological science in the heads of the scientists (B), so that science has to be understood in a biological framework. It is a new form of, and yet nothing but, survival skills, selected for. Science would be naturalized, Darwinized. It may also be that the history and logic of science (B) are illuminating the biological history (A), which is their precursor, and we come to see the incipient rationality of natural selection. The cognitive adventures that culminate in culture have already begun in nature. The biology (A) is a precursor, a developmental stage of science (B); this is the deepest sense of a really "genetic" explanation. We tell the story of the genesis of rationality, trying to be discriminating about both the continuities and the discontinuities.

There emerges more in the resultants B (science) than was present in the lesser antecedents A (biology). That does not demean the antecedents but places them earlier along a developing story line, necessary but not sufficient for the later events that unfold in such a way that the end transcends the beginning, whereupon the beginning has to be understood in terms of the ending. This appropriately evaluates these events, finds the values there, and sets them in their historical context. From this perspective, the evolutionary process, so far from being irrational is a partial prototype of the kind of rationality that we experience in ourselves, the kind so well exemplified in the scientists themselves.

The human brain/mind comes into being and matures under a heritage from the past, under genetic and also cultural cybernetic control. But the brain–mind too is, in its essential genius, a nonprogrammed learning center that openly and flexibly scans to see what else it can learn, and this was already foreshadowed in precerebral nature. The unity we seek may not lie in reduction of the more to the less but in development of the less to compose the more. Genesis has marvelously depended on genes, and the story comes also to include exodus and freedom of the spirit. We do want to unify the picture, if we can, but the unity is not that of reduction or of lawlike sameness. The unity is of developing story, not that of predictive argument. That there is transcendence need not mean that there are no precursors, and precursors are what they are in themselves, but they also are what they are in their story lines. Without the perspective of behind and before, we do not really know what is going on now.

Nor do the later, cultural – including the scientific – chapters make the biological precursors blind, inept, or less valuable, any more than adult life makes childhood to be pitied, or the third act of a play makes the first clumsy. The hand and the brain that evolved naturally are still more marvelous than anything these brains and hands have yet made. The discovery of language is as significant as any discovery subsequently made with language. The felt experience sponsored by the brain in the body has yet to be approached in any artifact, no matter how cleverly designed.

The story form is what justifies Earth history, which can only be evaluated as narrative. There are no premises from which one can deduce these conclusions. There is no argument from primeval Earth to Precambrian protozoans, to Cambrian trilobites, to Triassic dinosaurs, to Eocene mammals, Pliocene primates, eventuating in Pleistocene *Homo sapiens*. No theory can look at a protozoan and deduce an eye or a brain. There is no argument why this has happened, nor that it ought to have happened. There are no equations into which one can introduce amino acids, or microbes, or trilobites as initial conditions to specify the variables and then solve to produce dinosaurs, or mastodons, or persons. Earth history has been, rather, a splendid story – this saga of struggle, beginning with the simple and producing the complex; the increasing vitality, interdependence, community, freedom, individuality; the increasing power for action and cognition, for locomotion and cerebral power in the upper trophic rungs of the ecosystems, in which continue the previous "lower" forms as well, now in support of the adventures at the top.

Natural selection is not the transcending category, nor is rational selection, but historical selection integrates both. What gets selected enriches, that is, makes more valuable, the epistemic (which sometimes tragic) epic. All this is, and it ought to be. One great story in natural history is that of the evolution of rationality in freedom; now, in science, more than ever before, we have an emergent intelligence with the power of understanding who and where it is, a marvelous thing. If this is not valuable, why not? Ought this creativity not to be defended, conserved, shared? That value ultimately is the only argument of life, the only argument for life. Science is both evolution becoming conscious of itself and evolution transcending itself.

In terms of the alliteration that titles this book, our search for understanding must reckon with genes and their genesis, but this produces the human genius, which produces scientists, including geneticist geniuses – who themselves have genes. The creativity we are encountering is escalating, and, before we can evaluate it all, lurking still ahead, is the question of God.

Chapter 5

Ethics: Naturalized, Socialized, Evaluated

Once upon a time natural history produced moral agents, incrementally perhaps, but still appearing where there were none before. Since there isn't any ethics in nature, it may prove challenging to naturalize it. At least there is not much ethics in nature, even if some precursor primate behaviors are part of the story.[1] Ethics is distinctively a product of the human genius, a phenomenon in our social behavior. From a biological perspective, such an emergence of ethics is as remarkable as any other event we know; in some form or other ethics is pervasively present in every human culture, whether honored in the observance or the breach. So we must evaluate origins to discover the nature in, the nature of, our duties. One might know present duties and be ignorant of origins, of course, but a more complete account, if we can gain it, will be enlightening.

---

[1] Peter Singer's *Ethics* (1994) has a section "Common Themes in Primate Ethics," which includes the section "Chimpanzee Justice," and he wants to "abandon the assumption that ethics is uniquely human" (p. 6). But many of the behaviors examined (helping behavior, dominance structures) are more preethical than ethical; there is little or no sense of holding chimpanzees morally culpable or praiseworthy. After a careful survey of behavior, Helmet Kummer concludes, "It seems at present that morality has no specific functional equivalents among our animal relatives" (1980, p. 45). Frans de Waal finds precursors of morality but concludes: "Even if animals other than ourselves act in ways tantamount to moral behavior, their behavior does not necessarily rest on deliberations of the kind we engage in. It is hard to believe that animals weigh their own interests against the rights of others, that they develop a vision of the greater good of society, or that they feel lifelong guilt about something they should not have done. Members of some species may reach tacit consensus about what kind of behavior to tolerate or inhibit in their midst, but without language the principles behind such decisions cannot be conceptualized, let alone debated" (1996, p. 209).

There is no science in nature either, but precursors in natural cognition made some connections plausible. Scientific reasoning is perhaps easier to naturalize than ethics (which may prove easier than religion). Yet there must be some story to tell. We need a threefold account: first, how morality happened to appear in past evolutionary time out of precursors in animal behavior; second, what to make of the ethical heritage of cultures over the ages, that with the goal of understanding, third, how morality today can and ought to operate in society, especially (for our purposes) as this relates biology to ethics. Both biologists and ethicists are particularly challenged to give an account of the origin(s) of altruism, the genesis of generosity.

1. MORAL VALUE: LOVE, JUSTICE, AND RESPECT

(1) The Ethical Challenge

Although life has continued on Earth for several billion years, although human life (*Homo sapiens*) has continued for a hundred thousand years and more (the genus *Homo* for two million or so), there is a profound sense in which we humans in the twentieth century, in an age of science, turning the next millennium, know for the first time who and where we are. This has been Darwin's century, and we have more understanding than any people before us of the evolutionary natural history by which we arrived. We know the astronomical prehistory of Earth; we know its geological history. We know the natural history that lies behind us, around us, out of which we have come.

With this knowledge comes power. More than any people before, as a result of our technological prowess through science and industry, we humans today have the capacity to do good and evil, to make war or to feed others, to act in justice and in love. Nor is it only the human fate that lies in our hands. We are altering the natural history of the planet, threatening alike the future of life, the fauna and the flora, and human life. With such increasing knowledge and power comes increasing duty. Science demands conscience. Philosophers must join with scientists, theologians, political scientists, literary analysts, and others, to evaluate the origins and principles of ethics; more, philosophers and others, along with the scientists, have the challenge of evaluating its worth. Otherwise we lack a well-integrated account and, to that extent, do not know what we ought to do.

The origin of ethics is deeply ingrained in the archaic Genesis stories. Adam and Eve appear, and immediately there occurs the mysterious appearance of conscience, symbolized in the tree of knowledge of good and evil. The primal couple are given the responsibility of keeping the garden and held responsible for their presumptuous sin, and they are cast from the garden. Cain murders Abel in jealousy, refusing to be his brother's keeper, and becomes a social outcast. Even the ground cries out against the spilled blood. The story is of a primal "fall" of humanity into sin and of the need for an ethic, indeed for redemption. The genesis of responsibility – and culpability – these two are phenomena that need explanation and evaluation.

Those stories are quite archaic, or mythological, but (as already noted) one can hardly claim that modern science has figured out ethics, either its historical origins or a current evaluation. The more usual account is that ethics is not science, nor science ethics; the one is a descriptive discipline, what *is* (was, or will be) the case; the other an evaluative discipline, what *ought* to be. "Good and evil" (symbolized by the tree in the garden) are not categories that appear in science textbooks, which is not to say that scientists make no judgments about good and evil. Being human, they routinely must do so, both in ordinary life and in the pursuit and application of their science.

But science is never sufficient, and often not even necessary, for such judgments. This remains true despite the efforts (which we will here be examining) to naturalize, or Darwinize, ethics. Evaluating what has happened – for example, the richness of creation and the responsibility and culpability of the humans both toward each other and toward their world (as Genesis portrays these issues) – is hardly an archaic matter, but very much at the center of the agenda in a scientific and postscientific world.

So, although there is a profound sense in which we humans now know who and where we are, there is an equally deep puzzlement about what we ought to do, and the grounds of its justification. Science has made us increasingly competent in knowledge and power, but it has also left us decreasingly confident about right and wrong. The evolutionary past has not been easy to connect with the ethical future. There is no obvious route from biology to ethics – despite the fact that here we are, *Homo sapiens*, the wise species, lacking wisdom, troubled by ethical concerns, with our own future and that of our planet in our hands. The genesis of ethics is problematic.

These three features of our human life on Earth – knowledge, power, and duty – are especially puzzling. How does reason, the mind with its knowledge, fit into the biological picture? Does it produce only more survival power? Does not this human mind gain some new power, manifest in cumulative transmissible cultures, that changes the evolutionary story? Such knowledge and power, descriptive facts of modern life, seem inescapably to prescribe an ought. But the same science that demands a conscience has difficulty explaining and authorizing conscience, for we struggle to understand how amoral nature evolved the moral animal, how even now *Homo sapiens* has duties, humans to fellow humans, and humans to the community of life on Earth. The value questions in the twentieth century remain as sharp and as painful as ever in our history.

(2) The Domain and Focus of Ethics

Probing these fundamentals in ethics, we need to characterize ethics essentially and in the whole. We must look at the nature of ethics, so to speak, before we look at nature in ethics. There is a vigorous philosophical tradition in ethics, going back several thousand years, in cultures West and East, primeval, classical, and modern. Ethics is essentially about right and wrong, good and evil, asking what human actions produce the right and the good. Ethics has been, in many respects, plural and relative to cultures, shaped to optimize what this or that culture values. Ethics is, in other respects, pancultural, both in the sense that ethics is present in all cultures and that certain principles regularly recur (such as keeping promises, or not stealing or murdering, or being loyal to one's family). Or at least they ought to recur universally.

In view of the debates about altruism and its relation to genetics, we should take caution to recall that neither in deontological ethics nor in utilitarianism, the two main Western traditions, is altruism the pivotal principle. The moral agent does what is just, giving to each his or her due, and whether this due is to self or other is secondary. The question of fairness (justice) is not so much one of preferring self over other (I win; you lose), or other over self (you win; I lose), but of distributing benefits and losses equitably (summing wins and losses, we each get what we deserve). The agent does the greatest good for the greatest number, which might mean benefits to self and/or to other, depending upon options available.

The Golden Rule urges one to love neighbor as one does oneself, but this is not other love instead of self-love. "Do to others as you would have them do to you" seeks parallels in the self's doing for others with others' doing for the self, suggesting reciprocity as much as antithesis between self and other. The first and most wide-spread Hindu and Buddhist commandment is noninjury, *ahimsa*, whether or not the injury is to others or to self. The commandment enjoins self-defense as well as defense of others threatened with injury. Taoists seek to balance the yang and the yin, something quite different from egoism versus altruism. Aristotle recommended the golden mean, also a balancing of values. Doing the right, the good is a matter of optimizing values, which often indeed means sharing them, but this is never simply a question of benefiting others instead of oneself.

Socrates insists that virtue is its own reward and even claims to know with certainty that good people do not lose. "No evil can come to a good man" (*Apology*, 41d). Whoever wrongs another person always damages his or her own well-being more than the victim's. The only true harm that can befall self is to one's character; doing the wrong thing ruins character, the worst result imaginable. The right thing ennobles the person; beyond that there is nothing higher to be won. "Do we still hold, or do we not, that we should attach highest value not to living, but to living well?" "We do." "And that to live well is the same as to live honorably and justly; do we hold that too, or not? We do" (*Crito*, 48).

Doing the right is ipso facto such a great benefit that even if considerable other harms result, the just person never loses. For no accumulation of resulting harms can weigh negatively more than doing the right weighs positively. Doing right gains an *arete*, an excellence, and gaining such virtue more than compensates for other losses, such as one might have in business, political, or social affairs. Or, presumably also in the loss of possible benefits to one's offspring. In any case, Socrates' concern is amply for the self's doing well as the self does well by others. There is no egoism–altruism dichotomy pivotal to his ethics.

The Hebrews claimed that the righteous person is "like a tree, planted by streams of water, that yields its fruit in its season," by which the sages, prophets, and rabbis meant both good deeds and a prosperous family. Such a person is, in their idiom, "blessed" (benefited), and by contrast sinners "perish" (Psalm 1). The Hindus and Buddhists interpreted the value of virtue in terms of good karma, deeds that benefit others and self at once. Calculating whether the self wins or loses in a direct tradeoff with whether others gain or lose can hardly be said to be the principal axis of analysis of any ethical system in the classical past or contemporary present. The questions are more those of justice and love, or integrity and virtue, or honor, or optimal quality of life – that is, of good and evil, right and wrong.

Many dimensions of morality do not directly focus on altruism: questions of the rights of the minority, of capital punishment, the extent of free speech versus pornography, preferential hiring, abortion, euthanasia, and so on. Ethics is about optimizing and distributing moral and other values, about what sorts of values count morally, and what the moral agent ought to do to promote these values. This is a more comprehensive question than whether the self is preferred to others or vice versa.

Nevertheless, ethics involves altruism and constrains egoism. Altruism in the ethical sense applies where a moral agent consciously and optionally benefits a morally considerable other, without necessary reciprocation, motivated by a sense of love, justice, or other appropriate respect of value. Selfishness in the ethical sense is to be distinguished from self-defense and self-actualizing, both of which are commendable and necessary activities. Selfishness applies where a moral agent exceeds the bounds of legitimate self-interest and is so concerned with self that the appropriate motivation in love, justice, and respect for the interests of others fails.

Moral altruism is done because it is just or right, and done freely and generously under such intent. The motivation may at times be tacit; not all altruism or selfishness is calculating and consciously reflective on each particular occasion, but behavior must flow from formed character that, in the course of personal life, has come to be responsibly owned by the moral agent. A striking feature of such altruism, when culturally mature, has often been its universalism, and that will prove the ultimate test of any genetic, Darwinized ethics.

(3) Interhuman Ethics

Ethics emerges to protect values within culture, which, historically, has been its principal arena. As philosophers frequently model this, ethics is a feature of the human social contract. If ethics is in any sense "natural" to humans, it will be "in the nature" of this highly social species. If ethics is "rational" for humans, it will be that there are benefits to persons who live in the resulting kind of society, vital for the flourishing of this social species. Natural processes in the wild serve animal life well, but these are not processes in terms of which the values achieved in culture can be fully protected; one way or another, there emerges morally responsible agency to protect human life and its cultural values. Analogously, in the analysis of science (Chapter 4), those same genetic processes, superbly creative in biological evolution, were not processes in terms of which the values achieved in science could be fully interpreted, where deliberated rationality occurs.

In a general way, one can expect that peoples who have a functioning ethical code will do well.[2] In that sense, the emergence of ethics in a highly intelligent, highly social animal is not surprising. One can start to construct ethics by beginning with rational self-interest. I will try to arrange a society where others do not lie, steal, kill, and so on, because I lose when people do these deeds. A prevailing ethical system is my best shield against such troubles. This will be true for anybody in any society. Acting in my self-interest (I might first think) I myself will lie, kill, steal, as need be – but not they. But of course, when I realize that they too are actors thinking the same way, I will have to cooperate or they will not. There is no reason for them to tell me the truth, respect my life and property, and so on, unless I reciprocate. This, again, will be true for everybody. So it is in everybody's best interest to enter into this social contract. In this sense, every ethicist holds that ethics is good for people. The Hebrews put it this way: "If we are careful to do all this commandment" it will be "for our good always" (Deuteronomy 6.24–25).

All we had before was a concern for one's own advantage, but here "own advantage" has expanded to "shared advantage," "mutual advantage," not for all benefits but for many, which are shared

---

[2] A point elaborated by Allan Gibbard; see Section 2(5).

first in the genetic kin and second in those who are axiologically kin, that is, who share one's culturally acquired values. The benefits often diffuse into those that cannot be differentially enjoyed – such as public safety or the right to vote – since what makes society safe and democratic for me (life preservers at the pool, free elections) confers these benefits at once on others. With such benefits it is hard to be self-interested about them, at least in the individualistic and genetic senses of "self-interested" with which we began.

There will be tradeoffs, my good against yours, and hence arises the sense of justice (each his or her due), or fairness (equitable outcomes for each), or greatest good for the greatest number. Such standards can appeal to every actor, in whatever culture (even though the detailed content will to some extent be culturally specific), because on the whole this is the best bargain that can be struck, mindful of the required reciprocation. Often it may be hard to reach more than a truce between parties pressing their self-interests, enlarged as these may be into kin and reciprocating groups. But in such disputes issues of justice and fairness will arise.

A concern to behave fairly or justly is something more than a concern for self-interest, but at least those who press such self-interest publicly will have to do so in the name of fairness and justice, and they will learn how to argue fairness and justice for their own sake, and perhaps will learn to feel the force of the unfairness and unjust allegations should these be used against them by others, or even should they themselves realize (cheating when they can get by with it) that their conduct is unfair or unjust.

Further, there is considerable satisfaction both in being fairly and justly treated and in realizing that you keep your end of the bargain, even at some cost.[3] What one ought to do, in any place, at any time, whoever one is, is what optimizes fairly shared values, and this is generically good, for both the self and the other, who are in parallel positions. One way of envisioning this is the so-called "original position," in which one enters into contract, figuring out what is best for a person on average, oblivious to the specific circumstances of one's time and place, including one's genome or culture (Rawls 1971). This is a sort of self-ignorant self-interest, whereby one is ignorant of all the particulars of one's self, and thereby alerted to what would be generally in everybody's self-interest. This is where the

---

[3] The question of "cheaters" is addressed later.

sense of universality, or at least panculturalism, in morality has a plausible rational basis.

The result – though persons act in their generic self-interest – is to pull the focus of concern off self-center and bring into focus others in the community of persons. The single self must find a situated cultural fitness; a person ethically adapts to his or her neighbors. Beginning with a sense of one's own values to be defended, ethics requires becoming more "inclusive," recognizing that one's own self-values are widely paralleled, a kind of value that is "distributed" in myriad other selves. The defense of one's own values gets mixed, willy-nilly, with the defense of the values of others. Recalling terms used earlier to model the interconnections of genetic values (Chapter 1, Sections. 6–7), in this "contract," one has to "participate" or "share" in this larger community of valued and valuing agents. The self-defense of value gets "multiplied" and "divided" by this interactive network of connections.

Values must be recognized as widely "dispersed," "allocated"; as having extensively "proliferated" beyond oneself; now the protection of values has to be "shared," distributed in "portions," some in self and some in others. Now, however, contrary to what was true genetically, "share" can begin to take on moral tones, as can "selfish." So ethics develops into an effort to honor the intrinsic worth of persons, beginning with self and extending to others one encounters, and comes to require protecting them and what they value simultaneously with oneself. Toward their fellows, humans struggle with impressive, if also halting, success in an effort to evolve altruism in fit proportion to egoism.

Such human flourishing will require provision for human reproduction, of course, but nothing is here being said about maximizing or optimizing the numbers of one's offspring. In culture, people can, and often do, act in ways that decrease their fitness – if this is described biologically in the reproductive sense. Such acts can be understood in terms of conserving what the actors value, but the conservation of biological value underdetermines such events. The self is not simply biological and somatic but cultural and ideological. The self is expansive and finds an entwined destiny with many other persons, because what the self values can only be sustained if people act in concert, reproducing the gains of the past in the present and ensuring them for the future. Over generations, adequate biological reproduction of the human species is necessary for this, but cultural reproduction, conserving what one values in one's heritage, is equally required.

Though one enters into the social contract in enlightened self-interest, it does not follow that morality never rises to still more enlightened consideration of the interests of others. After the agent has interiorized his or her bonding to others in society, he or she may come more and more to identify with those with whom values are shared. An equitable distribution of benefits lies at the root of justice. But enlightened self-interest, supporting justice, is not the upper limit of moral development. Some persons, more than others, or all persons, some of the time more than other times, will move beyond such bargaining to envision a nobler humanity still to be gained in a more disinterested altruism that takes a deeper interest in others. Our sense of identity is enlarged, but this does not mean only selfishness enlarged. It cannot be that everything we desire, intend, or incline to do is acting in our self-interest, much less is selfish. We can consider and intend the interest of others, as part of our enlarged network of values in which the self is constituted.

This can motivate benevolence, beyond justice, where one acts to promote values respected in others, values both already there and facilitated by one's act of benevolence. One does not just fear loss from misbehaving others, but one is drawn to protect the benefits at stake in others by behaving morally toward them. This includes culturally transmitted values into which both the self and these others have been educated (such as "human rights," "the Christian faith," "the democratic tradition," or "French literature"), and one values the continuing success of enterprises larger than oneself. One ought to defend one's colleagues in such enterprises, both for their individual integrity and for the integrity of the values for which they labor. If this is an ideal not always real, even the failure to reach it attests its ultimate reality.

Such moral contracting is not possible for the other creatures. An oak tree is not endowed with the capacity to consider the welfare of other oaks, much less that of the squirrels who eat its acorns; nor can the squirrels consider the oak. Higher animals realize that the behavior of other animals can be altered, and they do what they can to shape such behavior. So relationships evolve that set behavioral patterns in animal societies – the dominance hierarchies, for example, or ostracism from a pack or troop. It seems, however, that the degree to which even the higher animals have options among which they can reflectively choose is minimal. Animals are capable of performative self-actualizing, but absent such considered options, they cannot choose either right or wrong.

Further, since animals lack the concept of other minds (Chapter 3, Section 1; Cheney and Seyfarth 1990) they do not reach the second-order intentionality required for contracting with other intentional selves. To become a reflective agent interacting with a society of similar reflective agents, knowing that other actors (if normal), like oneself, are able to choose between options and are responsible for their behavior – this is not within the animal capacity. Nor is there any cultural or ideological heritage to defend.

In this sense, animals are unaware that there are mental others, that is, other *minds*, who might be held responsible, to whom one might be held responsible. This precludes any critical sense of justice, of values that could and ought to be fairly shared because they are enjoyed by others who, like oneself, are existential subjects of their own lives. Even more, this precludes loving others in the morally responsible sense, although there are pair bonding, grooming, and the pleasure of the company of others. After her years of experience Jane Goodall writes: "I cannot conceive of chimpanzees developing emotions, one for the other, comparable in any way to the tenderness, the protectiveness, tolerance, and spiritual exhilaration that are the hallmarks of human love in its truest and deepest sense. Chimpanzees usually show a lack of consideration for each other's feelings which in some ways may represent the deepest part of the gulf between them and us" (van Lawick-Goodall 1971, p. 194).

In their private worlds, such consideration is not a possibility, nor is any morally binding social contract such as that in interhuman ethics. Yet all this, undeniably, has emerged within the human genius. In contrast to genes, which are not capable either of "sharing" or of being "selfish" in the deliberated, moral meanings, in humans we have a moral primate, where agents can and ought to share as well as to be self-affirming, and where such agents can and ought not to be "selfish."

(4) Environmental Ethics

Ethics is both interspecific, between humans, and intraspecific, in human relations with morally considerable nature. Though arising

within culture, ethics can and ought to return to encounter the creation and the creatures, the processes and products of nature's genesis. When humans do this, they transcend the spontaneous natural
environment because they have dutiful oversight of it. Animals have
the capacity to see only from their sector, their niche. But humans
also see others in their niches, and evaluate the biodiversity on our
planet and the evolutionary ecosystems that have generated and sustain such biodiversity. Humans study warblers to conserve them or
see Earth from space and set policy about global warming.

The originating context of ethics, human culture, has yielded a
sense of ethical priority toward humans, often ethical exclusivism.
Humans count most; indeed, some say, only humans count. Every
other living organism defends only its own kind; humans behave
that way too, maximizing their own kind – and justifying (defending) their position by claiming to be the only species with and of
moral concern. An environmental ethic that is in principle a resource
conservation ethic may be driven by self-regarding pragmatism; one
is helped or hurt by the condition of one's natural environment, no
less than by the condition of one's social environment. Analogously
to the way that humans in their cultures have an entwined destiny
with other humans (a social contract), humans are biological beings,
Earthlings, who have an entwined destiny with natural systems (a
natural contract, Serres 1995). They require photosynthesis, insects
for pollination, earthworms in the soil, and so on, for their flourishing. Such an environmental ethic may be driven by a respect for
other persons, including future generations, whom one regards altruistically and who are also helped or hurt by their environment. But,
unlike the social contract, nothing in nature is the direct object of
moral concern.

There also arises, however, environmental ethics in the deeper,
nonanthropocentric sense, where humans regard animals, plants, endangered species, and ecosystems with appropriate respect for their
intrinsic natural values. Unlike the social contract, in which other
moral agents may refuse cooperation, in environmental ethics the
objects of moral consideration are not reciprocating moral agents.
One protects the whales, but the whales cannot be expected to respond by protecting humans. Many of the endangered species protected (*Rhododendron chapmanii*, Chapman's rhododendron, rare and
endangered in swamps in Georgia and Florida) can hardly be said to

be important enough in ecosystems that human destiny is entwined with their flourishing. Many of them humans could do quite well without.

Nevertheless their value constrains human conduct. The human moral agent encounters value independently of human society and its contracts, value generated in spontaneous and self-actualizing nature. That value, which *is* present and threatened, *ought* to be preserved. That conviction arises from a love beyond self-love, although, as mentioned earlier, it need not follow that, in deeper senses, the moral agent is worse off for such acts.

Yes, so it seems – comes an objection – but the motivation can still be human-centered, and this is given away by noting that humans gain when they protect nature. At times they gain more than evident natural resource benefits: they gain an excellence of character, a virtue – recalling the virtue that Socrates so treasured. An interest in natural history ennobles persons. It stretches them out into bigger persons. Humans ought sometimes be admirers of nature, and that redounds to their excellence. A condition of human flourishing is that humans enjoy natural things in as much diversity as possible – and enjoy them at times because such creatures flourish in themselves. Humans can always gain excellence of character from acts of conservation.

If the human excellence really is the motivation, however, such an environmental ethicist is not especially seeking the good of nature, but rather seeking his or her own good – the real payoff. If ever it were the case that such a person could increase his or her welfare and harm nature, win–lose, there would be no restraint. Nature is only good as an enricher of persons. But this seems logically confused. If the virtue of human character really comes from appreciating other, nonhuman forms of life, then surely that is also to discover intrinsic value in these others. The human virtue is, at this point, axiologically tributary to that. How could humans gain much virtue by valuing and conserving something that has no value in itself? When a human donates money for the preservation of the whales, this is not covertly the cultivation of human excellences; the life of the whales is the overt value defended. An enriched humanity results, with values in the whales and values in persons compounded – but only if the loci of value are not confounded.

In environmental ethics, in the deeper sense, the human mind has grown toward the realization of its excellences by appropriate respect for nonhuman values in nature. Here humans are higher than whales only as and because humans, moving outside their own immediate sector of interest, can and ought to be morally concerned for whales, whereas whales have no moral capacities to care for humans and can neither cognitively entertain a concept of humans nor evaluate their worth.

Notice, too, in view of the question of naturalizing, or Darwinizing, ethics toward which we are headed, that maximizing the moral agent's offspring is not here of concern. When I donate money to the fund for the whales, I need not even know that I have genes, or if I do, the genes be damned, so long as the whales are saved. I may do it "for my children," but I may donate even if I have no children. I do want to convert other persons to my conservationist ideology, but their genetic relationship to me is immaterial. I enjoy knowing that the whales are safe in their marine ecosystems. Label that a "selfish" motivation if you must, but that enjoyment does nothing to increase my fertility. John Muir and David Brower, if anything, will have fewer, not more offspring in the next generation as a result of their effort spent in protecting Hetch Hetchy and Glen Canyon. I do not expect whales, warblers, or grizzlies, much less forests and canyons, to reciprocate with some reproductive benefit; the animals can do nothing to assist me (or any other humans) somatically or genetically.

Insist perhaps that what I am really doing is identifying my "self" with the ecosystemic whole, or preserving my life support system, or whatever; this does not aggrandize the self or its genetic line so much as it stretches the "self" and expands its line beyond recognition into the community it inhabits. Insist that I do not really have a concern for the whales, warblers, gorillas, and pristine forests. I am only protecting my recreational opportunities. Perhaps, on an even less plausible account, I am only self-deceived and parading my beneficence so that other humans will laud me and assist my offspring, since I am an environmentalist hero. But isn't the simplest account not to argue away what evidently seems to be going on, and to recognize that the "self" has been elevated once more into genuine morality, now regarding nonhuman others, where the self can detect values outside itself, and come to embrace these in freedom and love because it is right to do so.

(5) Ideal and Real: Moral Failure

Morality is an ideal, only partially attained. There is a gap between the ideal and the real; the human moral status is ambiguous. A great many thinkers have concluded that humans are born selfish, and some influential ones have concluded that there is not much, if any, possibility of altruism in the deeper sense, short of some kind of redemption or transformation of the nature that humans inherit biologically. These views have sometimes claimed to be scientific, as with Freud in psychoanalysis, Skinner and his behaviorism, or Durkheim and his sociology, but before the rise of science they were just as intensely advocated by pre-Darwinians such as Luther, Calvin, Aquinas, Augustine, Saint Paul, Jesus, or Gautama Buddha. Human *concupiscence* or *cupiditas* (selfish desire) has long been recognized and lamented, *eros* (self-love) produces vices as readily as virtues, and the need for moral reformation is no new claim. What might be new is that, now in biological science, the cause has, for the first time, been found in genetic determinants.

Moral failure appears with the possibility of "cheating" on the social contract. Yes, one ought to cooperate; yes, that is in fact what produces the greatest good for the greatest number. Yes, if everybody else cheated, one would suffer. If one is caught, one is penalized, is ostracized, and loses. But one might not be caught, and so could gain at the expense of others. Left to one's calculating self-interest, "cheating" seems promising, if risky. To this one may well be inclined, if all that is operating is one's animal heritage, since animals are impelled to their own self-defense, outside any reflective social contract. The selfishness that is deplorable arises when persons are unable to rise from the defense of life proper at the animal level to a moral level proper to human destiny.

Self-actualization remains appropriate, a heritage of the biological past and requisite still for conserving the values of personhood. But this becomes selfishness where the human career is "nothing but" such self-defense, failing to rise to the moral perspective that others too have lives worthy of our defense. The historic route into such moral development involves reciprocating cooperation, and there is nothing amiss about enlightened self-interest cooperatively joining with similarly moral others. Of course also, on the immoral side, the self, perhaps in a dominant social class with other selves, can take advantage of opportunities to exploit human tendencies to be moral, and such selves, in their social classes, can use morality to enforce their positions of power, as well as their biases and prejudices (perhaps with property laws reinforced by commandments against stealing, or legitimate patriotism enforced by the divine right of kings, or commandments to chastity used by dominant males to control subordinate women).

2. NATURALIZED ETHICS? EMERGENT, SOCIALIZED MORALITY

How far can ethics be naturalized, or Darwinized? Perhaps there are ways in which it can be naturalized, though not Darwinized (Nitecki and Nitecki 1993; Thompson 1995; Bradie 1994; Farber 1994; Rolston 1995; Sesardic 1995; Kaye 1986; Breuer 1982). By some accounts, ethics *genuinely emerges* out of animal cooperation (Section 2). There does appear more out of less, an ideal mixed with, guiding the real. By other accounts the emergence of ethics is *illusory*; the real forces are still those of natural selection (Section 3). What appears to be more is really less than appears. Or, though natural selection is at work, the emergence may be *epiphenomenal*.

(1) From *Is* to *Ought*: Emergent Morality

How does what is not possible for animals become possible for humans? Human mental and cultural development somehow generates a possibility space for ethics, which emerges where none was before. This account of the origins of ethics will not be by implication. We cannot posit flowers, squirrels, or chimpanzees and infer moral agents, any more than we can infer scientists or saints. We will have to narrate the evolutionary adventure into conscience, finding the plausible routes, which may make a certain sense, even though there is no formal implication. The challenges will be to unify movements that ought to be unified and to keep separate things that ought to be kept distinct. Without discrimination we will have a muddle. Too much unity too soon (natural selection as the real determinant of all events in nature and in culture) is likely to be simplistic. Too much disconnected plurality (nature at random drift; nature severed from culture; plural ethical codes with nothing in common; a blooming, buzzing confusion) is likely to miss the storied development of nature evolving into this emergent in culture.

An account of historical origins is not the same as an account of contemporary logical relationships; to think so is the genetic fallacy. If evolutionary theory is the only available explanatory category, then scientists can be expected to make a determined effort to cut all evidence to fit it, including ethical evidence. The ethical evidence might, in fact, be counterevidence to any extrapolation of the theory into this range of human behavior. Such morality is a phenomenon of culture, where it emerges as a higher level of the defense of value, rising out of animal cooperation. This is one of the marks of our humanity, along with the capacity to construct religious, philosophical, and scientific worldviews. There are precursor animal roots, but few will claim that morality is "nothing but" genetically determined animal behavior.

G. G. Simpson, after surveying evolutionary history, insists:

> As applied to man the "nothing but" fallacy is more thorough-going than in application to any other sort of animal, because man is an entirely new sort of animal in ways altogether fundamental for understanding of his nature. . . . The human species has properties unique to itself among all forms of life. . . . Man's intellectual, social, and spiritual nature are altogether exceptional among animals in degree, but they arose by organic evolution. They usher in a new phase of evolution, and not a new phase merely but also a new kind. . . . Man is a moral animal. (1967, pp. 284 and 293–295)

Ernst Mayr agrees, "Genuine human ethics emerged from the inclusive fitness of our primate ancestors" (1988, p. 77). Biology still, in a general way, frames these moral decisions, for humans no less than other animals are somatic organisms and must reproduce genetically. At the same time, in the metaphor of Peter Singer, morality forms an "expanding circle" where our moral concerns exceed our biological interests (1981).

One would expect, therefore, both continuity and emergence. For example, one plausible route by which individual animal self-defense could have evolved into animal cooperation with unrelated others is as follows: Where there are memory and a capacity to discriminate between individuals, remembering who reciprocates and who does not (even though there is not yet any knowledge of other minds, only of other behavior), a strategy (dubbed TIT FOR TAT) can evolve, which involves cooperating initially, never thereafter refusing to cooperate if the other does, refusing to cooperate when and so long as the other refuses to cooperate, and restoring cooperation at once if the other ventures it (Axelrod and Hamilton 1981; Axelrod 1984; Nowak, May, and Sigmund 1995; Crowley and Sargent 1996). At least such a strategy can appear on computer simulations.

The benefited other may (or may not) be kin; what counts is that the other be a reciprocator. The number of future interactions with a known reciprocator needs to be frequent and indeterminate, that is, an ongoing small community (Axelrod 1984; Axelrod and Dion 1988). Such a strategy can get established in a (computer-modeled) population, remain established, and resist invasion by various other strategies, particularly by noncooperation.[4] Dawkins remarks, interestingly, that here, their selfish genes notwithstanding, "Nice guys finish first" (1989, p. 202). Though initiated at the nonmoral level in animals, that is not an immoral strategy, were a moral agent deliberately to continue it. It is a sort of operational version of the Golden Rule, doing to others as you would have them do to you, while refusing to be taken advantage of.

Axelrod explains:

> What accounts for TIT FOR TAT's robust success is its combination of being nice, retaliatory, forgiving, and clear. Its niceness prevents it from getting into unnecessary trouble. Its retaliation discourages the other side from persisting whenever defection is tried. Its forgiveness helps restore mutual cooperation. And its clarity makes it intelligible to the other player, thereby eliciting long-term cooperation. (1984, p. 54)

In such kinds of community, this strategy will displace another one (dubbed ALWAYS DEFECT), in which cooperation is withheld in favor of self-benefit, whenever the actor tries to get by with it. In terms of our axiological model, however, such a strategy really means always defend your own values (ALWAYS SELF-PROTECT),

---

[4] Such behavior patterns might be maintained once well established but be difficult to get started. The mutant "altruist" who ventures cooperation in a society of "selfish" others will always lose because others will take advantage of the novice altruist. Mutant altruists will be culled out. There would be a trajectory problem getting up to high enough frequency of reciprocators, crossing a critical threshold where reciprocal altruism can begin to work. Nevertheless, at least on computer models, such patterns are regularly originated and established (Crowley and Sargent 1996).

and this is the only strategy possible to lower life-forms. There is nothing improper about this there. There is also something impressive when in higher animals this evolves into, If possible, always cooperate in defending your values, but refuse to be a pushover for noncooperators, because this destabilizes the cooperative system. As before, values thereby become entwined in community.

With the TIT FOR TAT strategy as a starting point, especially in heterogeneous systems in which there are uncertainties, probabilities, and errors, some "fuzziness," a further strategy will evolve in which an actor is more "forgiving." If confronted with a defector, one retains a "tendency to cooperate" and responds in kind less frequently, perhaps once in three defections, or only after several defections by the same party. This strategy is dubbed GENEROUS TIT FOR TAT, again run on computer simulations but now thought more nearly to resemble what life is like in the real world (Nowak et al. 1995; Nowak and Sigmund 1992).

There are other variations. If there is too much tolerance of defectors, the reciprocating system can grow unstable and collapse, so variations often insist on penalizing defectors; they can include penalizing those who too generously (or naively) forgive defectors and thus do not enforce cooperation. This results in a strategy called FIRM BUT FAIR, for example. All the successful variants initiate and stabilize the required reciprocation and, where it fails, simultaneously penalize failure and restore cooperation when opportunity arises.

Such strategies are mathematical game models, run on computers, which are neither biological nor ethical. They are mostly one player facing another player, not groups, and may not model social systems.[5] In groups, ostracizing (refusing to play) with noncooperators also arises. The cooperators find ways of segregating themselves into subgroups, which outperform the others. Real animal life is in social groups where the opportunity for likelihood of reciprocation is uncertain and variable. Estimating how probable it is requires powers of recognition and evaluation of others (their reputations!) that animals (including humans!) may or may not have. One can model

---

[5] Robert Axelrod, seminal in the development of TIT FOR TAT strategy, is a game theorist and political scientist, not a biologist. The application was first to nation states, international relations, law, corporate business, and markets; application to biology was secondary and derivative.

some of these features in probabilistic modes on computers, but it is hard to say which conditions, if any, model the rather messy real world adequately. The computer models do not require foresight, or intention, much less consciousness, only memory, and operate with rather formally structured and overly simplified rules. Whether they map anything in the real world of genetics and ecosystems is arguable. TIT FOR TAT and its various modifications are not particularly well verified in the animal world, although there is, for example, food sharing among chimpanzees presumably because individual chimps do better, on average (de Waal 1989).

Even if the computer simulations did map onto the biological world, these are not yet moral systems. This is animal behavior, arising where animals reach powers of memory and cognitive discretion sufficient to permit such reciprocity. But the point is that even here the logic of the system generates cooperation. The values defended by an individual, even when still operating in its self-interest (as such animals can only do), interlock the individual into cooperative reciprocity. Willy-nilly, the individual is webworked into accommodating the values of other individuals. Values have to be "distributed," "multiplied," "divided," "collaborated," "portioned out," or "shared" – as surely as can they be "selfishly" defended. The whole point of these reciprocating strategies is not to beat your opponent – those who try that fail – but to cooperate to win. To win, two must act as a team, self and other. Three, four, five, a clan, a tribe may be still better. Winning must be done in concert.

Whether the models apply to personality and morality needs much more discussion (Sesardic 1995, pp. 149–155; Dugatkin 1997a; 1997b). The outcomes simply result from rules of the game, with randomly ventured heritable trials, interpreted as "strategies" for gaining benefits (offspring). "Cooperate," "defect," and "generous" are metaphorical interpretations of such computer simulations, as are "selfish" and "sharing," but if such simulations do reflect events in natural history, the evolution of such reciprocity processes will be of philosophical interest.

One possibility is that in *Homo sapiens*, such a strategy, which had evolved in precursor animal life, is continued and becomes the pivot point from which there emerges a more expanded, reflectively self-conscious social contract. A moral community is superposed on what before was only a reciprocating biological community. The moral agent comes to live in an ambience of values in which it is simulta-

neously in one's self-interest and in the interest of others to participate and to share. In human affairs, the emotions come into play, as
well as self-interested rationality. One develops a concern, a caring,
for those with whom one has been so regularly reciprocating. Commitments develop (Frank 1988). This may not yet be altruism, but it
is ethical. One ought, both (whether) for one's own sake and (or) for
the sake of others, to cooperate in such reciprocity. Thus morality
gains a foothold, by which the step to genuine altruism could then
be taken.

In the analysis to follow, we recall (Section 2[2]) the evolution of
animal cooperation and inclusive fitness, and (3) enlarge this into
human reciprocal altruism, (4) further expanded into an indirect, socialized altruism, all this setting the stage for the emergence of (5) an
ethics naturalized, then socialized, and even (6) universal altruism.
But any such seeming altruism – so these biologists will now be
claiming – is to be explained as illusory, naturalized, or Darwinized
(Section 3). The deeper motive is always genetic self-interest, whether
in inclusive fitness or reciprocal or indirect altruisms, or (Section 3[1])
self-deceived altruism, or (Section 3[2]) induced and inflated altruism. By still other accounts, morality is not so much illusory as epiphenomenal (Section 3[3]). The result (Section 3[4]) is a dilemma how
any moral actor, and especially these scientists themselves, can get
from *is* to *ought*.

(2) Animal Cooperation: Inclusive Fitness and Reciprocal Altruism

Animals evidently sometimes cooperate, extensively so (Dugatkin
1997a; 1997b; Harcourt and de Waal 1992; de Waal 1989). It seems
plausible that the human capacity to be ethical arose out of animal
cooperation, although one should be circumspect. Animals might behave cooperatively as a result of biological and ecological causes,
whereas humans behave cooperatively for cultural, philosophical,
and religious reasons. The sets of causes and reasons might partially
overlap; causes might evolve into reasons. Human language is highly
developed, permitting reasoning levels unavailable to animals, and
this introduces capacities for evaluating others critically, as well as
evaluating one's relationship to them. Animals are unable to address
the problems posed by morality (such as fairness, justice, benevolence, rights, equality, impartiality, responsible guilt and merit,

moral intentions in other minds), but they do interact in their societies and have their behavioral rules, which must have been where
ethics got launched.

Such an emergence of ethics seems generally plausible. And yet, if
one supposes selfish genes producing a selfish organism, there also
seems, initially, little place for helping others, especially unrelated
others. "The central theoretical problem of sociobiology [is]: how can
altruism, which by definition reduces personal [individual] fitness,
possibly evolve by natural selection?" (Wilson 1975a, p. 3). The problem is that much helping of others will prevent such altruists from
reproducing themselves. They will be culled out of the population.
Or so it first seems, but then again it seems obvious that, sometimes
at least, helping others can help oneself. Certainly this is true of
helping related others.

Cooperation with kin can be understood in terms of "inclusive
fitness" ([Hamilton 1964; Chapter 1, Section 6; Chapter 2, Section
3[1]). In troops of chacma baboons (*Papio ursinus*), the dominant male
places himself in an exposed location as a lookout while the other
troop members forage (Wilson 1975a, p. 121). He is at some risk
while on sentry duty. Predators will see him first and he is not getting anything to eat. He is providing a benefit to other troop members. Without supposing that the dominant male is a moral agent,
"an entity, such as a baboon, is said to be altruistic if it behaves in
such a way as to increase another entity's welfare at the expense of
its own. . . . 'Welfare' is defined as 'chances of survival' " (Dawkins
1989, p. 4).

But look more closely. At the genetic level, the dominant male is
the father of many juveniles and has nieces, nephews, and many
relatives in the troop. Summing up his inclusive fitness, adjusting for
risks and probabilities, the "selfish" benefits distributed to relatives
exceed any losses and risk to the baboon himself. He is really defending his enlarged, reproductive self when he risks his individual organismic self.[6] Meanwhile, he also conveys some benefits on unrelated others; these will be accidental to the actual determinants of his
behavior.

Humans evolved with this animal heritage, and therefore one can
use the same theory to interpret "altruistic" acts when a family member risks danger to protect his kindred. Alexander states, with em-

---

[6] The case is more complex than appears. See Kitcher (1985, pp. 159–193).

phasis, the fundamental principle of both animal and human behavior: *"Lifetimes have evolved so as to promote survival of the individual's genetic materials, through individuals producing and aiding offspring and, in some species, aiding other descendants and some nondescendant relatives as well"* (1987, p. 37; 1993).

Since the baboon, though on sentry duty, is not a moral agent, he has no moral duty. We have already doubted whether "selfishness" is the appropriate interpretive category here for wild nature, labeling it pejoratively with a term borrowed from more complex human moral failure in culture. The behavior is self-defense, self-actualizing proper to animal life, a defense not only of somatic self but of familial and specific forms of life. The "self" cannot be isolated and singularly preserved; it must be integrated, socialized, redistributed, mingled with other "selves," likewise shuffled. If human cooperation originates here, this much alone could be a promising origin of ethics in values already shared in premoral animal behavior. "Inclusive fitness," whereby "my" becomes "our," is a welcome precursor to ethics, although we must be clear about what additionally emerges with its elevation into altruistic moral concern.

At the next level, we complicate the inclusive fitness picture with "reciprocal altruism" (Trivers 1971; 1985, pp. 386–389). Animals may serve their self-interest by helping each other out oblivious to close kinship. There are certain things it is difficult or inconvenient for a baboon to do for itself (backscratching) and that others can easily and conveniently do for it, and it can reciprocate (scratch their backs). Genetic relationships make no difference; a foreign backscratcher will do as well as a brother, subject only to the likelihood that the second will reciprocate later when the first gets an itch. Partners do not have to be kin. Reciprocal altruism underlies the success of the TIT FOR TAT strategy, examined earlier.

In a cooperative society, animals can lower their risks. A vervet monkey will give an alarm call and identify danger by the type of call – leopard, eagle, or snake (Seyfarth, Cheney, and Marler 1980; Cheney and Seyfarth 1990). Any other monkey, related or not, can benefit from the call, while the caller puts himself at some risk by identifying his location to the predator. On a later occasion, if the caller himself is unaware of an nearby predator and is alerted by some more distant monkey, perhaps one outside his family line, his life is saved. The monkey in danger is at high risk of losing everything; the caller alerting others to a predator spied at a distance is at comparatively low risk. Because of this asymmetrical risk factor, a lot of help given at a little cost, both parties can, overall, lower their risks. Each gains and is more likely to live to reproduce than if neither calls.

The monkeys are not moral agents, though they seem to intend to change others' behavior. They need not even intend anything; there may just be natural selection favoring those who instinctively respond so, eliminating those who do not. Everything depends on reliable reciprocation in such a simian society. Those who first benefit could later "cheat," decline to reciprocate, with all gain and no loss. To prevent this there will evolve ways to keep would-be cheaters in line, perhaps by remembering which ones they are and refusing to cooperate with them, perhaps by intimidating them, ostracizing them, not mating with them, and so on. That too is part of the TIT FOR TAT strategy.

When reciprocal altruism is working well, there are no losers on long-term average, although there are short-term losers. Generally each gains more than is lost, although benefits and losses may, on statistically rare occasions, be maldistributed. In a win-win situation, when one "self" has an interest that coincides with that of another, the mutual parties are each acting in their self-interest, self-actualizing, but this ought not to be called "selfish" with any implications for censurable moral selfishness. All this could be said to be enlarged "selfishness." But in the same way that "inclusive" fitness is not a very selfish kind of fitness, reciprocal altruism is not so "selfish" as alleged. The "self" is getting coupled up to other selves willy-nilly. At the same time reciprocal "altruism" is also a misnomer, since, by the genetic definition of altruism, neither of the reciprocating parties is sacrificing any fitness.

One approach is resolutely to hang on to the central model of "selfishness" and see all these others as being exploited by the original self. But it is just as plausible to see the self as being distributed further into the reciprocating system and to transpose to a communitarian paradigm. The backscratcher or alarm caller is getting socially entwined with the lives of others, somewhat analogously to the way in which, earlier, the system embedded the fate of any one gene with the collective fates of myriad others copresent in the genome of the integrated organism. That organism was in turn embedded in a family, its genes spread out over kindred, and all these genes were interlocked sexually with mates.

Beyond that, here, in social systems, the self is again being expanded, past those who are kindred, to all those of like kind with whom one interacts. An animal's genetic identity (Chapter 2, Section 1) is taking on further dimensions of social identity. We can return to a diagram used earlier (Figure 2.2) and add these interacting reciprocators (Figure 5.1). Before, "selfish" had to be stretched to cover benefits to father, mother, niece, nephew, cousin, children, aunts, uncles, and so on; here it must be stretched to cover benefits made to reciprocating nonkindred others. The "my" that once seemed located from the skin in has been so much the further reallocated into a broadly scoped "our." The evolutionary adventure here is becoming still less private or individualistic, still more social and communal. The picture we are getting is of benefits dispersed as much as of benefits hoarded. In terms of the paradigm we prefer, values are conserved only as they are also shared (reciprocated). The selfish gene hypothesis, which has already been enlarged through the interconnections of somatic and inclusive fitness with related others, is now enlarged by yet more extensive reciprocating interconnections, now with unrelated others.

Reciprocal altruism (so-called) is present even in animals, but animal relationships are usually not sufficiently complex, enduring, or remembered to permit its elaboration (Wilson 1975a, p. 120). Most organisms, living in rather local environments and narrow niches, are incapable of much reciprocity. "Each organism exists at the center of its own little eddy of inclusive fitness in a very shallow sea of reciprocity" (Hull 1988, p. 433). The "very shallow sea" (a pejorative term) really only means that animals do not have much capacity to act or interact outside their own immediate sector of residence (what can a warbler do to help a grizzly bear, or vice versa?); nor, even within that sector, do they have much capacity to learn deeper reciprocal relationships (neither warblers nor grizzlies can do much backscratching). One does not want to fault them for not being more than they are.

(3) Human Reciprocal Altruism

In humans, by contrast, reciprocal cooperation becomes widespread. In ancient and classical cultures, people did not help just their blood relations; they helped other members of their tribe. Today, persons

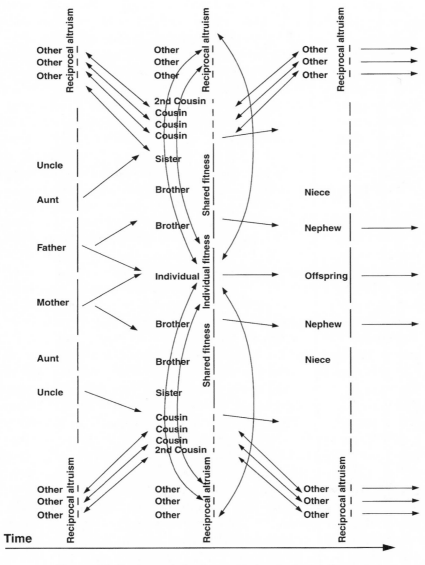

Figure 5.1. Reciprocal altruism added to inclusive fitness.

237

cooperate at work, in politics, at school, in business, and so on, with hundreds of others with whom they have no known kinship. In modern nations, with trade by truck, mail, and telephone, they may never even see or know the names of these people. The "shallow sea of reciprocity" of the animal world becomes a national and international network of cooperation. In this kind of behavior, judgments of kinship are irrelevant. This embedding of individuals in society involves neural information transmission superposed on genetic cybernetic systems. It involves language, artifacts, markets, computers, oil tankers, and jet planes.

Out-group cooperation can be just as beneficial as in-group cooperation. Brothers and cousins are nearby and can often help but are not likely to possess goods to which I do not myself also have some access. Foreigners have access to goods and skills I may need – and this is in fact what has happened in the modern world. The local self eats his or her breakfast (coffee, orange juice, bananas) with resources drawn from ten thousand miles away and brought to him through the reciprocal cooperation of ten thousand persons (all those who had anything to do with getting his breakfast there). Then he or she drives to work in a car made in Japan. Much of the prosperity of modern peoples depends on specialization of labor, and that increases the necessity and extent of reciprocating, made possible by increased powers of transport and communication. People, who still have genes, depend on global markets and international treaties. Boyd and Richerson conclude: "The evidence regarding the scale of cooperation . . . is difficult to reconcile with any model, including sociobiological models, which portrays human behavior as self-interested in the inclusive fitness sense" (1985, p. 287).

But it could still be grand-scale reciprocation, much elaborated "backscratching" with the additional complication that now we have to make judgments of the likelihood of reciprocation over great distances and time spans. To some extent, there may be just natural selection favoring those who instinctively respond to such culturally elaborated backscratching, eliminating those who do not. But there is also considerable judging going on consciously in our heads, since humans frequently do decide whether or not to help others out. There is reciprocating, but, again, this "helping each other out" is misnamed "altruism" either by the genetic definition, since agents do not have fewer offspring in result, or by the moral definition, since agents do not intend to benefit others at cost to themselves.

In many market transactions there are no losers; buyer and seller both gain simultaneously at the moment of sale. Even if the delivery of goods is later, cooperation is a win–win situation. But sometimes, there are short-term losers, persons who act with immediate loss and for the benefit of another, in the expectation of long-term beneficial reciprocation. Again, in longer focus, there are no losers (unless people make mistakes). Again too, there is nothing "selfish" about helping each other out to the mutual advantage of both. No ethical system, nor any religion, has ever lamented cooperation in which both partners gain.

Alexander claims, "That people are in general following what they perceive to be their own interests is, I believe, the most general principle of human behavior" (1987, p. 34; 1993). We expect rational persons to do what is best for them – choose the bargain, store for the future, protect their property, seek medical help. There is nothing particularly amiss about assuming a world of rational, self-interested agents. Economists and politicians do this and work out theories of how markets will operate or how citizens will vote. People are not fools all the time, not even most of the time, when their perceived self-interests are at stake. They want as much money as they can get; they want as much as they can get with their money. So sellers charge what the market will bear, and buyers buy as cheap as they can. The voters elected the Democratic congresswoman because she favored an increase in social security benefits to be offset by elimination of waste in military spending. All this is appropriate behavior in the world, and no one denies it.

Prudence is a first principle of intelligent action. If humans can help themselves in partnership, well and good; there is increased community. After all, the Second Great Commandment urges us to love others as we do ourselves, and that presumes self-love as an unquestioned principle of human behavior and urges us to combine this with loving others. If we can do this with overall loss to none, so much the better. We need not always love others instead of ourselves to fulfill this commandment. So one welcomes mutually beneficial reciprocity, wherever this can be found or arranged. This is seemingly as far as one can get, operating with current biological theory. Matt Ridley, searching for "the origins of virtue," concludes, "The argument of this book" is that "people are . . . calculating machines intricately designed to find co-operative strategies only when they assist with their enlightened self-interest" (1997, p. 214).

The most we can ever do is generate virtues out of a genetic legacy of self-interest, by modifying incentives so that persons, acting in their own interests, simultaneously act cooperatively.

But is this all there is to say about the determinants of behavior? Will it do enough explanatory work?

Consider the case when there is loss to one and benefit to another in the presence of some uncertainty about the future. One is inclined to think of this more altruistically. Someone is drowning and an unrelated friend jumps in to save him. He does so impulsively, yet he acts in character with this kind of policy, or at least with this result: the rescuer takes a risk, the drowning person benefits. The rescuer, perhaps a good swimmer, does not know whether he will ever need rescue; he does not know whether his friend will be nearby if and when he needs rescue, nor whether the friend will be inclined to reciprocate. So the act is an altruistic occasion and whether there will be reciprocation is unknown. Nevertheless, the rescuer expects continued, strengthened interpersonal relationships with various benefits; his other friends will admire him and help him out in turn, and he will in the long term probably benefit more than he loses. His altruism builds a reputation that is to his advantage, when his behavior patterns spread widely and become the ambience of his society.

As with the monkeys and their alarm calls, when drowning we are at high risk of losing everything; when rescuing another we are at comparatively low risk. Both persons can, overall, lower their risks by helping each other out (Trivers 1971; 1985, pp. 386–389). Both will, probably, leave more offspring as a result. "Doesn't even reciprocal altruism turn out to be just a more subtle and complex form of selfish behavior?" "Yes. Over the long pull, natural selection will always favor the genes of individuals who, by their behavior, have increased the reproductive success of themselves and their relatives" (DeVore and Morris 1977, pp. 51 and 84).

To keep this system going, reciprocation, though always uncertain, needs to be made as reliable as possible. There is evidence that the larger the group, the more complex arrangements become, and, in result, the more defectors can go undetected or unpunished. "Altruistic cooperation in large groups of unrelated individuals is unlikely to evolve" (Boyd and Richerson 1985, pp. 230–231). Nevertheless people do cooperate in very large groups (nations and markets). Various social devices evolve to prevent cheating, including laws, courts, police, and the human conscience, producing a sense of guilt, keeping cheaters in line.

An average benefit does not guarantee a benefit gained on every particular occasion, and so calculating individuals will find occasions to break the rules (if they can go undetected) that are to their long-term advantage. Morality helps to prevent such cheating, rather curiously encouraging would-be offenders to act in their long-term self-interest (since on average they will be detected and sanctioned), when they are tempted to cheat. But now it becomes difficult to say whether the agent, when influenced so by moral injunction, loses or wins. In the end, honesty is the best policy.

(4) Indirect, Social Altruism

Expanding the circle, there is indirect reciprocity. The drowning person may be rescued by a total stranger accidentally passing by. There is only a minuscule possibility that the saved person will ever be in a position directly to reciprocate. But this does not matter so long as people in society at large are so disposed to rescue others. This makes it probable that if ever I need rescue, some other total stranger will rescue me. "Rewards from society at large, or from other than the actual recipient of beneficence, may be termed indirect reciprocity" (Alexander 1987, p. 153). The term "indirect," however, fails to register the pervasively "social" character of such altruism.

Where there is a stable enough social order to set up a climate of expected return, I may not even know who the beneficiaries of my altruism are. I just have "to appreciate the pervasiveness and the consequences of indirect reciprocity" (Alexander 1987, p. 95). After rescuing a drowning person, I may return to install life preservers around the lake. If people in general are safety-minded, perhaps installing fire extinguishers elsewhere or promoting seat belts and highway safety, I can expect to benefit overall more than I lose. There is a feedback loop from single persons to "society at large," the "unselfish" act of any particular individual benefits unspecified beneficiaries, and this common good promoted redounds to the benefit of the individual self. That, once more, entwines the "self" with the community at large, and, as remarked on several previous occasions, there is nothing problematic about finding that self-interest is sometimes interlocked with the common good. We might not want to call such behavior pure altruism, but it is certainly not pure selfishness. Why not say that in certain areas, like public safety, there are shared values?

By now, human behavior, though perhaps "naturalized" in that the self acts (as is "natural") in his or her self-interest, has been also extensively "socialized": the self's well-being integrated with a reciprocating community. We may no longer be so sure whether the determinants are "Darwinized," in the sense of leaving the most offspring in the next generation, but such cooperation certainly contributes to human well-being, and this will include positive effects in human reproduction. By now also, an ethical dimension is beginning to emerge, for, although those entering into such a "social contract" stand to gain on average, they also acquire obligations to support this contract.

The "original position" appealed to here may include such a self's concerns about reproduction; it does include figuring out what is best for a person on average, oblivious to specific circumstances of one's time and place, genome or culture. But just this reflective element rationalizes (makes reasonable) and universalizes the recommended behavior. The altruism is "indirect" in that there is no one-to-one benefactor-to-benefited reciprocal exchange, but the altruism is quite "inclusive" just because of this indirectness. The concern is all the more widely "shared" or "distributed." One helps out in general, and one expects to be helped out in a society of reciprocating helpers. To think of this as pure "altruism" remains a misnomer, since, on the genetic definition, the agent does not lose genetic fitness, nor, on the moral definition, does the agent expect to lose when aiding others. But to think of it as pure "selfishness" is equally inaccurate, since, on the moral account, others do get aided by intention of the agent, who also benefits, and, on the genetic account, all parties remain well positioned for future offspring.

One problem with trying to be selfish with indirect benefits like this is that they are too pervasive. They loop back to the agent himself or herself, but they loop back to everybody else, nasty or nice, with about the same probability. "Everyone gains by the presence of beneficent people and the possibility of interacting with them" (Alexander 1993, p. 188). So they do not proportionately benefit the agent because they are benefiting both community and self. If so, natural selection cannot "see" the benefit to select any particular person's genome, because the agent, so far as that benefit is concerned, is not differentially benefited in producing more offspring.

(5) A Naturalized, Socialized Ethics

Allan Gibbard, in *Wise Choices, Apt Feelings*, offers "a theory of normative judgment" for both science and ethics, based on "a naturalistic view of ourselves" (1990, p. 254). Moral judgments are based on sentiments that evolved to promote the survival and welfare of human societies. "Human moral propensities were shaped by . . . multiplying one's own genes among later generations" (p. 327).

> How, then, are we to understand our normative life as part of our nature? The key must be that human beings live socially; we are, in effect, designed for social life. Our normative capacities are part of the design . . . not from design literally, but from that remarkable surrogate for design, genetic variation and natural selection. (p. 26)

To say that *x* is "moral" (as it is for "rational") is to express a sentiment that one approves norms of judgment that result in coordinating social interaction with, in turn, resulting survival benefits (pp. 7 and 46).

The problem with the usual explanations of ethics offered by sociobiologists and behavioral psychologists is that they are too "direct," and therefore "simplistic," as though persons behaved intending to leave more genes in the next generation, or as though we now lived in the hunter-gatherer societies in which such genetic dispositions evolved. "Good evolutionary treatments of human life will be indirect" (pp. 27 and 29). People seldom want to have the most children possible. "The point is not . . . that a person's sole goal is to maximize his reproduction; few if any people have that as a goal at all" (p. 67). Rather, generally in result, there is this effect:

> Human cooperation, and coordination more broadly, has always rested on a refined network of kinds of human rapport, supported by emotion and thought. A person sustains and develops this network, draws advantages from it, and on occasion keeps his distance from aspects of it. . . . We are evolved animals, and so biological evolution must account for our potentialities

[pp. 26–27]. . . . Propensities well-coordinated with the propensities of others would have been fitness-enhancing. (p. 67)

This is reminiscent of the "indirect reciprocity" of Alexander. Also, such broad, indirect results make the tie to genetic survival somewhat "loose" or connect genes and ethics with a very long leash (recalling Wilson's metaphor 1978, p. 167; Chapter 3, Section 3).

A prominent reason for the broad, indirect, looser connection results from language. Humans have unique capacities to talk about these things. In a society, various persons express their approval of such norms and carry on a "conversation" about these; they communicate with each other and move toward a "consensus" (pp. 7 and 46). When a person says, "*x* is moral," he and others discuss it. Those societies do best where, in result of this conversation, there is some agreement; they are the better coordinated, with survival benefits. Gibbard continues:

> Normative discussion might coordinate acts and feelings if two things hold. First normative discussion tends toward consensus. The mechanisms here . . . are two: mutual influence and a responsiveness to demands for consistency. Second, the consensus must move people to do or feel accordingly. . . . We evaluate in community. (p. 73)

Consensus, of course, is not always achieved, but there are pressures toward it, at least toward the minimal morality necessary to keep the society coordinated. Societies are held together by their mores. Morality functions with this result, said to be a Darwinian survival benefit. The character of the morality is not preset in the genes, however, but appears in the "conversation" of the culture into which one is educated and in which one takes part. So the claim that this is a Darwinian genetic perspective is getting stretched over into a more generic utilitarianism. Ethics is good for people, producing the greatest good for the greatest number; those kinds of societies flourish. This is group selection, and the groups are selected not simply on the basis of their genes but on that of their cumulative conversations, transmitted culturally.

So one has to look more closely at this "conversation" with its need for "mutual influence" and "consistency" as a member of society develops a "system of norms" (pp. 91 and 153) worthy of adherence. Embedded in the social system, one needs to be "systematic" and to press "hard inquiry" (p. 326). One makes "conversational demands" that one's audience accept one's norms, and they do likewise (pp. 172–176). Interacting with others, that surely will mean recognizing the needs, wishes, desires, pains, pleasures, rights of others, parallel with my own. These come to influence each other mutually. That is part of the consistency. Consistency is egalitarian, fair, evenhanded. Using normative judgment, it does not make sense, it is not rational, neither plausible nor consistent, not to allow the influence of their goods. I know they will demand this of me; I will demand likewise of them. "Morality . . . concerns the moral emotions it makes sense to have from a standpoint of full and impartial engagement" (p. 128). Double standards would soon break down the desired consensus and social coordination. The "indirectness" is becoming quite "inclusive," or, as Gibbard puts it, "interpersonally valid." There are "objective pretensions in our normative talk" (p. 155).

This need for competent judgment, using generic properties consistently applied to all, without bias to special features of oneself or one's group, much less one's genes, combining with a tendency toward consensus, will move to produce a universalism in ethics. "None may give special weight to his own judgment simply as his own" (p. 182); that is the problem of "parochialism" (pp. 205–208). Still, Gibbard worries that these social interactions are for the most part local, in the community in which one is a functioning member. Globally, there is no consensus: ethics is too "fuzzy"; the most one can hope for is relative consensus "among neighbors and other conversants" (pp. 233, 199 and 211). International, widespread agreements are likely to be only on "narrow topics" (p. 250). "Mutual influence" and "consistency" nevertheless tend to produce something like the Golden Rule in ethics. We have to reciprocate to others as we would be done to. One has to be "objective" about that (pp. 153–170) but to give up "grandiose objectivity" (pp. 199–201).

If humans are really going to make "wise choices" on the basis of "apt feelings," they are going to need critical capacities that educate their feelings, evaluating which innate biological tendencies to cultivate and which to curb. Gibbard concludes:

> Human moral propensities were shaped by something it would be foolish to value in itself, namely multiplying one's own genes in later generations. Still, the kinds of coordination that helped our ancestors pass down their genes to form us are worth wanting – for better reasons. . . . Darwinian forces shaped the concerns and feelings we know, and some of these are broadly

moral. . . . Having those concerns promotes a good we can recognize . . . and we can try to do better. (p. 327)

Amen. But how are we to gain these better reasons; how are we to try to do better? What standards, whether global or local, universal or relative, are we to use in our conversations that mutually influence each other and in our search for consistency, unless we have capacities that transcend because they emerge out of the genetic creativity that once shaped us. After we get the evolutionary story of why we have such moral feelings, that they were "adapted" for their survival value, what are the "apt feelings" from here forward? "To pass from why we *do* care . . . to why *to* care means assessing feelings. We must assess which feelings, if any, to take as guides. . . . We need to settle what norms to accept as governing these feelings" (p. 254).

"Adapted" does not mean "apt," though it might be of course that some of our "adapted feelings" coincide with what our "apt feelings" ought to be, others not. Hidden in this normative "apt" is a kind of blend of "adapted" and "appropriate." "Adapted" in the usual biological meaning of "more fit, so as to leave more genes in the next generation," need not mean the same thing at all as "appropriate," since this is not "fitting" or "apt," not an appropriate behavior in an overpopulated world (as Gibbard fully realizes). "Adapted" in more general sense of "promoting social consensus and coordination with the resulting society flourishing" need not be "apt" in the sense of "fitting the norms of justice" at all, since a society might have a consensus about the divine right of kings or the legitimacy of slaves or the role of women and be well coordinated. Prophets, radicals, and feminists arise to challenge this consensus. Novel ethical insights can be upsetting; they can bring civil war.

Doing ethics, making "wise choices," we are going to need new "information," with no source of it in the genes. Here are new "possibilities for moral system" opening up (p. 254), almost (as we will later worry) possibilities floating in from nowhere. Those inclined to be religious about developing such a better ethic might even welcome some "inspiration" enlarging these possibilities. Gibbard himself, to return to our ad hominem argument, in his search for wise choices and apt feelings, "find(s) plausible . . . feelings with a broadly moral import, such as benevolence, respect, fairness, a sense of worth, reverence, moral inspiration" (p. 291), and thereby illustrates a mind that critically examines and evaluates what is going on. His very taking part in this conversation seeks to advance ethics beyond anything in the legacy of the genes. Perhaps ethics starts with a picture of ourselves as natural animals, perhaps it socializes us, but it becomes increasingly clear, when we grow sophisticated at critiquing our origins and natures with norms that are not there in our genetics, that ethics demands and generates novel powers of both analysis and feeling. We begin to wonder whether this latter ethics is any longer even "indirectly" naturalized, much less Darwinized; in our genesis of ethics, assist us to survive though this may, humans are a quite exceptional species.

(6) Universal Altruism

People start out acting in their self-interest and continue to act for a society in which such selves can flourish, but there is increasingly required reciprocal interaction with others, finding that what the others value is entwined with what we value. Now also people come under a sense of obligation to keep such a reciprocating community functioning well. From this point, it hardly seems a great stretch of thought (nothing beyond the human genius) to entertain both a generalized, universal obligation toward others and a particular respect for other individual persons whom one encounters in the circumstances of one's own career, whether or not one's self-benefit or self-cost has been figured. After one has carried on by arguments a defense of his or her self-interest in the company of others arguing their self-interests, one does become enlightened enough to see that a civil society is inconceivable without obligations to reciprocate, to be fair and just, and to trust others and ways to police this arrangement against the noncooperators. Nor is this just logic. Commitments continue to develop. One develops an affection for, an emotional attachment to, a community of such cooperating agents; one becomes committed to this kind of world. One votes for it to be so if it can.

The motivation for behavior moves from a grudging consideration of others to a rationally endorsed coordination, and then to a genuine caring for such others. Reasoned cooperation can be "just" self-enlightened cooperation, but it can sometimes become justified cooperation, or even *just,* cooperation in justice, or even benevolent cooperation, where the self identifies with values located in others. There takes place the genesis, or emergence, of altruistic ethics, now in a genuinely altruistic sense. Such a person acts, on the moral account, intending to benefit others at cost to himself or herself, and, on the genetic account, increasing the likelihood of the aided person's having offspring over one's having them. Not every person may develop such ethics – there are stages in moral development (Kohlberg 1981) – but some, who will serve as moral role models and ideals, can and will.

A notable example of such moral behavior is the Good Samaritan, celebrated for his expansive vision of who counts as a "neighbor," a model example of helping another at cost to one (Luke 10.29–37). The Samaritan – and this is important for our case – is not genetically related to the Jewish victim whom he aids. "Jews do not share things in common with Samaritans" (John 4.9).[7] The story is commonly regarded as a parable, though, since nothing in the text specifically identifies it so, it may report an actual contemporary event that Jesus used to illustrate the scope of loving neighbors. In any case here is a recommended ideal, influential across two millennia of ethical history. Parallel models can be found in other traditions, as the widespread presence of variants of the Golden Rule illustrates.

The Good Samaritan did what he intended: spent time, energy, and money helping an alien (nonkindred) genetic line, a victim that his ethics valued as a neighbor. Framing the deed in our axiological paradigm, the Samaritan is not simply maximizing his personal self-interest (biological somatic value) nor maximizing his offspring (biological genetic value); he as moral agent is defending an other of his kind altruistically, maximizing value in *Homo sapiens*, his species, if you like, and this by maximizing cultural and ethical value instanced in this victimized person, who, so assisted, would be more likely to have offspring. A society with many such persons is likely to do well in competition with societies from which such behavior is absent.

The determinant here is an "idea" (helping a neighbor, with sympathetic compassion) that is not just subservient to but superposed on the genetics. Such an "idea" can be transmitted nongenetically, as has indeed happened in this case, since the story has been widely retold and praised as a model by persons in other cultures who are

---

[7] *The New Testament and Psalms: An Inclusive Version* (New York: Oxford University Press, 1995).

neither Jews nor Samaritans. In ethics, persons regularly persuade others and are themselves persuaded to adopt ethical creeds.

Such persuasion contains both rational and emotive elements. The Golden Rule appeals to a parity of reasoning. It is irrational to hold that one's own self-interest is all that is present; the other's self-interest is equally present. It is irrational (at least, unreasonable) to expect others to respect your self-interest if you are not going to respect theirs. There is a compassionate element, a com(mon) passion; the pain and pleasure in oneself are paralleled by pain and pleasure in the other self. Sympathy is appropriate. Animals, we should notice, have minimal capacities for such reflection. Their reciprocal behaviors are naturally selected, but in human social behaviors there appears rational selection of the more persuasive ideas.

There are present both an ideal and the real; persons fail to form ethical creeds, fail to act on the creeds they do form; there is moral selfishness. There are thieves as well as Samaritans, exploiters as well as missionaries, assassins as well as prophets. But such failure is proof, not disproof, of the norm – an ethics that holds that one ought to help others individually – that will also maximize the general sense of "neighborliness" pervading a culture. Neighbors are whomever one encounters whom one is in a position to help. The Samaritan respects life not his own: that is, he can value life outside his own self-sector, and there arises the conviction that he ought to do so. That value, which *is* threatened in this injured person, *ought* to be preserved.

3. NATURALIZED ETHICS? ILLUSORY, DARWINIZED MORALITY

E. O. Wilson begins and ends his *Sociobiology* with a "biologicized" ethics: "What . . . made the hypothalamus and the limbic system? They evolved by natural selection. That simple biological statement must be pursued to explain ethics and ethical philosophers, if not epistemology and epistemologists, at all depths." "The time has come for ethics to be removed temporarily from the hands of philosophers and biologicized" (1975a, pp. 3 and 562). "Human behavior . . . is the circuitous technique by which human genetic material has been and will be kept intact. Morality has no other demonstrable ultimate function" (1978, p. 167). However circuitous the cultural variations, the ultimate connection is, after all, pretty direct (pace Gibbard's worries that this is too simplistic). A quarter century after first making these claims, Wilson is as insistent as ever: "Causal explanations of brain activity and evolution, while imperfect, already cover most facts known about behavior we term 'moral' " (Wilson 1998, p. 54).

Michael Ruse, a philosopher, joins Wilson: "Morality, or more strictly, our belief in morality, is merely an adaptation put in place to further our reproductive ends. Hence the basis of ethics does not lie in God's will . . . or any other part of the framework of the Universe. In an important sense, ethics . . . is an illusion fobbed off on us by our genes to get us to cooperate" (Ruse and Wilson 1985, pp. 51–52). "Morality is a biological adaptation no less than are hands and feet and teeth" (Ruse 1994, p. 15; 1986, p. 222). Bluntly put, ethics results in fertility; that is its deepest explanation.[8]

A morality that conserves human genetic material is welcome enough. But this also brings deeper trouble. More bluntly put, evolution produces this fertility through a radical selfishness incompatible with ethics. George Williams claims, "Natural selection . . . can honestly be described as a process for maximizing short-sighted selfishness" (1988, p. 385). Richard Dawkins summarizes: "The logic . . . is this: Humans and baboons have evolved by natural selection. . . . Anything that has evolved by natural selection should be selfish. Therefore we must expect that when we go and look at the behaviour of baboons, humans, and all other living creatures, we will find it to be selfish" (1989, p. 4).

Michael Ghiselin concludes his scientific analysis with memorable rhetoric:

> No hint of genuine charity ameliorates our vision of society, once sentimentalism has been laid aside. What passes for co-operation turns out to be a mixture of opportunism and exploitation. . . . Given a full chance to act in his own interest, nothing but expediency will restrain [a person] from brutalizing, from maiming,

---

[8] Ruse can, with little sense of contradiction, say in almost the same breath that science "soars into the cultural realm, transcending its biological origin. In the case of ethics, the Darwinian urges a similar position. Human moral thought has [biological] constraints . . . [yet] leads to moral codes, soaring from biology into culture" (1986, p. 223). Such a soaring ethics seems unleashed from natural selection for maximal offspring, hardly "an illusion fobbed off on us by our genes."

from murdering – his brother, his mate, his parent, or his child. Scratch an "altruist" and watch a "hypocrite" bleed. (1974, p. 247)

If all ethics must be Darwinized, this is revolutionary – "the greatest intellectual revolution of the century" – because the implication is that there is no actual altruism, as classically understood. All that natural selection permits is forms of quasi altruism that are actually self-interest, more or less enlightened or disguised forms of selfishness. Richard Alexander concludes:

> I suspect that nearly all humans believe it is a normal part of the functioning of every human individual now and then to assist someone else in the realization of that person's own interests to the actual net expense of the altruist. What this "greatest intellectual revolution of the century" tells us is that, despite our intuitions, there is not a shred of evidence to support this view of beneficence, and a great deal of convincing theory suggests that any such view will eventually be judged false. This implies that we will have to start all over again to describe and understand ourselves, in terms alien to our intuitions, and in one way or another different from every discussion of this topic across the whole of human history. (1987, p. 3; 1993)

Dawkins claims that with the Darwinian revolution begun in *The Origin of Species* (1859) culminating in his theory of selfish genes, all the old answers to the question about how humans ought to live and act are discredited. "The point I want to make now is that all attempts to answer that question before 1859 are worthless and that we will be better off if we ignore them completely" (1989, p. 1).[9] Challenged about this, Dawkins insists: "There is such a thing as being just plain wrong, and that is what, before 1859, all answers to those questions were" (1989, p. 267). These are not modest claims. They are "models designed to take the altruism out of altruism" (Trivers 1971, p. 35).

Returning, then, to the Good Samaritan, let us see whether we can take the altruism out and find Jesus' answer worthless and plain wrong. Let us start all over and describe his behavior in terms alien

---

[9] Citing G. G. Simpson, though Simpson's insistence on the uniqueness of humans, especially in ethics (see earlier discussion), might have given Dawkins some pause whether Darwinism makes all previous efforts in understanding human nature so worthless.

to our intuitions. Let us scratch this altruist and see whether a hypocrite bleeds.

(1) Self-Deceived Altruism

Is it still possible to describe the Samaritan's behavior as being "selfish" in the Darwinized sense. Yes, we must – insist some sociobiologists and evolutionary psychologists – for enlarged self-interest is really all there is to morality. Genetics allows only one explanatory framework for any and all human (or animal) behavior, and so the Good Samaritan must be fitted into that explanatory box. Alexander concludes, "This means that whether or not we know it when we speak favorably to our children about Good Samaritanism, we are telling them about a behavior that has a strong likelihood of being reproductively profitable." Conscience is a "still small voice that tells us how far we can go in serving our own interests without incurring intolerable risks" (1987, p. 102). "The main reward is reputation, and all the benefits that high moral reputation may yield. Reputation as an altruist pays" (1993, p. 188). Even the Bible enjoins, "Cast your bread upon the waters, for you will find it after many days" (Ecclesiastes 11.1).

The Good Samaritan – so the theory holds – is constitutionally (= genetically) unable to act for the victim's sake. And so, all appearances to the contrary, there cannot be real altruism here (helping another at one's own genetic expense); there must be a self-interested account. Of course the Good Samaritan did not think of himself as increasing the likely number of his offspring. He had compassion for the victim. He thought of himself as a good neighbor; he did not even know he had any genes. He knew the difference between crass self-interest and concern for others; thieves had robbed this hapless fellow, and he by contrast was trying to help him.

But this concern for others, apparent to him, was only apparent. What the Samaritan intends is not what is resulting. Despite the intended altruism, the Samaritan's act promotes his own genetic interest. The fact that it is some sort of appearance even to him is explained this way: the whole transaction works better if persons are self-deceived when they act as moral agents. Not only do they not know about their genes; they do not know they are really acting in their self-interest. The Good Samaritan gets these results by indirection. He has to want what he doesn't really want to get what he really wants. Alexander explains, "I mean that such information is not a part of their conscious knowledge, and that if you ask people what they *think* their interests are they would usually give wrong answers" (1987, p. 36).

The apparent sincerity guarantees the reciprocity. If the victim knew the Samaritan's real motives (putting genes in the next generation), he would be disinclined later to reciprocate, had he such opportunity. If even the Samaritan knew his real motives, he would be a bad actor and his insincerity would leak out. So the Samaritan has to be blind to his own deepest motives, blind to the genetic impulses that fundamentally frame his behavior; he has to appear convincingly concerned, if the reciprocity is to go through. "If the theory is correct humans could not have *evolved* to know it, and to act directly and consciously in respect to it" (1987, p. 38).

Ruse and Wilson put it this way:

> Human beings function better if they are deceived by their genes into thinking that there is a disinterested objective morality binding upon them, which all should obey. We help others because it is "right" to help them and because we know that they are inwardly compelled to reciprocate in equal measure. What Darwinian evolutionary theory shows is that this sense of "right" and the corresponding sense of "wrong," feelings we take to be above individual desire and in some fashion outside biology, are in fact brought about by ultimately biological processes. (1986, p. 179)

Remember, "Ethics . . . is an illusion fobbed off on us by our genes to get us to cooperate" (1985; earlier discussion). The Good Samaritan is operating with an "ideal" that one ought to aid neighbors, but this is his delusion, his hidden reputation seeking. The Good Samaritan (a half-breed himself, part Jew, part Gentile) really assisted the luckless victim on the Jericho road in order to leave more genes in the next generation. What a hypocrite! That selfish bastard!

He doesn't know this, but we can allow no disconfirming or confirming evidence from people's verbal reports. Their conscious motivations are superstructural, epiphenomenal; their deep genetic determinants are not available to them. Genes are microscopic and humans historically knew no more about their genes than do monkeys today. "Genes remained outside the range of our senses in all respects until the twentieth century" (Alexander 1987, pp. 38–39).

Humans, however, have long known what it means to be self-interested, and they have had to create an illusion of altruistic morality for the reciprocity to work.

This even means that scientists can expect this theory of ethics to be rejected by critics, continuing to deceive themselves. "Natural selection . . . appears to have designed human motivation in social matters as to cause its understanding to be resisted powerfully." This is why "evolutionary biologists who attempt to explicate human behavior are ignored or maligned" (1993, pp. 192 and 189). Here genes are going to work both to create an illusory ethics and to make such hypocrites that it becomes difficult for good science to reveal what is going on. We may be headed toward a dilemma both in ethics and in science (Section 3[4]).

There is a presumption here that takes the biological level to be final. If x can be shown to be biological, then no further explanation is permitted or required. There is also a presumed discovery that takes the biological processes to be deceptive. We are programmed to believe what is not so. Explanatory schemes are difficult to deal with when they make an end run around our capacity to reason, when they tamper with our capacity to think. There is, of course, a great deal of rationalizing (unconsciously pretended reasons, hypocrisy) in human behavior, as well as much selfishness, and both do undermine our capacity to think. Psychologists and biologists were not the first to discover either tendency; ethicists and theologians had been lamenting it for centuries – if we can trust those verbal reports.

Even if we can get ourselves freed from this selfish rationalizing enough to examine the scientific claims here, matters are going to be tricky to disentangle. The fundamental claim is that selfish persons outreproduce unselfish ones, but superimposed on that is the claim that (really) selfish persons who are self-deceived into thinking they are unselfish outreproduce selfish persons who know their own selfishness. Really, those damned thieves will leave fewer offspring in the next generation. Neither the priest nor the Levite will do well either. Initially, the claim to be tested seemed simply that cooperative persons outcompete combative ones. Good Samaritans outreproduce thieves. (Is there any evidence that theft is declining over generations, that Good Samaritans are increasing? If so, is the cause of this genetic?)

Later, the claim to be tested is that pseudoaltruism (altruism, really self-interest) outreproduces unenlightened selfishness. Self-deceived Good Samaritans outreproduce thieves. Later still, the claim is also that tacit pseudoaltruism (altruism, really self-interest, but unawares) outreproduces even enlightened selfishness (persons made explicitly aware of their self-interest in reciprocal altruism). Deluded Good Samaritans outreproduce nondeceived, wised-up Good Samaritans.

But there is no evidence even that altruistic persons are increasing in the genetic pool over selfish ones, or vice versa. Meanwhile, one hardly needs evidence that cooperators do well in society. If there were some evidence of the increasing genetic frequency of altruists, it might be difficult to say whether it was supporting cooperation over combativeness, or genuine altruism over unenlightened selfishness, or pseudoaltruism over enlightened selfishness. Nor is there any evidence that altruists who are deceived about their motives are, over the centuries, outreproducing altruists who are introspective enough to realize the benefits of mutual cooperation. The difficulty of interpreting whatever behavioral patterns we find is going to be compounded by the fact that all verbal reports of motives have to be dismissed as unreliable. Since psychological, ethical, and experiential evidence is inadmissible, we could find it difficult to reach the conclusion that the biological determinants are underdetermining the outcome.

A biologist inhabiting this paradigm is going to have trouble accepting counterevidence. Lawrence Kohlberg discovered six stages of moral growth: (1) In the juvenile stage there is an egocentric obedience to authority, avoiding punishment. (2) Later, the right is serving each other's needs and making fair deals. (3) Then the right is being concerned about other persons and their feelings, keeping loyalty and trust. (4) In the next stage, the right is doing one's duty in society, upholding the social order and maintaining the welfare of society. (5) Later still, the right is upholding the basic rights, values, and legal contracts of a society, even when they conflict with legalistic rules. (6) In the last stage one is guided by universal ethical principles that all humanity should follow, seeking to respect the equality of human rights and the dignity of human persons as individuals (1981).

Of course it was impossible for Kohlberg to discover these stages without some reliance on (unreliable) verbal reports, and Alexander, dismissing those, sees only increasing levels of self-deception in the upper levels where one supposes oneself to be guided less and less by self-interests and more by what is universally right. "I see these final 'stages' of moral 'development' as being just as self-serving (in reproductive terms) as the first three stages" (1987, p. 134).

These are stages in the lifetimes of individual persons, reaching from childhood through maturity, some persons developing further morally, some less than others. Most do not reach the higher stages until later in life (females are perhaps past menopause by then) or do not reach the higher stages at all. Further, such persons may or may not reproduce and care for their children during these decades of growth that cross several moral stages. All this is going to make it troublesome to check this claim, compounded again by the fact that we too will be unable to trust the verbal reports of those we study. So it may be difficult to find empirical evidence that stage five persons do or do not outcompete stage four, or stage one, persons.

More complications follow.

(2) Induced and Inflated Altruism

Humans on average get selected for the most fertile mix of reciprocal altruism, including the right levels of deception. Past this beneficial self-deception – so continue these ingenious sociobiological accounts – there is harmful deception when a moral agent gets tricked into edging past the point of diminishing returns and moves over into what is in fact real altruism, benefiting the other at cost to the self. Here the actor not only thinks he or she is an altruist, but indeed is an altruist, and the advantage passes over to the person aided. Truly altruistic acts cannot be favored by selection, but here is selection for "the ability to induce others to behave altruistically" (Williams 1988, p. 400).

In such "induced altruism" an individual is favored who can trick others into believing that altruism is the right quality to have, this coupling up with the moralist's own native, naive self-deceptions about his or her duties. "We, therefore, would expect the evolution of abilities and tendencies to deceive potential altruists into serving inadvertently the interests of others" (Alexander 1987, p. 114; 1993). Alexander is forthright, claiming a "general theory of behavior":

> Society is based on lies. . . . "Thou shalt love thy neighbor as thyself." But this admirable goal is clearly contrary to a tendency to behave in a reproductively selfish manner. "Thou shalt give the

impression that thou lovest thy neighbor as thyself" might be closer to the truth. (1975, p. 96)

The hoodwinked altruist's kind will be reduced, and the trickster's tribe increases. So trick prevails over truth.

If we think of a spectrum with total selfishness at one end and total altruism at the other, it is advantageous to move along this scale just so far as one remains on the portion of the spectrum that is only apparent altruism, combining self-interest with helping others. It is fatal to edge over any further. A study of blood donors concludes, from their verbal reports and behaviors, that many of them give blood anonymously with seemingly altruistic intent. But Williams concludes that they are "victims of manipulation." "Anyone who makes an anonymous donation of money or blood or other resource, as a result of some public appeal, is biologically just as much a victim of manipulation as someone whose self-sacrifice serves the interests of a tyrant." In such altruistic behaviors donors lose and those who get the transfusions of blood or money gain. An ethicist who takes philanthropy as authentic "misses the role of manipulation in philanthropy" (1988, p. 400). These donors are really losers. The only philanthropy that wins, though unaware of doing so, is really self-seeking and results in actual gain to the donor. Meanwhile philanthropy that knowingly realizes that it seeks its own interest is not convincing enough to succeed.

Super Good Samaritans are suckers, outcompeted by self-deceived but successful Good Samaritans, who in turn outcompete wised-up Good Samaritans. Always look for the subtler self-interested motive. If you do not find it, look again. It must be there because the theory demands it. If you cannot find it, there must be a mistake, either yours in not detecting where the genetic self-interest is present, or a mistake of the actors, who fail in acting in their self-interest. "I do not doubt that occasional individuals lead lives that are truly altruistic and self-sacrificing. However admirable and desirable such behavior may be from others' points of view, it represents an evolutionary mistake for the individual showing it" (Alexander 1987, p. 191).

The Good Samaritan must not edge past the point of his or her own self-interests, not allow the groans of the wounded man to con him into too much risk, not promise to pay at the inn any more money than he is likely to gain benefits from in return. He should not offer a blood donation. He must resist induced altruism. But further, a supersmart Good Samaritan can himself become a trickster. In the struggle between trick and countertrick, he can con the victim into thinking that his rescuer is more of a Good Samaritan than he really is. "Individuals are expected to parade the idea of much beneficence, and even of indiscriminate altruism as beneficial, so as to encourage people in general to engage in increasing amounts of social investment whether or not it is beneficial to their interests" (Alexander 1987, p. 103; 1993). This is "inflated altruism."

Though the Good Samaritan must not actually let himself be induced into being a Super Good Samaritan, if he can manage to appear this way to the victim, then the victim (or other admirers) will be all the more disposed to reciprocate with benefits, benefits to the Good Samaritan that now exceed the advantage conveyed by the Good Samaritan to the victim. The Good Samaritan, first found to be only apparently a loser in favor of the victim, is, at this second level of deception, found out to be inflating this appearance even more, so that he can win bigger still. That is why he told the innkeeper he would pay more, if need be, on his return trip. He wasn't being tricked into extra altruism; he was parading his beneficence for future gains: image building. The victim is twice victimized, once by the thieves and a second time by the Samaritan, who inflates his already only apparent altruism and suckers the victim into over-reciprocating later on. That selfish bastard is at it again!

Alexander concludes, summarizing both induced and inflated altruism, "The long-term existence of complex patterns of indirect reciprocity, then, seems to favor the evolution of keen abilities to (1) make one's self seem more beneficent than is the case; and (2) influence others to be beneficent in such fashions as to be deleterious to themselves and beneficial to the moralizer, e.g., to lead others to (a) invest too much, (b) invest wrongly in the moralizer or his relatives and friends, or (c) invest indiscriminately on a larger scale than would otherwise be the case" (1987, p. 103). "Now biologists realize that the conflicts of interests that exist because of histories of genetic difference imply . . . that nearly all communicative signals, human or otherwise, should be expected to involve significant deceit" (1987 p. 73; 1993). Mind initially evolves to know enough truth about the world to be able to cope, to find a way through the world. But later it further evolves to deceive others, and in such a way that it is self-deceived while doing so (1987, pp. 114–117).

Perhaps. But first one ought to make sure there is no mistake in the core theory, and one ought to notice that Alexander insists that biology says "nothing whatsoever" about what humans ought to be doing (1979, p. 276), leaving him in a dilemma to which we return later. A critic will be better advised first to check the logical structure of such evolutionary psychology applied to ethics. We may only be dealing with a blik, that is, a paradigm grown arrogant, interpreting and reinterpreting all evidence in its favor. The empirical facts, which seem to be frequently examined, may in fact make little difference. The theory absorbs the evidence into its position. Perhaps we hardly need bother to bring any further moral behavior into the court of evidence. Alexander knows before he looks that all human behavior, however apparently moral, is selfish (apart from that of anomalous misfits), just as he knows before he looks that the fittest survive (the misfits soon go extinct).

If one's categories are limited to the merely biological ones, one will have to call Good Samaritan behavior some kind of a mistake, dismissing the actor's altruistic accounts of his behavior because they are anomalous to one's interpretive categories. There *must* be deception here somewhere. The theory demands it, and phenomena cannot gainsay the theory. But the deception could be in the theory, not the phenomena, which is disposing us to interpret as an illusion the altruism that is in fact taking place before our very eyes. So far from understanding what is going on, one will miss a critical new turning point: the emergence of these "ideas" become "ideals" – altruistic love, justice, and freedom.

Natural selection is relaxed in favor of *ethical* selection, analogously to the way it was earlier, in science, relaxed in favor of *rational* selection, although neither ethical nor scientific societies do poorly in competition with other societies. Something is selecting the more ethical theory and behavior; universal altruism is winning out over group selfishness and xenophobia. "Love your enemies; do good to those who hate you" (Luke 6.27) – that might result in peaceable societies that flourish and leave more offspring in the next generation, but it is certainly not evidently reciprocal altruism or indirect reciprocity or enlightened self-interest.

Curiously, the double deception (deceiving others and being deceived about the fact that one is deceiving others) forces positing a double negative to substitute for an apparent positive (that one helps another altruistically). The appearance of this as a real positive cannot be allowed because the theory does not allow such emergents.

Therefore, these seemingly altruistic events must be mere appearance, not virtue but something virtual.

At this point, one begins to wonder just who is being deceived: the moralist who acts with these altruistic intentions, or the reductionist scientist whose theory forces a double negation of a positive emergent? The induced blindness as to what is really going on could lie in either place. There is no particular cause to see ethical advocacy as so much fluff over unconscious genetic determinants.

(3) Epiphenomenal Altruism

In what we term "epiphenomenal" accounts, ethics arises as an anomaly associated with rationality. Rationality has survival advantages; ethics appears because, although in fact a little unreasonable, it is quite pragmatic for the human mind. Herbert Simon (1990) proposes this alternative origin of altruism: it is quite advantageous to individual humans to gain the skills that are transmitted culturally, and they can do this only if they are teachable, or "docile." To a considerable extent, the more docile they are, the more children they have, since they gain skills that help them rear their children successfully. Thus being docile increases fitness. But people are not smart enough to be able to evaluate everything they learn with a view to how much this or that recommended practice makes them better able to reproduce children; they are not that discriminating; human rationality is limited. So they take the social heritage more or less as a package.

Societies whose members cooperate in this way outcompete societies with less cooperation. Much cooperation is based on kin selection or reciprocal "altruism." But there is more, a place for genuine altruism. If a society evolves so as to slip into the social heritage that it bequeaths its docile members a limited amount of really altruistic behavior, those docile members still gain considerable reproductive success from their docility, even though they are also now losing a little reproductive success as a result of this real altruism. Docile persons benefit the society as a whole greatly enough (that is, contribute enough benefits to other members of the society, aggregated) that such a society outcompetes other societies where such real altruism is not so present. These docile individuals lack the capacity to discriminate against moral teaching that reduces their individual fitness; their disposition is an efficient rule of thumb, true enough but not quite true. This has the felicitous result that their kind of society survives; so, indirectly, they benefit from their behavior in this larger context.

Hence societies that have docile members survive, and those that have docile members who accept moral teaching exhorting a (limited amount of) genuine altruism survive even better. Simon calls this extra, nonreciprocal, real altruism a "tax" that society can impose. This will require, perhaps not unreasonably for *Homo sapiens*, the social species, a form of group selection, selection of those cultures that can impose this "tax" on individuals, such groups outcompeting other cultures that impose no such tax.

Like the TIT FOR TAT models, Simon's is mathematical; it can be computer-modeled with arbitrarily assigned numbers. But Simon does not offer any actual genetic, psychological, or social measures with which to test such a theory, either between individuals within a society or between competing societies. There are no measures of docility in excess of reproductive advantage, of this "tax," or of the proportionate numbers of the docile, or the excessively docile, in one society against another.

Even if one could find out that this account were true, one would be left quite puzzled about what such docile persons, once they wised up to what is going on, would or ought to do. If a person's sense of identity was sufficiently formed by this social heritage, now exceeding the sense of genetic identity, he or she might well conclude that such a society was right after all.

Boyd and Richerson suppose, somewhat similarly, "a conformist effect": "Conformist transmission may be favored . . . because it provides a simple, general rule that increases the probability of acquiring behaviors favored in the local habitat. It is plausible that, averaged over many traits and many societies, this effect could compensate for what is, from the genes' 'point of view,' the excessive cooperation that may also result from conformist transmission" (1985, p. 236). Conformists are more easily persuaded to be altruists, acting for the good of these groups, and such groups do outcompete other groups. Hence ethics arises. So Boyd and Richerson find widespread "a general tendency of humans to behave altruistically toward members of various groups of which they are members" (p. 205). Again, ethics is not so much reasonable, as an economic strategy that substitutes conveniently for more careful reasoning. Again too, one has to wonder what a wised-up conformist ought to do.

Such scenarios suggest how the benefits of cooperation can produce various sorts of feed-back loops that will further strengthen cooperation, in nonmoral forms at first, but out of which the ethical sense might once have emerged, at least an interhuman ethics within local societies. Ethics is epiphenomenal to these other benefits. But these accounts do little to suggest how either a more universal interhuman ethics or an environmental ethics might have appeared. Neither the "taxed" nor the "conformist" social members are allowed to be reflective or self-critical about their behaviors. Yet the history of ethics supplies centuries of just such critical reflection.

Francisco J. Ayala holds that ethics is a by-product of selection for intelligence in the hominoid line. His account makes ethics fully rational, but still epiphenomenal. With increasing intelligence, reaching the large brains of *Homo sapiens* there arise "(a) the ability to anticipate the consequences of one's own actions; (b) the ability to make value judgments; and (c) the ability to choose between alternative courses of action" (1995, p. 118). These features contribute to survival; if I can evaluate, choose, anticipate consequences, and then act, I shall more likely survive than a competitor who cannot do these things or does them less well.

It just so happens that these three gifts of general intelligence are exactly "the three necessary, and jointly sufficient, conditions for ethical behavior" (1995, p. 118). This is curious biologically, although it is logically necessary. Exactly the same factors that are required for general intelligence are required for conscience. Intelligence is the target of selection and conscience the "by-product." There is no cause to think that conscience would contribute to survival; there is much cause to think that other activities of intelligence do. "Ethical behavior came about in evolution not because it is adaptive in itself, but as a necessary consequence of man's eminent intellectual abilities, which are an attribute directly promoted by natural selection." With ethics disconnected from survival, Ayala is free to hold that the normative content of ethics is culturally based, not biologically driven. "Moral norms are products of cultural evolution, not of biological evolution" (1995, p. 118).

Now we seem to have moved to the other end of a spectrum. The accounts examined in the previous section claimed that ethics was always and only a survival tool, and that seemed excessive, but, against Ayala, one may be reluctant to conclude that ethics makes no contribution at all to survival. It seems quite plausible that ethics is adaptive, both in the sense that groups that cooperate do well and in the sense that individuals within those groups do well. Various ethical systems are urged because they yield "the greatest good for the greatest number," which presumably includes prosperity and success in child rearing. In any case, Ayala, faced with such epiphenomenal ethics, concludes, "Biology is insufficient for determining which moral codes are, or should be, accepted" (1995, p. 134), and that leaves him too facing the dilemma to which we next turn.

(4) From *Is* to *Ought:* A Dilemma

How does one move from the DNA code that *is* to a moral code that *ought* to be? Any adequate account of human ethical behavior has, at this point, to appraise what these scientists themselves recommend and how they behave. Ruse identifies the *is* that has resulted from natural selection with the *ought* of moral life: "The good is simply that which evolution through selection has led us to regard as good" (1984, p. 93). In humans, whatever norms and values have been selected for are ipso facto good. Ruse is reasonably comfortable with such an ethic, but many biologists find this equation to be incorrect.

Richard Alexander, although insisting on what moral philosophers do not yet know – that morality as routinely practiced has evolved to serve human reproductive selfishness – turns in the end to join moral philosophers and ask, "What does evolution have to say about normative ethics, or defining what people *ought* to be doing?" His answer: "Nothing whatsoever" (1979, p. 276). The toughest assignment for a Darwinized (or any naturalized) ethic is going to be how to find an answer to that latter question, especially if science gives inadequate help, or none whatsoever, in answering it. John Maynard Smith is blunt: "A scientific theory – Darwinism or any other – has nothing to say about the value of a human being. . . . Scientific theories say nothing about what is right but only about what is possible, and we need some other source of values" (1984, pp. 11 and 24).

Alexander has been "accounting for the altruism of moral behavior in genetically selfish systems" (1987, p. 93), but he also thinks that we can break away from our genetics. "To say that we are *evolved* to serve the interests of our genes in no way suggests that we are *obliged* to serve them. . . . Evolution is surely most deterministic for those still unaware of it" (1987, p. 40). Finding out what is going on enables us to stop it, and we can afterward do what we ought to do, not what we are evolved to do. But what ought we to do? Finding out what we are evolved to do breaks the leash, but just being unleashed does not give any positive direction. We have wised up the Good Samaritan; we have been cautioned against overinflating one's image, cautioned that others may dupe us with an overinflated image. After all that is said, ought we still to be Good Samaritans or not? Alexander does say, despite his "general theory of behavior," which discovers how we act always selfishly, that "Thou shalt love thy neighbor as thyself" is "an admirable goal" (1975, p. 96), but just how did he find this out and how does he know whether he himself is being edged over into induced altruism? How will he encourage latter-day Samaritans to act so admirably (assuring them that he is not inflating his altruism and inducing theirs)?

The details are left unspecified, but ideally the ethic will include love, justice, and freedom. Meanwhile, notice that this ethic, whatever it is, will command its *ought*'s having been freed from the prior evolutionary determinism. At least one thing that a scientist morally ought to do, as Alexander himself is doing, is to investigate "the biology of moral systems"[10] in order to free humans from that biology. But even this *ought*, which launches us into an ethic as yet open-ended, cannot be evaluated until what it offers has already happened. Alexander will first have to cut the leash and then bring in by skyhook these nonevolved ethics, with their outside authority that has "nothing whatsoever" to do with genetic determinism. Alexander has figured out how biology drives morality, a discovery, a good piece of true science, he thinks, not just a pragmatic behavior that helps him reproduce. Not only that, he has further gotten himself an imperative that he ought to use this science to help others (who are unrelated readers) to escape their biology.

Other scientists regularly join Alexander. Williams is intense about the need for humans to overturn their biology. "An unremitting effort is required to expand the circle of sympathy for others. This effort is in opposition to much of human nature." "Natural selection . . . can honestly be described as a process for maximizing

---

[10] The title of his book.

short-sighted selfishness." "Brought before the tribunal of ethics, the cosmos stands condemned. The conscience of man must revolt against the gross immorality of nature" (1988, pp. 437 and 384–385). Not only must humans find an ethic from outside their biology, they must defeat their biology with it. When he urges expanding the circle of sympathy for others, Williams seems to be finding something of value in the classical, pre-Darwinian insights of philosophical and religious ethicists, as though somehow, somewhere, some persons have already managed to escape their biological legacy.

When Dawkins reaches the conclusion "Let us try to *teach* generosity and altruism because we are born selfish" (1989, p. 3), he certainly intends a moral use of "generosity," here mixed confusedly with a contrasting moral and also biological use of "selfish," where the genetics spills over into morality, disclaimers to the contrary. Dawkins laments this bad human nature. But Dawkins can do this only because he has found some vantage point from which he can recommend that altruism be taught in culture, educating us out of our beastly nature. If so, then he himself has reached a more comprehensive ethics – one in which he has escaped, or at least knows that he ought to escape, the biological legacy. He has found genuine altruism, ideal if not yet as real as he wishes.

It is puzzling to say where he found this, since he has dismissed all ethics prior to Darwin as worthless, and all he can find in Darwinism is a disposition to selfishness, which is the wrong answer. Generosity, nevertheless, seems to be a recommendation he has gotten from somewhere. He needs, in our idiom, to generate generosity, urging us to do so. Perhaps he cannot reach this biologically with his selfish genes, but philosophically he has the ideal, and, philosophically, we have to give an account of how such generosity is generated, of which he himself is an instance. Meanwhile, he can take himself seriously as an ethicist, and we can take seriously what he morally advocates, only if he and we are exempt from the theory that he is advocating as a scientist about how ethics works.

Wilson claims, "Our societies are based on the mammalian plan: the individual strives for personal reproductive success foremost and that of his immediate kin secondarily; further grudging cooperation represents a compromise struck in order to enjoy the benefits of group membership." But in the same breath he can urge as the three primary principles of interhuman ethics that (1) one ought to protect "the cardinal value of the survival of the human genes in the form of a common pool over generations," (2) one ought to "favor diversity in the gene pool as a cardinal value," and (3) one ought to regard "universal human rights . . . as a third primary value" (1978, pp. 197–199). That doesn't sound like grudging cooperation; it sounds like appreciating shared values. The behavior of at least one human conflicts with his conclusions, namely, Wilson's own behavior. Wilson is neither behaving so as to maximize his own offspring nor recommending that others do so.

Not one of these values that Wilson has reached can be obtained by operating with his selfish genes. "Of all the evils of the twentieth century, the loss of genetic diversity ranks as the most serious in the long run." Wilson fears a tragic loss of "the variety of human genes out of which endless new combinations can be drawn for the attainment of genius and further genetic evolution" (1980a, pp. 61–62). The one thing selfish genes do not do is promote diversity not their own. His vision of universal human rights nowhere is derived from his biology; it is borrowed from the philosophers whose illusory ethics he has been undermining.

Wilson warns, "The naturalistic fallacy has not been erased by improved biological knowledge, which still describes the 'is' of life but cannot prescribe the 'ought' of moral action" (1980b, pp. 430–431). One version of this is that the human nature with which we once evolved, and that is still genetically coded within us, does not any longer incline us to do the right thing. People must evaluate and correct their inherited nature and do otherwise.

> The trap is the naturalistic fallacy of ethics, which uncritically concludes that what is should be. The "what is" in human nature is to a large extent the heritage of a Pleistocene hunter-gatherer existence. When any genetic bias is demonstrated, it cannot be used to justify continuing practice in present and future societies. . . . For example, the tendency under certain conditions to conduct warfare against competing groups might well be in our genes, having been advantageous to our Neolithic ancestors, but it could lead to global suicide now. (1975b, p. 50)

This Paleolithic tendency ought to be replaced by a more peaceful, universal altruism. "Human nature can adapt to more encompassing forms of altruism and social justice. Genetic biases can be trespassed, passions averted or redirected, and ethics altered" (Wilson 1975b, p. 50). Ethical persons can and ought to defend universal human rights in place of self-defensive xenophobia, morality overriding genetics.

But then just where is Wilson getting these *oughts* that cannot be derived from biology, unless from the insights of ethicists (or theologians) that transcend biology? This no longer sounds like a biologist biologicizing ethics and philosophy. It sounds like a biologist philosophizing without acknowledging his sources. Perhaps there is more than one kind of trespassing here. The genes are being trespassed (= transcended?) by ethics, and biology is trespassing into ethics without either resources or authority for the nonbiological, nongenetic norms it preaches.

Turning to environmental ethics, Wilson is notable for his ardor. In *Biophilia: The Human Bond with Other Species,* he urges "an advance in moral reasoning . . . to create a deeper and more enduring conservation ethic." No one is going to say here that Wilson is parading his beneficence, inducing readers to benefit him reproductively, or making an evolutionary mistake. One cannot dismiss his moving appeal as an unreliable verbal report. Rather we can only join his sincere efforts as Wilson struggles both to keep and to break out of a selfish conservation ethic. "The only way to make a conservation ethic work is to ground it in ultimately selfish reasoning – but the premises must be of a new and more potent kind." He worries about only "a surface ethic," and continues, "it is time to invent moral reasoning of a new and more powerful kind . . . a deep conservation ethic [based on] biophilia." We have selfish genes through and through, but "our predatory actions toward each other and the environment are obsolete, unreliable, and destructive." They are "prevailing myths." "The more the mind is fathomed in its own right, as an organ of survival, the greater will be the reverence for life for purely rational reasons" (1984, pp. 119, 131, and 138–140). In sum: "To the degree that we come to understand other organisms, we will place a greater value on them, and on ourselves" (p. 2). "Love the organisms for themselves first" (1994, p. 191). "Wilderness has virtue unto itself and needs no extraneous justification" (1992, p. 303). That certainly sounds like widely distributed and shared values.

Wilson is finding it difficult to get biophilia out of selfish genes. The welfare of the self is being stretched over to a nobility of character that comes from "generosity beyond expedience" (1984, p. 131) that Wilson wishes to embrace but cannot really reach on the basis of his theory. "Generosity beyond expedience" certainly sounds like "altruism beyond self-interest." So Wilson, like Dawkins, has the problem of generating generosity. Selfish genes are never generous beyond expedience; that is the core of sociobiological theory. He is clearly wrestling with how humans ought to behave morally, not simply how they are driven to behave biologically. He concludes that there ought to be a respect for life in which we value other forms of life as we do our own, a sort of Golden Rule in environmental ethics. The self-interest that an environmental ethic serves cannot be of the usual backscratching kind; the ants that Wilson wishes to protect are unlikely reciprocators. Rather for those humans who appreciate them, "splendor awaits in minute proportions" (1984, p. 139). That, if one insists, is an enrichment of human welfare, but it has nothing to do with fertility. None of this inquiry can be undertaken without being released from an ethics that is nothing but selection for maximum production of human offspring.

To the contrary, Wilson asks, "What event likely to happen during the next few years will our descendants most regret?" His answer: "The one process now going on that will take millions of years to correct is the loss of genetic and species diversity by the destruction of natural habitats. This is the folly our descendants are least likely to forgive us" (1984, p. 121). Why ought this catastrophe not happen? No doubt these descendants will suffer losses in those species that do not survive. Their human quality of life may be at stake, but maximum reproductive success, the largest human population possible on Earth, is no criterion of this ethic.

Quite the contrary again, it is antithetical to it. "Genetic biases can be trespassed" here too; indeed our human reproductive instincts must and ought to be replaced by biophilia and concern for environmental integrity. "To rear as many healthy children as possible was long the road to security, yet with the population of the world brimming over, it is now the way to environmental disaster" (1975b, p. 50). One morally ought to limit family size. Meanwhile, again, selfish genes do not promote diversity or integrity not their own, much less morally limit family size. Wilson does not escape the naturalistic fallacy; he falls into "the naturalistic paradox" (1980a, p. 70).

We have various assurances that, though this is the way human evolved mental and behavioral dispositions *are*, nothing follows about what *ought* to be. But it does follow that humans, moving from their psychological and behavioral *is* to what they *ought* to do, will need to have enough mental ability to evaluate what *is* and decide what *ought* to be, and enough discipline of their emotions and motivations to do otherwise, should their decisions about what ought to be run counter to their innate dispositions. If so, however, it *is* the case that (at least some) humans have such mental capacities and self-control of their behavior; they have to get this ability from somewhere, and nothing in these scientists' theories provides or allows such ability. We inherit these selfish genes, but from somewhere too we inherit genes that prompt us to sympathy, to mutual care, and to cooperation, and from somewhere we (some of us at least) get enough mental power to reflect over our evolutionary genesis and to generate an ethic about what *ought* to be in the light of this *is*. We may need to correct the moral slippage in our evolutionary natural history. That was the challenge so intensely faced by Gibbard in his search for "wise choices, apt feelings" (1990; Section 2[5]).

This reach toward more comprehensive ethics is not simply argued away by sociobiologists, when it upsets their categories of interpretation. They themselves appeal to and argue for this ethics. It is not simply that persons like the Good Samaritan are anomalies to their theory; they are their own anomalies. What they value, what they preach, how they behave, can neither be judged nor explained by their own accounts. They are, by those accounts, extremely unlikely creatures; the more they are able to reason about either the science or the ethics of human behavior, the less likely their theory is to be true.

So long as "altruistic" behavior produces and coincides with fertility, one cannot be sure whether the determinant is moral or genetic. One needs to escape the theory long enough to examine it. Before, presented with the theory that science serves selfish fertility, we could not examine it intelligently unless we assumed that during our investigation it did not. Presented now with the theory that ethics *does* serve selfish fertility, we cannot examine whether this further scientific theory is in fact true without presuming on the occasion at hand that it is not. Further, if we ask whether ethics *ought* to serve fertility, we cannot presume to answer unless we are, for the duration of inquiry, free from having all our ethical judgments determined by genetics. So far as they recommend that we modify behavior that has previously been genetically determined, both we and they are henceforth going to have to be exempt both from the scientific theory and its ethical bondage. We have to be able to give a genuine answer, unselfishly.

If Wilson, Ruse, Alexander, and others are recommending their (allegedly) *scientific* account of the origin and operation of ethics because this serves their reproductive interests, that is reason to distrust it. If they are recommending whatever *normative* account of ethics they recommend because this serves their reproductive interests, that is, again, reason to distrust it. If they deny they have any such reproductive interests at stake or in mind, their theory warns that we cannot trust verbal reports. Let us be on guard against their moral self-inflation. They might be trying to impress us with their magnanimity, and they might be psychologically suppressing even to themselves that they are doing this.

In fact, however, these sociobiologists appear to be genuinely interested in helping other humans to understand the human evolutionary history and genetic makeup. Indeed, they seem eager to help us to break away from our biological legacy, or at least to deal with it as effectively and as morally as we can. These verbal reports (their books are full of words) may only be long-winded attempts at inducing our altruism or at seeming inflation of their own. But, if we can set aside at least that part of the theory and take their verbal reports seriously, they seem to want to help. None of these scientists thinks that we humans are here to be selfish, enlightened exploiters of each other, or enlightened exploiters of nature, each seeking to maximize his or her offspring in the world. They witness to some larger meanings found in life; found for humans with their capacities for ethics, science, and religion; and also found for life in its biodiversity.

They do seem to urge us to accept and act on principles that are the right ones, that can be justified, and not simply to do what maximizes our offspring. But we have been given no resources from within the theory itself with which to do this; indeed, the theory denies that we are free and able to do this. The theory has to be false, not only when we evaluate whether it is true, but – should we discover that the theory is true – when afterward we evaluate what we ought to do (Barnett 1988, pp. 134–140).

What we ought to do is becoming an ever more pressing a question with the rapidly growing possibilities of genetically regenerating our human species (Kitcher 1996; Peters 1997). Genetic engineering redoubles the need for capacity in the human mind to think critically – in science, in ethics, and in religion. Humans might be genetically disposed to grow bald, fear strangers, avoid incest; men to dominate women; both to want many children or to be selfish. But how should we modify these traits in our genetically improved future? To make men less promiscuous, or women more aggressive than they now are by nature? To make both wish fewer children?

Ruse, claiming that "the good is simply that which evolution through selection has led us to regard as good," will wish no changes. Alexander and Maynard Smith, finding "nothing whatsoever" in science to help answer, will indeed need "some other source of value." Williams and Dawkins, having from somewhere learned already to "condemn" evolution, will welcome genetic engineering in their "unremitting effort to expand the circle of sympathy for others," "trying to teach generosity and altruism because we are born selfish." Wilson will no doubt wish to engineer more "biophilia" into future generations, more love of environmental conservation, to prevent the folly and catastrophe he fears, but he will need minds that can evaluate their selfish genes. Deliberated genetics will be genesis redoubled, though one ought not to forget that the most complicated part of the process, the brain of the deliberator, was genetically "engineered" by nature.

Ruse remains absolutely certain of the truth of Darwinian science as it controls ethics. "I grasp a truth that others have not" (1989, p. 8). So Ruse's genes will let him know what nobody else's genes will, that morality is an illusion, and he is trying to convince others of this, contrary to their genetic dispositions. But Ruse becomes increasingly doubtful whether he can find any *ought* at all, even a locally relative one, and eventually finds that he must part company with Wilson.

> Many of my fellow evolutionary naturalists . . . believe they can take evolutionary theory and use it to justify claims, not only about the physical world, but also about the moral world. [They] think that the Darwinian theory of evolution justifies claims about the moral obligation to care for members of one's fellow species. I deny this absolutely. Evolution explains why we believe that we should love our fellow beings. However, evolution demonstrates that such beliefs have no foundation. One should not therefore give up morality and go on an immoral rampage. . . . The person who tries to step outside morality soon feels severe personal inner contradictions. (1989, p. 7)

So Ruse is ethical, not because ethics is true or has any foundation in fact or logic. It is just the way his genes have shaped him to feel; he feels guilty if he abandons morality and he doesn't like those feelings. His position, he thinks, is the final truth in the sense that it is scientifically true, but in the ethical sense it fails theoretically because there really isn't any truth in it. In such Darwinized epistemol-

271

ogy and ethics, skepticism doubles back on itself and leaves as much reason to be skeptical about the skepticism as to be skeptical about anything else. The sad contradiction here is that the moral values that humans have gained over the millennia of ethical struggle to rise to higher ground are no longer being conserved. Moral vision lapses back into the bondage of unredeemed self-interest. If we take him at his word, unable to step outside his self-interest, Ruse acts biologically so as to minimize his personal pain.

But, reading between the lines, that pain indicates much more: that Ruse, like the others, operates with a genuine moral concern considerably beyond that to which his theory entitles him. Looking for the justification for such moral concern, is there any cause to think that these scientists-cum-prophets can redeem persons into ideals of love, justice, and freedom hitherto unattainable or envisioned only in delusion? Perhaps they can supply insights about the genetic basis of selfishness. But it may well be that philosophy, ethics, and religion (as well as literature and other humanities) have been at work humanizing persons for millennia.

(5) Altruism and Selfishness Defined and Confused

The contrast of self and "other" (*altrus*) seems clear at the start, but matters become complicated because the self is not an isolated self, but instantiates a family, a population, a species, and, in the human case, a society and a culture. The identity problems examined earlier at the level of genes and organisms return now from the perspective of morality, with added confusions in the crisscross of levels extrapolating moral vocabulary into domains where there are no moral agents. "Selfishness" and "altruism" are fundamentally moral terms, and we may prove to need another, fundamental category, that of value, adequately to explain what is going on in both animal and human behavior.

Multiple layers of meaning are confusing as much as clarifying issues, despite seemingly careful definitions.[11] Wilson says, "When a

---

[11] Just how fluid meanings are is revealed when one tries a set of technical definitions to isolate the various components:

> 1. Biological somatic altruism, altruism$^b{}_s$, altruism from the perspective of the self as organism, is found when in some transaction the organism loses (or risks loss), an other gains. A baboon stands on sentry. A

person (or animal) increases the fitness of another at the expense of his own fitness, he can be said to have performed an act of *altruism*. Self-sacrifice for the benefit of offspring is altruism in the conventional but not in the strict genetic sense. . . . In contrast, a person [or

monkey gives an alarm call. A parent feeds a child. A passerby rescues a drowning person. To ascertain whether such altruism is present, we need only observe behavior and its results. There may but need not be intention; inner states are irrelevant. Some somatic identity gives away some increment of benefit. The opposite is biological somatic selfishness, selfishness^b_s.

Reproduction involves altruism^b_s, parent to offspring, since one somatic self spends time and effort to bring other somatic selves into the world, half-copies of oneself genetically. Even an offspring is an "other" (*altrus*), a different individual from the parent. The identity presumed is somatic identity (Chapter 2, Section 1[2]). Parent and offspring are two selves somatically, though genetically only half-different.

2. Biological genetic altruism, altruism^b_g, altruism from the perspective of the self, genetically is found when the genetic self loses, and another genetic self wins. What is won or lost has shifted; now it is the genetic self, present proportionately wherever in descendants or nondescendant relatives there are partial copies of "my genes." The identity switches to genetic type identity (Chapter 2, Section 1[5]). The (only apparent) opposite is biological genetic selfishness, selfishness^b_g. Calling this "altruism" is a misnomer. Real (nonapparent) altruism^b_g does not exist either in nature or in culture (it is claimed), except by mistake (induced reciprocal altruism, which will tend to become extinct, since it is selected against). When parent feeds child, this is altruism^b_s, but it is not altruism^b_g. When a passerby rescues a drowning person, this is altruism^b_s, apparently altruism^b_g, but actually (since the rescuer gains reciprocal benefits increasing his probability of reproduction) it is selfishness^b_g, not altruism^b_g. Again, since one is dealing only in biological categories, behavioral patterns are required; intentions are irrelevant.

3. Biological kin altruism, altruism^b_k, is found when selves help their kin, as regularly happens. The somatic self loses; the kindred self gains. The identity is kinship identity (Chapter 2, Section 1[3]), which is, at the phenotypic level, the outcome of genetic type identity. The seeming opposite is biological kin selfishness, selfishness^b_k, but in fact there is identity: altruism^b_k = selfishness^b_k. Kin altruism is, again, a misnomer. When a baboon stands on sentry or a monkey gives an alarm call, some of the sentry and alarm benefits are received by kin, evidently altruism^b_k, but some of the benefits are received by nonkin (see next), which when we include reciprocation and the increased fertility therefrom (altruism^b_r, later), is really also altruism^b_k.

4. Biological reciprocal altruism, altruism^b_r, is found when there is action

animal] who raises his own fitness by lowering that of others is engaged in *selfishness*" (1975a, p. 117). There is reference only to fitness effects, not yet to any moral behavior. Wilson can inquire about "an altruistic bacterium" (p. 116), for instance, to find that only selfish bacteria exist. Still, in conclusion he wants to "biologicize ethics," seemingly referring to some as-yet-unspecified mixture of genetic and moral altruism. "In biology, as in everyday life, altruism is defined as self-destructive behavior for the benefit of others" (Wilson

> resulting in benefit to nonrelatives with a result also of benefit to one's offspring and relatives, or to oneself as assisted to support one's offspring and relatives. This mostly takes place in the human world, when people help each other out generally, and in the Good Samaritan case, when they think they are behaving altruistically. The opposite is biological reciprocal selfishness, but this is an only apparent opposite, since, seen for what it really is, altruism^b_r is in fact selfishness^b_r, and, again, misnamed.
>
> All of the preceding, with the superscript b, biological, involve biological consequences only, not intent. Intent may not even be present; altruistic intent is rarely, if ever, present in animal behavior. Intent is often present in human behavior, but (it may be claimed) this is not the real determinant of behavior. Intent may sometimes be present contrary to the consequence. Moral praise and censure are irrelevant.
>
> 5. Moral somatic altruism, altruism^m_s, is found in the action of a somatic self, who is a moral agent, when such an agent intentionally benefits another somatic self, which may or may not be genetically related. A human parent aids a child. A stranger rescues a drowning person. Such moral agents have options, make decisions, and form intentions that determine what they do. Altruism^s_m is commendable; its opposite is selfishness^m_s and is to be censured, although acting in one's self-interest is not ipso facto to be censured. We all must and ought frequently to act in our self-interest. Selfishness is excess self-interest. Altruism^m_s is the (self-deceived) Good Samaritan's intent, but (it is claimed) the consequence is other than his intent; it is selfishness^b_s and selfishness^b_k.
>
> 6. Moral genetic altruism, altruism^m_g, is found when a self, who is a moral agent, intentionally benefits another self, who is genetically unrelated. This is not present when a parent aids a child; it is present when a stranger rescues a drowning person, additionally to the altruism^m_s also present in the act. Again, the reciprocity involved means that the alleged altruism^m_g is really selfishness^b_k, though it may not be selfishness^b_s. The stranger may in fact die in the rescue, but his offspring and relatives will prosper in the kind of world in which such rescues are the prevailing expectation and practice.

274

et al. 1977, pp. 458–459). His theory of selfish genes and kin selection "has taken most of the good will out of altruism" (Wilson 1975a, p. 120).

Dawkins cautions that it is important to realize that his "definitions of altruism and selfishness are *behavioral*, not subjective" (1989, p. 4). Behavioral altruism does not exist in nature. But, as we have heard, in culture, Dawkins urges, "Let us try to *teach* generosity and altruism, because we are born selfish" (p. 3). That is urging moral altruism and lamenting what? Behavioral selfishness devoid of moral intention? Or moral selfishness with censurable intent? It muddles analysis to protest that one does not mean anything moral, or even intentional, about being "selfish" in an allegedly careful definition based on fitness effects, and then to preach, a few sentences later, that we, as actors with motives, should overcome a censurable selfishness inherited at birth. Perhaps human nature inclines us to be not only self-interested but selfish, and we must teach altruism to offset this, but such selfishness will have to be accompanied by subjective awareness and personal responsibility before we can condemn any of these inborn behaviors.

Trivers gives this definition: "Altruistic behavior can be defined as behavior that benefits another organism, not closely related, while being apparently detrimental to the organism performing the behavior, benefit and detriment being defined in terms of contribution to inclusive fitness." That is without reference to anything moral, but the sentence follows an abstract that proposes to explain human "friendship, dislike, moralistic aggression, gratitude, sympathy, trust, suspicion, trustworthiness, aspects of guilt, and some forms of dishonesty and hypocrisy," as "important adaptations to regulate the altruistic system" (1971, p. 35). Evidently the distance from nonmoral to moral altruism is rather short, only one sentence.

Williams, urging human morality, says, "We need all the help we can get to overcome billions of years of selection for selfishness" (1988, p. 401). Really the first help we need is to get clear whether the alleged selfishness back there in the trilobites and dinosaurs is really the same thing as the moral selfishness we need so much help overcoming. The distinction between biological and ethical altruism has been made, but has a disconnection been made? There may be an indiscriminate glossing over from an admirable biological vitality and proper self-defense to a censurable human selfishness. The word "selfish" has become an accordion narrowed to one use one moment and expanded to another use a few seconds later. It shuttles back and forth between moral meanings and technical biological meanings. We may need rescue because we are lost in a semantic morass.[12]

---

[12] Independently and without reference to genetics, psychologists have been interested in altruistic motivation. A standard claim is that persons aroused by the plight of others often act to aid them, but that the real goal is to reduce unpleasant arousal, not directly to help others. Helping is instrumental to a defense of the ego, troubled by the distress of others or afraid of feeling guilty or of suffering social disapproval. One acts so as to incur the least costs and most benefits in terms of feeling good about oneself and having others think well of oneself. "Adult altruism," claims Cialdini, "is a form of hedonism" (Baumann, Cialdini, and Kenrick 1981, p. 1039). What the Good Samaritan was really doing was repairing his own good mood, interrupted when he came on the ugly scene. Actual altruism, if there is any, requires a benefactor with intentions directed toward the end state of increasing the other's welfare.

Since helping others is obviously often accompanied by feeling good about one's success in so doing, it is difficult to isolate these. The usual case, in which altruism coincides with positive self-regard, does not serve to determine which is cause and which effect, and the debate often becomes an exercise in accommodating verbal reports and behavioral observations to whichever theory one prefers. Could the motivations be isolated, it might prove difficult to say which was primary, especially if one distrusts verbal reports. Even if one can find behaviors that help without the good feeling, these may be so atypical that the main hypothesis (that altruism is really just ego defense) is not tested under the anomalous circumstances.

In the years between 1962 and 1982, there were over one thousand empirical studies of altruism, and a review of them leaves the matter unsettled (Dovidio 1984) but does not eliminate the possibility that on occasion moral altruism is the primary determinant. One hardly wants or expects to find many occasions on which the altruist feels bad about his altruism, or feels indifferent, or is censured by others, and the debate almost becomes mute. Of course the Good Samaritan was troubled by the scene he encountered; whether what he wanted most was the victim aided or his own discomfort relieved is rather like arguing whether a glass is half full or half empty: it turns on perspectives preferred in description of the same event.

In studies over the last decade designed to isolate these perspectives, however, the egoistic hypothesis that seeming altruism is in fact done to reduce negative arousal has met serious empirical challenge. There is "impressive support for the empathy–altruism hypothesis," "that empathic emotion evokes truly altruistic motivation, motivation with an ultimate goal of benefiting not the self but the person for whom empathy is felt," and this is important "for our understanding of human nature" (Batson and Shaw 1991, p. 107; Monroe 1996).

None of these studies asks, much less answers, the question whether such

(6) Analogy and Category Mistakes

These two words, "altruism" and "selfishness," are borrowed from culture and problematically redefined to mean something different in nature;[13] but then, for all that, they slip back to culture and resume some of their conventional meanings. The same word is being used to describe widely different forms of life, from bacteria to Good Samaritans, at levels from molecular genes to public moral education. Are these words throughout the debate meaning anything similar enough to have a common intension or connotation, much less anything common in their extension or denotation? Denials of subjectivity notwithstanding, there is something anthropomorphic, almost animistic, about thinking that bacteria, or rats, or even baboons can be selfish. Or altruistic. In result, their behavior is "like" the decision of a moral agent. Biologists, complains Gunther Stent, need "terminological hygiene," terms that mean what they mean with precision; to say that genes are selfish is quite unhygienic: it only confuses (1980, p. 16). Others think that stretched metaphor can be productive (Hefner 1987). Words are regularly getting stretched to new meanings, willy-nilly, and sometimes we can do this creatively.

In conceptual *analysis* one asks what terms have meant and continue to mean in contemporary thought. Perhaps, to this date, altruism and selfishness have had only moral meanings. In conceptual *development* one asks how terms might be employed for breakthrough to new understanding of what is going on. Sometimes one narrows the meaning of a term, makes it more precise. This happened to *aqua*, water, when chemists began to define H_2O, water. Sometimes one widens the meaning of a term, as happened to "memory" when scientists built computers.

Still, words do break when stretched too far. There is attraction (psychological) between human lover and beloved; there is attraction (biochemical) between flower and insect pollinator; there is attraction (electromagnetic) between magnet and filings. But if one says that

---

good-feeling seeming altruists outreproduce their competitors, that is, whether their behavior enhances their inclusive fitness. Also, there is nothing amiss if indeed those who help others feel good about it. It would be rather fortunate if we had this tendency in our genes.

[13] J. B. S. Haldane seems to have been the first to introduce the word "altruism" into biological discussion (Haldane 1932, pp. 207–210).

the magnet "loves" the filings, or that the flower "loves" the insect (or vice versa), we begin to worry about identity and survival of concepts, and we fear equivocation. Love is an emergent psychological phenomenon, not yet present in electromagnetism, not yet present in botanical life, and probably not yet present even in insect life. To speak of a "selfish magnet" is unlikely to bring insight into either morals or electromagnetic theory; it is just to make a category mistake and speak nonsense. If morality exceeds genetics, trying to use biology to understand altruism may be like using valence (chemical attraction, exchanging of electrons) to understand animal sexuality (biological attraction, exchanging of genetic information). The efforts will fail because biology exceeds chemistry and because ethics exceeds biology.

One seeks fruitful but not misleading analogies. A child who grabs all the cookies may be selfish, but what of a chicken that grabs all the grain that the farmer scatters in the barnyard? Or an insect that grabs all the pollen? Perhaps even the magnet that grabs the filings is selfish! Or the cation that grabs the electrons. Perhaps magnets and ions can neither grab nor love nor be selfish. Perhaps chickens and insects can grab but not be selfish in any moral sense. An analogy is likely to mislead if it reduces, interpreting the more in terms of something less, but concealing this by too simplistic an analogy.

Something like that happened when physicists discovered relativity, and people said, "Everything is relative," and jumped from space and time to ethics, philosophy, and religion. Relativity in physics has only doubtful connections with cultural relativism. It is perfectly possible, even plausible, that the Golden Rule is an absolute in every inertial reference frame that is inhabited by moral agents, though the time of day shifts from time zone to zone even on Earth. The analogy only confuses, though it has, unfortunately, confused thousands of persons, some trying to understand physics and some trying to do ethics, philosophy, and theology.

An analogy is fruitful if it unifies, showing how the less evolves into the more, how the earlier parts of the story relate to the latter. But analogy is not yet story, and analogy too may not do enough explanatory work, even at its richest. Magnets attract, flowers attract, lovely women attract, but nothing follows about there being any covering law that explains electromagnetism, insect pollination, and sexual charm. Even if there were, this would not narrate the evolutionary history that starts with physical phenomena (including electromagnetism), continues through biological phenomena (the rise of insects and dicots), reaches at length psychological phenomena (love between man and woman), and unfolds this in concrete history (Antony and Cleopatra, their love affecting Roman and Egyptian history). There may not be any single idea, such as adaptive fitness for reproduction, that can illuminate everything that is going on in nature and in culture.

Sometimes it is fruitful to extrapolate, but often it is not. Statisticians know that it is risky to extrapolate very far. We have to ask what is the relevant domain of the regularity. Failing an adequate answer, we will extrapolate "altruism" or "selfishness" and get absurdity. There is the problem of knowing when to stop, before science transposes to poetry (or nonsense), fact to fiction, illumination to confusion. Are we enlightened, unifying the world with a better science? Or have we just slipped up with slippery terms? Elliott Sober concludes, "The psychological concepts of altruism and selfishness are quite independent of the evolutionary concepts that go by the same names" (1993b, p. 206).

When a bee flies to the hive and does the waggle dance, it communicates correctly. The bee is not to be commended for telling the truth; it does not have the capacity to lie, any more than a thermometer does. Ants are not to be censured for their castes, which are genetically based, though human caste systems ought to be censured, because they are not genetic. If either bee, ant, or thermometer were to fail, it would simply be broken. We might say that the newly informed bees, flying out to the flowers, "expect" to find nectar, perhaps that they are "confused" if they fail, even "satisfied" if they succeed. But one should not try to evaluate the bee as a "selfish" moral agent. That would strain the words so that we really undermine the point we are trying to make, because the words no longer carry the meanings we are trying to put on them. We might here be moving a term from culture into biology and faulting the biology we thereby misdescribe, only to reimport such faulty biology back into culture, thereby misunderstanding culture, thinking it to be determined by biological genetic selfishness when in fact in culture we can and ought to choose between moral selfishness and altruism.

To interpret events in terms of biology, where these are events in culture, is to fail to see that there is an emergent chapter in the story. It is an archaic interpretation. To interpret events in biology in terms drawn from cultural phenomena, which emerge novel to biology, is anachronism, a misplaced interpretation historically. It is a mistake both to see ourselves in fur and feathers and to see ourselves as nothing but fur and feathers. Try as we may to redefine the terms borrowed from one domain for use in the other, we fail this way when we label the behavior of bacteria, bees, and baboons as selfish and then find that human altruistic behavior must be more of the same. At best, such analogy is but one of various metaphors that provoke some novel insights and then outlive their usefulness and become dogmas, straitjackets that are obstacles to further creative thought.

4. EVALUATING ETHICS: VALUES DEFENDED AND SHARED

The model we advocate maps the same phenomena from the domain of value. Ethics is still our focus; there are two realms of concern: humans and nonhuman nature. In a comprehensive account, one needs value naturalized as well as ethics humanized; then ethics will require appropriate respect for value, whether human or nonhuman. Classical, interhuman ethics arises to defend and to share human values, as these arise in human cultures over the millennia. Such ethics arises out of evolutionary natural history, in which values have already been arising, being defended and shared over the epochs of life on Earth. Becoming aware of such genesis of value, environmental ethics arises, recognizing the human destiny entwined with valuable nature. Such genesis of ethics, distinctive to the human genius, testifies both to human uniqueness, emergent from natural history, and to the creative power evidenced in the spontaneous genetics, the primal source now transcended with the appearance of genuine and universal caring and altruism.

(1) Moral, Valuable, and Evaluating Persons

The human self achieves the novel possibility of defending values as a moral agent. Such a self can recognize and make its own concern intrinsic values outside its own local sector. The self can take an interest that ennobles and enlarges itself. The person can and ought to defend values that are more comprehensive than those of self-love; the person can love the other as it does itself. Failure to rise to this possible humanity, impossible for the animals as well as inappropriate for them, is selfishness, now censurable. The shadow of moral possibility is moral failure, indeed, moral tragedy.

None of this puts culture at odds with nature; nor does it find in nature any human norm for social ethics; rather it insists that culture be superposed on nature, with its novel integrity and yet integrated with the global story. Such defense of value will still include human fertility, for humans remain biological and must reproduce over generations, but the values defended will be the vital values of culture as well, values whose transmission and defense are no longer genetic. Even in human society, the self-actualizing inherited from animal life remains foundational and appropriate, although the character of the "self" so formed is elevated from the animal to the personal self. The person reaches *Existenz*, deeper than animal existence. In this axiological model, the self *must* defend its values, a *description* of the essential life process, and the self *ought* to defend its values, a *prescription* for ethics. In that sense, self-love is proper and appropriate, the presumption of life. "Selves" are one of the wonders of creation, and there is nothing unsatisfactory, per se, about interest satisfaction.

In culture, one can gain enlarged interests and so an enlarged sense of identity. The first cultural unit is the nuclear family, where there is also genetic identity, kinship identity. But the cultural self, like the biological self, lives the life of myriad interconnections, extending far beyond the family. One works for a business firm, serves on a town council, is a volunteer at the hospital, spends time in military service, makes a donation to the college of which he or she is an alumnus or alumna, supports a scientific research project, teaches a class of students with kindred interests, joins a conservation society, leaves a will with a bequest not only to children but to those institutions he or she wishes to see continue after death. Almost everything that the self cares about has to be cared about in concert with others, and all these others have their myriad connections in turn. The cultural self comes to transcend, even to replace, in part, the biological self. What one wishes to survive is one's ideas, ones values, or, more accurately, those ideas and values into which one comes to be educated and in which one meaningfully and critically participates.

This sharing is of cultural beyond genetic information. One can insist if one wishes that this is just enlarged selfishness, reciprocating with all these values now located in persons outside oneself, in the institutions of the self's heritage, or in the natural world with which one is environed. But it is far more plausible to face the epistemic crisis and move to the better paradigm. The self has entwined its identity with others: that is, it blends its own self-defense with that of others; the self becomes an altruist, more and less, because this optimizes the sharing of values.

The bold hypothesis of selfish genes dies the death of a thousand qualifications once again, because these genes live the life of ten thousand cultural interconnections, beyond the ten thousand genetic interconnections found before. Human altruism, with its genuinely emergent properties, "takes over" the biology; it "takes off" from precedent phenomena, in an elevated but analogous way to that in which a living organism "takes over" the electronic bondings of chemistry – ionic, covalent, van der Waals, hydrogen, and the like – superimposing biochemistry on chemistry, biological functions onto physicochemical laws and states. The organism "takes off" into life. Much later, the human animal "takes off" into ethics.

Socrates drinking the hemlock, loving Athens and protesting its injustice; Jesus at the Last Supper, with his vision of a realm, a kingdom, of the love of God and neighbor; Buddha postponing nirvana until he could enlighten others; John Stuart Mill advocating the greatest good for the greatest number; Immanuel Kant urging his categorical imperatives; the Good Samaritan helping the victim of thieves; or even Richard Alexander writing his book and hoping to release us from our evolutionary biology – these are not individuals exhibiting behaviors that one can profitably understand by asking about somatic versus genetic forms of self-interest, about their results in offspring, any more than one can understand them by asking about biochemical movements in which electrons are transferred from this atom to that, resulting in energy dissipated and reconfigured biomolecules with an altered set of synaptic neural connections.

One has to understand social facts as social facts, ethical convictions as ethical convictions. Else we do not really narrate what is developing in the story. One falls back to lower, inadequate categories, admirable ones in their own domain but juvenile and even tragic if not superseded, because this robs us of our humanity. We fail to rise to all our possibilities. There are twin truths: nature is a womb that humans really never leave, and so ethics does have to be "naturalized," to fit human biology, including human reproductive needs. Yet there is an exodus out of nature into the freedom of spirit in cultural life, superimposed on biological life. We never become free from nature, but we do become free within nature.

Many precepts in our moral system will be specific for *Homo sapiens*. After all this is an ethic that human persons, an earthy species, must use to evaluate each other and to defend their kinds of value. There is nothing undesirable about having morality applied to the human species; one does not want an ethic of no Earthly use. A species-blind moral system would be inadequate. Ethics needs to be situated where the moral agents live, internal to the agents' conditions – physical, environmental, social, cultural.

Ruse and Wilson are eager to deny that there can be in cultural ethics any breakthrough past biology, and this is where their claim becomes problematic. Their theory of ethics gets stalled in the preethical world: "Ethical premises are the peculiar products of genetic history, and they can be understood solely as mechanisms that are adaptive for the species that possess them. It follows that the ethical code of one species cannot be translated into that of another. No abstract moral principles exist outside the particular nature of individual species. . . . Morality is rooted in contingent human nature, through and through" (1986, p. 186).

All that really follows is that human ethics will have evolved to suit human biology and to defend the sorts of values that persons can instantiate, enjoy, or lose and need to protect. If resources are in short supply, there will need to be an ethic about stealing. If killing is possible, there will need to be an ethic about murder. If the moral agent is sexed, there will need to be an ethics of sexuality. If humans have lusts, there will need to be a command not to covet. Ethical principles will need to fit human sociology and psychology, as this is superimposed on our biology. Ethical principles will also need to fit human cultural institutions – such as contracts of marriage, or business dealings, or citizenship in states – many of which have little precedent in nature. Ethics will need to protect, both by defending and by sharing, the multiple capacities that humans have for enjoying values.

And so it is. Ethics is as undeniably present, ideal and real, as are genes, and just as much among the wonders of creation.

(2) Amoral, Valuable Nature

In this creation, undeniably present, are myriad living organisms, resulting from the evolutionary genesis. Every such organismic life must be defended, and defense of somatic identity (Chapter 2, Section 1[2]) is the vital (=valuable) condition of all life. Each organism inherits life as a given from its past. Self-actualizing, each thrusts this life forward to the future. This requires short-range preservation of material identity (the stuff of one's body) and its long-range replacement through dynamic resource input and utilization, and waste output – materials turnover and energy throughput. Alleged biological somatic "selfishness" is this admirable biochemistry, metabolism, physiology, the proper life of the organism, all that it "owns." Animals, though not moral agents, have their norms to defend; they can do no other and survive. An animal preserves for its own sake the good-of-its-kind that it instantiates. This is the conservation of intrinsic value.

In most of the biological world, animals have limited capacity to help each other out, although they are interrelated in ecosystems and can sometimes reciprocate. Oxpeckers and rhinoceroses cooperate (coact) for mutual benefit. But warblers cannot aid grizzly bears, or vice versa. In social animals, where reciprocal "altruism" develops (the monkey giving an alarm), one life can and does aid another, and the result (averaged over the population) is increased conservation of somatic and genetic value. There is nothing killjoy about this explanation (no "altruism" reduced to "selfishness") when we find that self-defense combines with other-(*alter*)-defense to result in maximal protection of values held in common by the animals involved. It would be a mistake for one animal to lose where this did not bring high enough gains in kindred lives bound with it in community. That would result (on average) in the loss of intrinsic value. Natural selection selects against such behavior.

Somatic identity is short-lived; death comes soon. The organismic values can be preserved only if reinstatiated in "others," offspring. This involves making more of one's kind, more others, not only replacements but additionals, which increases subsequent somatic value in oncoming generations. The only kind that an organism can make is its own kind. It is comedy again to think that warblers can breed grizzly bears. An animal's defense of its genetic line is proper to it, the only reproductive power it has. Natural history is not of somatic individuals with everlasting life; it is a history of inheritance, life to life. That is, it is genetic. Intrinsic value is perpetually perishing, perpetually regenerated, perpetually conserved. This is kinship identity (Chapter 2, Section 1[3]).

Animals can defend only the alleles they own, not some others they do not have, but this results also in a defense of species identity (Chapter 2, Section 1[4]), in that the animal breeds after its own kind, shares large numbers of genes common to the species, and outbreeds sexually with other members of the species. There is rapid turnover of gene tokens, with relatively fleeting identity; all that can be preserved is genetic type identity (Chapter 2, Section 1[5]). What is thereby in fact preserved is a genetic cybernetic identity, that is, the information that codes a species, the valuable know-how for this form of life (Chapter 2, Section 1[8]). Any such life-form is *what* it is *where* it is, within an ecosystem, networked into a community of life (Chapter 2, Section 1[7]). This is the defense of a good kind in a good kind of place.

Now we can see the sense in which genetic "altruism" does not exist in nature because it cannot. Setting aside any borrowed moralistic terms, and more objectively interpreted, reproduction is the transmission of intrinsic value, instantiated in the organism as somatic value and transmitted as genetic value. The animal parent disseminates what value it owns to its offspring. It will only introduce confusion to think this might be genetic altruism and then be disappointed to find this an alias for genetic selfishness. One corrects the confusion by interpreting the event as the distributing in parts (sharing) of the intrinsic value of animal life.

In the defense of life, somatic and genetic, there is much value capture. All heterotrophic life depends on this. When a predator eats prey or feeds its young, when a grazer eats grass and gives milk to nurse young, the eaten genetic "self" loses, and the eating "self" wins. But one should not interpret any and all self-defense of a somatic self or a genetic line as selfishness, any more than one should interpret the captured prey or the eaten grass as behaving altruistically. That broaches animism and anthropomorphism. The losers here are sacrificed for the welfare of others; such exchange belongs in the trophic pyramids of ecosystems and makes possible the evolution of highly mobile, perceptually acute forms in the top trophic rungs. This too is the redistributing of value throughout the interactive system, instrumental values tributary to intrinsic values. There is nothing morally culpable here, nor ought there to be actions that are morally laudable. All such supposition is a category mistake.

Real genetic altruism, if this existed in nature, would involve one animal by programmatic genetic disposition behaving in such a way as to result in genetic loss for itself exchanged for genetic gain by another animal. Animals are eaten by others; they lose, and their nutritional and energetic resources are cycled instead through these others. Animals aid others in reciprocal "altruism"; on statistical average they all win, though on occasion some individual animals lose. But no animal has the power to sacrifice itself genetically for others as an inherent component in its DNA coding; that would amount to innate self-destruction. Interpreted from the axiological paradigm, that would negate the conservation of intrinsic value. Far from some censurable genetic selfishness, there is present rather the ability to conserve value. Genes can code for, organisms can defend, only the vitality they have, not some "other" (*alter*) they do not. Any organism coded for degrading the values it instantiates would be an evolutionary mistake; such alleles will soon go extinct.

The selfish/altruistic conceptual scheme, found in sociobiology and evolutionary ecology, is set up so that nothing will count as altruism in any real, nonselfish sense unless it decreases fitness. This is tantamount to saying that the only way to be genuinely altruistic is to fail. Such failure may be partial, but inevitably, by definition, is proportional to the altruism. In this scheme, altruism is failure, so among the survivors, one cannot expect to find it. Indeed, "selfishness" is survival, and we know before we look that, one way or another, the survivors have been "selfish."

In the axiological model, by contrast, value is conserved, and what we may expect is that one generation transmits to another its valued skills and achievements, its know-how. There is a contest of values in ecosystems, value capture, loss, and gain, but on average over evolutionary history and in ecosystemic dynamism, by self-defense and reproduction of kinds, value remains in the series of its replacements. Indeed, in the larger history value develops and diversifies dramatically. This account removes any moral overtones from nature while conserving value present over the millennia of natural history.

Morality is not intrinsic to natural systems. In fact, there are no moral agents in wild nature. Nature is amoral, but that is not to disparage it. That is to set aside irrelevant categories for its interpretation. Amoral nature is fundamentally and radically the ground, the root out of which arise all the particular values manifest in organisms and ecosystems. This includes all human values, even though, when they come, human values rise higher than their precedents in spontaneous nature.

In fact in nature, there is systemic process, profoundly but partially described by evolutionary theory, a historical genesis during which spectacular values are achieved. At the core, the critical category is *value*, revealed in the term "survival value," valuable information for "living on and on" (*sur-vival*), coded genetically, apt for coping, by which life persists in the midst of its perpetual perishing.

(3) Global and Universal Morality

Humans have expanded their territories all over the globe; they have been quite successful reproducing. They are also capable of asking – and are now asking – how ought humans to live as they reside in their cultural communities on this Earth we have so much occupied. For each person to maximize the number of his or her offspring? Each family? Each tribe? For *Homo sapiens* as a species to maximize the number of offspring? For each to maximize his or her own self-interest, reciprocating as need be to accomplish this? For humans to maximize the high intrinsic value of our kind? The most convincing answers rather urge a more global, a more generous defense of value.

Humans can get "let in on" more value than any other kind of life. In interhuman ethics there is already a striking novelty, unprecedented in prior natural history, even though there are more and less plausible accounts of how this might have emerged. Humans can enter into an ethical contract with other humans, the principles of which are oblivious to the specific circumstances of time and place, genome or culture. Ethics – at least in ideal, if not in real – has universal intent. The Golden Rule, the categorical imperative, the greatest good for the greatest number, the Ten Commandments are for all humans, panculturally, pangenetically. Ethics takes up an "original position." Ethics becomes globally inclusive – without denying that there may be differing duties to family, friends, community, business, heritage, ethnic or interest group, nation state, and humanity at large. Human values are widely distributed and shared, as witnessed, for example, in the Universal Declaration of Human Rights.

In environmental ethics, rather than using mind and morals as survival tools for defending the human form of life, mind forms an intelligible view of the whole and defends the varieties of life in all their forms. Persons have their excellences, or genius, and one way they excel is in this capacity for overview. Such an ethic, although rather neglected in the modern West, is in the archaic memory. Adam and Eve were set on Earth as keepers of the garden, even before Cain was charged with being "his brother's keeper." The covenant (contract) that the Hebrews claimed they had with God included other humans (a social contract) and also the wild animals (a natural contract): "Keep them alive with you" (Genesis 6.19). In Hindu and Buddhist traditions, noninjury, *ahimsa*, was fully applicable to all living things. These ancient motifs are now resurrected in a scientific era when reevaluating nature has taken on a new urgency.

The novelty in such an environmental ethics is class altruism emerging to coexist with class self-interest, sentiments directed not simply at one's own species but at other species fitted into biological communities. Humans ought to think from an ecological analogue of what ethicists, as noted, call the original position, a global position that sees Earth objectively as an evolutionary ecosystem. Interhuman ethics has spent several millennia waking up to human dignity. Environmental ethics invites awakening to the greater story of which humans are a consummate part, a drama of which classical and primal peoples sometimes had a better intuitive sense than have we modern humans lately in the scientific West – and this despite the fact that we today, thanks to science, have more objective knowledge about who and where we are than did they.

Environmental ethics, in this sense, is the most altruistic, global, generous, comprehensive ethic of all, demanding the most expansive capacity to see others, and this now especially distinguishes humans. This is not naturalized ethics in the reductionist sense; it is naturalized ethics in the comprehensive sense, humans acting out of moral conviction for the benefit of nonhuman others. There is a widening sense of shared values, including values produced in the evolutionary genesis. Restricting morality to the species interests of *Homo sapiens* would be rather like a nation's defining their foreign policy only as their national self-interest. They minimally ought to recognize that other peoples in other nations have their interests, which must be recognized, in any reciprocity of defended national interests. People maximally can take an interest transnationally, respecting human values wherever they occur. But so too in environmental ethics, at still deeper levels of community, ethics moves outside the human sector.

Ruse and Wilson are especially eager to conclude that there is nothing absolute or permanent about these ethical commandments; they fit our human biology and nothing more. A naturalized, Darwinized ethic will be Earth-bound. "A conclusion of central importance to philosophy [is] that there can be no genuinely objective external ethical premises" (1986, p. 186). Drawing a contrast between somatic and extrasomatic moral truths, Ruse and Wilson insist, "Everything we know about the evolutionary process indicates that no such extrasomatic guides exist." Ethics is "idiosyncratic" to the biology of a species (1986, pp. 186 and 173).

There is much ambiguity here. That ethics is idiosyncratic to species is one claim; that ethics is selfish is another. An ethic that defends a species line is already rather "extrasomatic," if the *soma* (body) referred to is that of the individual pressed to maximize the numbers of its offspring. Does everything we know about natural history indicate that no "genuinely objective external ethical premises exist"? One thing we know, for example, is that Wilson in his studies of natural history finds values present in the biodiversity generated there that he desires to protect with a "generosity beyond expedience" (1984, p. 131); another is that he urges us to "favor diversity in the gene pool as a cardinal value" and to regard "universal human rights" as a "primary value" (1978, pp. 197–199). All this sounds quite genuine, objective, and external – external at least in the sense that the moral agent recognizes values outside the self deserving of appropriate respect. A comprehensive global ethic might be objective and external enough, as humans do not at present have any opportunity to be ethical elsewhere (astronauts excepted).

The evolutionary process does not produce extraterrestrial guides, because all the species achieved are environmentally situated, well adapted for their environments. Even these species, of course, must have extrasomatic information about "outside-body" affairs and how to survive among them; such knowledge is recorded microscopically within. Information about behaving ethically where one does not reside is irrelevant. Humans too receive this fortunate biological legacy, but are elevated into a cultural process superimposed on the evolutionary one. One might suppose that perhaps extrasomatic guides exist, but that extrasocial ones do not. Certainly ethics is social and informs us how to make a way through both the social and the natural worlds. We know only one ethical species – humans – and it is difficult to universalize from only one known case. Perhaps humans do not need any ethics for larger realms than those in which they actually reside.

But the human mind sometimes seems to be able to reach truths about realms that it does not inhabit, extrapolating and reasoning from the realms it does. We learn this even in biology. Life is Earthbound, but on Earth humans reach outside their own sector to study warblers, viruses, and dinosaurs. Without leaving what is true on Earth, humans become still more universal in physics and chemistry, learning about the microworlds of elementary waves and particles, about the astronomical worlds of outer space and millennia past, about truths in other inertial reference frames. Science is rooted in human nature, employs biologically evolved perceptual and conceptual faculties, is a social construct, but, for all that, it sometimes flowers to discover objective truths – such as relativity theory or the atomic table, which are true universally, that is, all over our universe. Perhaps we can ask whether the human mind can reach ethical principles that may transcend our somatic embodiment, that are "objective" to our humanity. "Absolute" is a forbidding word, but "universal" is not. Histories, such as Earth history, are particular and idiographic; does an ethic for a world history need to be absolute, or even universal?

Some insights in our human moral systems may be transhuman. Keep promises. Tell the truth. Do not steal. Respect property. Do to others as you would have them do to you. Love your enemies; do good to those who hate you. Such commandments may be imperatives on other planets where there are no humans, but rather where alien species of moral agents inhabit inertial reference frames that have no contact with ours. Wherever there are moral agents living in a culture that has been elevated above natural selection, one can hope that there are love, justice, and freedom, although we cannot specify what content these activities will take in their forms of life. It certainly does not follow that nothing generally true can appear in human morality because it emerges while humans are in residence on Earth. There is nothing particularly Earth-bound about "Do to others as you would have them do to you." That could be true whether or not the moral agents are *Homo sapiens*.

In environmental ethics, if visitors from outer space were to come here and wish to set up a space station that required destroying a rich tropical forest ecosystem, filled with endangered species, would not Wilson urge his "generosity beyond expedience" upon this non-Earthen species, rather as he would urge "universal human rights" upon the Martians, should they try to capture humans as slaves to build their space station? The ethics that humans have reached, although certainly appropriate to our Earthen residence, discovers values that are objective enough to urge on moral agents of whatever extraterrestrial origin. There would be something censurable about moral agents anywhere, anytime, who lied, cheated, stole, hated, or were unjust – or caused unwarranted pain in animals or destroyed endangered species without adequate justification. One can plausibly venture the claim that if there are moral agents anywhere, anytime, they have not matured until they have reached the capacity for altruism, indeed, as we will hear next from religious humans, the capacity for suffering love.

The surprising point is that we humans have enough wit to do this. And the wit to do this either has to come in by skyhook (revelation, prophecy, supernatural knowledge) or to come by these inherited traditions and such creative human inspirations and breakthroughs that rise to new levels of overseeing, understanding, and compassion. Far from a killjoy reduction of ethics to nothing but biology, we have discovered that ethics is naturalized only at the start by way of anticipation and launching; afterward it is conceived and socialized in culture. As it matures, we are left wondering whether it does not even move beyond, glimpsing universals. The genesis of ethics, especially in the genesis of generosity,[14] distinctive to the human genius, continuing but exceeding the genesis in the genes, reveals transcendent powers come to expression point on Earth.

[14] Remarkably, "generous" (like "genius") goes back to the same root as "gene," which, as noted earlier, is from the root for "nature," "giving birth." The "generous" are those of "noble birth."

Chapter 6

Religion: Naturalized, Socialized, Evaluated

Since there isn't any religion in nature, it may be difficult to naturalize it, even harder than with ethics, where there is animal reciprocal cooperation as a precursor. Religion is without antecedents in wild nature. Still, the capacity in persons to be religious somehow evolved within, or emerged out of, natural systems, where before there was no such capacity. Religion too, like ethics and science, is eminent in the human genius. Human societies, historically, always produce religion. Again, there must be some story to tell – this time of the genesis of religion. Now too, however, one must assess such accounts of the genesis of religion as these are set alongside religious accounts of the genesis of nature and culture, analogously to the way one needs an account of the genesis, in culture, of science, set alongside science's account of such genesis of nature and culture.

One should be wary. The question is of the logic as well as of the origin of religion. Religion today may be something quite different from what earlier religion and its precursors initially were. The monotheism widespread in the West, which has interacted with science for several centuries, is quite different from aboriginal animisms. The many religions may not have common origins or any common logic; their origins and operations may differ. Most of what the earliest humans thought is lost in the mists of the past; any psychoprehistory is speculative. One would commit the genetic fallacy if one overlooked ways that religions have matured, reformed, and transformed over the millennia of cumulative and critical transmissible cultures. Religions change, as much as does science. We know much more about what religion now is, and, failing knowledge of the routes traveled in the remote past, it could be a mistake to be so sure we know the determinants along the way.

Still, origins contribute to explanations in a deeply historical world. In evaluating religious accounts of origins, we will be wiser if we can discover how humans came to be religious, especially since religious capacities are a unique mark of *Homo sapiens*. Two questions are entwined: Is there a plausible religious account of genes and their genesis? Is there a plausible account of the genesis of religion, relating its origin to genes? Neither account will be by way of implication, whether deductive or inductive. There are no covering laws (such as natural selection) plus initial conditions (such as primates) from which one can infer religion (persons who are priests), any more than one can assume microbes as a premise and deduce primates in conclusion. Perhaps we cannot even predict what universal religions in advanced societies will be like, if we know what early religions in simple societies were. That development too may be more historical or culturally contingent than logically or biologically necessary.

The best explanation available will be a "how-possibly" explanation, not a "why-necessarily" explanation (Hempel 1965, p. 428); that is, it will trace a pathway along which religion might have appeared. It will also be true that there appears on Earth later on something, religion, of which there was exactly none before. Ideas may also appear within later religions that have little or no precedent in previous religions. Religious experience transforms animal experience of the environment. Experience of nature takes on a dimension of depth, the experience of the sacred. Nature becomes sacramental, as also do events in culture. This could be illusory mythology that is successfully functional. It could be an epiphenomenal anomaly, like dreams that have little to do with the real world. But it might be, like science and ethics, the achievement of new levels of insight. The fact that our perceptual and conceptual faculties have evolved does not mean that nothing true appears in them, nor that nothing new can ever appear in them.

Although origins contribute to understanding, the genesis of religion, unique to humans, is unlikely to be something one can extrapolate from earlier explanations in biology. A good rule facing the future is to stick with explanations previously tested; the sun will rise tomorrow because it rose yesterday and yesteryear. One will be right almost all the time, for the future is regularly like the past. One will be wrong every time the event under consideration is an advent making a critical difference introducing what is genuinely novel. One will miss every occasion of originating genesis. All one's explanations will be anachronistic. In the continuing creation, the future is never like the past: life appeared where none was before, exoskeletons and endoskeletons arose, photosynthesis evolved, so did sexuality and warm blood. So did vertebrates, sentience, pain, hiding, smelling, alarm calls, courting of mates, aggressive displays, learned behavior. So did fire building, tool making, language, writing, money, internal combustion engines, computers, rockets. So did ethics, science. And religion. What are we to make of sacrifice, prayer, altars, sacraments, cultus, shamans, priests, prophets, saviors, preachers?

It is tempting to dismiss novel appearances at first as "apparent" anomalies, because lawlike explanation dislikes counterexamples that defy the law. Some way is found by which these are "nothing but" appearances, and the old account, extrapolated, holds despite appearances. But when the appearances continue to mount, the law diminishes in its logical appeal. The developing appearances, the anomalies, are sometimes recompounded into history. When religion appears, can one subsume it under yesterday's explanatory categories? Or is this a deepening of the plot?

1. THE DIVINE EPIC OF LIFE

Religion is generated confronting nature – the sunset, the midnight sky, the wind and the rain, the forest primeval, birth and death, life renewed in the midst of its perpetual perishing. Though religion arises only within human societies and notably helps humans to manage within such societies, coupling neighbors and God, it will not suffice to get religion socialized (Sections 2, 3, and 4). One must also get religion naturalized, not so much in the sense of explaining it (away) naturalistically, as of explaining the numinous encounter with manifest nature. Biology does generate religion: the phenomenon of life evokes a religious response whether or not a functional human society is at issue, whether or not one is being altruistic or evangelistic toward others. Nature is the first mystery to be encountered, and society comes later, much later, after one learns evolutionary history. Surveying paleontological history, Loren Eiseley exclaims, "Nature itself is one vast miracle transcending the reality of night and nothingness" (1960, p. 171).

Religion begins also in physics and chemistry, matter and energy; in cosmology. Why is there something rather than nothing? Why is there something of a kind that spins this surprising kind of universe? But the most startling results are on Earth, not in the heavens. Our native-range life world stands about midway between the infinitesimal and the immense on the natural scale. The size of a planet is near the geometric mean of the size of the known universe and the size of the atom. The mass of a human being is the geometric mean of the mass of Earth and the mass of a proton. Astronomical nature and micronature, profound as they are, are nature in the simple. At both ends of the spectrum of size, nature lacks the complexity that it demonstrates at the mesolevels, found in Earthen ecosystems, or at psychological levels in human persons in their societies. Humans do not live at the range of the infinitely small, nor at that of the infinitely large, but we may well live at the range of the infinitely complex.

There is in a typical handful of humus, which may have ten billion organisms in it, a richness of structure, a volume of information (trillions of "bits"), resulting from evolutionary processes across a billion years of history, greatly advanced over anything in myriad galaxies, or even, so far as we know, in all of them. The human being is the most sophisticated of known natural products. In our hundred and fifty pounds of protoplasm, in our three pounds of brain, there may be more operational organization than in the whole of the Andromeda galaxy. The number of possible associations among the trillion neurons of a human brain, where each cell can "talk" to as many as a thousand other cells, may exceed the number of atoms in the universe. On a gross cosmic scale, Earth is insignificant and humans are minuscule atoms. But on scales of prolific genesis, Earth is quite significant, and mind is a most impressive creation. The brain is so curiously a microcosm of this macrocosm, since the mind can contain so much of nature within thought and thus mirror the world. We might live at the center of the most genesis.

As far as we can gain it, we, who have such minds, need a unified account, one that narrates the whole Earth story and locates ourselves in it. Call such worldviews "myths" if you wish; they must now be couched in scientific mythology; afterward one can see whether such accounts of the genesis that has taken place here remain congenial to any of the classical religious myths.

(1) The Prolific Earth

One thing is right about the fertility hypothesis as the key to under-standing life: humans reside on a fertile Earth. Evolutionary history has been fruitful, prolific. This is no myth; it is among the best estab-lished facts. But what hypothesis best explains this fact? From the dawn of religious impulses, in the only animal capable of such reflec-tion, this vitality has been experienced as sacred. Such experience has been often fragmentary and confused, as has every other form of knowledge that humans have struggled to gain, but at its core this insight developed that religion was about an abundant life, about life in its abundance. Classical monotheism developed (evolved) into a fertile (widely reproducing) hypothesis that claims – to take the He-brew form of it – that the divine Spirit, Wind (Greek: *pneuma*), breathes the breath of life into the dust of the Earth and animates it to generate swarms of living beings (Genesis 2.7). Eastern forms can be significantly different – *maya* spun over Brahman, or *samsara* over *sunyata* – but they too detect the sacred in, with, and under the pro-fuse phenomena.

In that sense, the fact that religious conviction cherishes, con-serves, and celebrates this fertility is no reason to think religion sus-pect; to the contrary, it is reason to think it profound. If this be animal faith, we still need to ask whether the animal in which such faith emerges, *Homo sapiens*, is coping now because it is detecting the truth: there is a divine will for life to continue. Genes and their gen-esis do lie behind the genesis of religion – but not (we will be claim-ing) in the way typically alleged by behavioral psychologists and sociobiologists. Rather, the genesis in natural history, when humans discover and reflect over this, generates religious responses. "Fertil-ity" is precisely what evokes religious belief. The prolific Earthen "fertility," "fecundity," or generative capacity is what most needs to be explained in the spectacular display of life in which we find our-selves immersed.

"Fertility" is literally used of the fauna and flora, though perhaps we are metaphorically extending it to evolutionary ecosystems and the global biosphere. "Nature" (we recall) has, as root idea, "giving birth." If we must use metaphors, after Darwin, the Earth is as much like a womb in these gestating powers as it is, after Newton, a clockwork ma-chine, or, after Einstein, energy and matter bubbling up out of a space-time matrix. The genesis is widely distributed over the planetary space and long continuing over evolutionary time, evidenced by the biodiversity so well documented in the biological sciences.

This genesis is hard fact. No one doubts that these myriad species, including *Homo sapiens*, are here. No one doubts that there lies behind us some sort of genesis, and few readers of this book doubt that evolutionary natural history is a key to this genesis. But a self-generating nature is not self-explanatory. One needs an account of the setup, an account of the generating processes; of how possibilities get actualized, of how possibility spaces come to be; of the depth sources of the creativity. In this genesis, "more" regularly comes from "less." Something comes, if not from nothing, at least where nothing like that was present before. Information does appear, superimposed on matter and energy, a key to the vital generation of life. This is a pregnant Earth. But we know what pregnant means with females giving birth, the vital information transferred in DNA from one generation to the next, and we must puzzle over where and how such information originates on Earth (Section 5[2]).

Such hard fact is hard to explain without some sort of generative principles before which many persons are inclined, one way or another, to become religious. Ernst Mayr, one of the most eminent living biologists, concludes, "Virtually all biologists are religious, in the deeper sense of the word, even though it may be a religion without revelation. . . . The unknown and maybe unknowable instills in us a sense of humility and awe" (1982, p. 81). We detect something sublime in the awe-inspiring sense because there is something sublime in the etymological sense of that word, something that takes us to the limits of our understanding, and mysteriously beyond.

Viewing Earthrise from the moon, the astronaut Edgar Mitchell was entranced:

> Suddenly from behind the rim of the moon, in long, slow-motion moments of immense majesty, there emerges a sparkling blue and white jewel, a light, delicate sky-blue sphere laced with slowly swirling veils of white, rising gradually like a small pearl in a thick sea of black mystery. It takes more than a moment to fully realize this is Earth . . . home. (Kelley 1988, at photographs 42–45)

The astronaut Michael Collins recalled being Earth-struck:

> The more we see of other planets, the better this one looks. When I traveled to the Moon, it wasn't my proximity to that battered

rockpile I remember so vividly, but rather what I saw when I looked back at my fragile home – a glistening, inviting beacon, delicate blue and white, a tiny outpost suspended in the black infinity. Earth is to be treasured and nurtured, something precious that *must* endure. (1980, p. 6)

Ernst Mayr's thoughtful biologist not only has religious humility, but a respect for nature. "And if one is a truly thinking biologist, one has a feeling of responsibility for nature, as reflected by much of the conservation movement" (1985b, p. 60).

The Earth is a pearl in a sea of black mystery. That is metaphor, but metaphor witnessing to the eventful genesis on Earth and witnessing to the power of such genesis, when scientifically known, to generate convictions of value present, to generate religious wonder. Whatever may be said of the rest of the universe, Earth is a prolific place, a pro-life place. That is the testimony of science, as well as a religious conviction. To use a weighted term, the *telos*, ending, heading, of the Earth process is "fertility," generativity, as evidenced in the *telos* (lives defended as ends-in-themselves) of the organisms that are its myriad products. Say if you like that there is a bias for self-organizing or autopoiesis in the process (Kauffman 1993; 1995; Maturana and Varela 1980) that explains the remarkable results. That may be good science, but now we are in a religious or metaphysical mode and need to explain this remarkable bias. Nature has been generously fertile.

(2) Nature and Spirit (*Geist*)

The story is nowhere more fantastic than in the evolution of spirit within and out of nature. Molecules, trillions of them, spin round in complicated ways and generate the unified, focused experience of mind. This too is among the established facts. For this appearance of "genius" (Latin: spirit) scientists can, as yet, hardly imagine a theory, though if ever such a theory appears, we shall welcome it as the most ingenious theory of all. Meanwhile, putting together molecular parts does not really explain how inwardness comes out of outwardness, how felt experience arises where before there was none.

At this point scientists, no less than religious persons, believe what they do not understand: that the output exceeds the input, that the results outrun the causes, that there evolves, incrementally and yet ex nihilo too, something in kind (subjects) where, if one looks rearward far enough, nothing of that kind existed before among the Earth objects. The human Geist is especially fertile in its generation of cumulative transmissible cultures, something novel in kind again, now in only one species, for, if one looks rearward in any of the other several billion species over evolutionary time, there is nothing of this kind. Persons too are among the established facts with which we must deal. There is personal narrative as an ego travels through the world.

The real surprise is that the human intelligence can be religious and philosophical; we nowhere approach that elsewhere in animal life. That is why it is so hard to get religion naturalized; it is quite unprecedented. Human spirits have *Existenz*. They anticipate death; they sense their finitude. They face the limit questions, sense the sacred, worry about communion with the ultimate or atonement of their sins. They know guilt, forgiveness, shame, remorse, glory, pride. They suffer angst and alienation. They build symbols with which they interpret their place and role in their world. They create ideologies, affirm creeds, and debate them. They are capable of faith and need salvation. They worship God. All of this is summed up in the one word: "spirit" (*Geist*).

Out of physical premises one derives biological conclusions, and, taking these as premises in turn, one derives psychological conclusions, which, recompounded again, yield spiritual conclusions. This kind of logic seems more story than argument; the form of argument is not so much rational as, to use a religious word, incarnational, since each step has to be embodied. Story is a better category than unfolding law, much less random drift, or selfish defense of life, when one wants to get more out of less. If one tries to interpret the world as law plus initial conditions, there is little plot. If one tries to interpret the world as statistical probabilities, there is little story. But when we tell the story of suffering through to something higher, over the millennia of microbes, and trilobites, dinosaurs and primates, persons who are scientists and saints, we have enough bite for a dramatic story.

(3) Nature and Sin

Humans forge their cultural history beyond biology, but this is not particularly to praise humans and belittle beasts. Part of the human genius is the genesis of sin. Humans have a superiority of opportunity, capacities unattained in animal life. Alas, however, the human capacity is but brokenly attained. Much of the history that humans have made is checkered enough. There are noble achievements, but humans repeatedly stand condemned because they could and ought to have made for themselves better history than they did. Religion has tried to face this fact full on, cognitively, existentially, and redemptively. All classical religions find the human condition to be deeply flawed; humans need salvation. "Our civilizations were jerrybuilt around the [human] biogram," Wilson laments (1975a, p. 548). But to discover that the world is in a troubled condition is no new revelation to religious sages; to the contrary, it is what they have regularly taught. In Judeo-Christian monotheism, the central category is that of "sin," missing the mark; in Eastern faiths the category is that of "ignorance," *avidya*. Islam uses both evaluations.

There is something "original" about sin, something in human origins that produces sin perennially, something in human biology, in the flesh, that makes it inevitable for humans to lapse into sin. At this point biology and theology are well within dialogue; indeed they can seem to be saying almost the same thing. The innate biological "selfishness" concurs with what classical religions have been teaching for millennia. But this congruence of biology and religion will have to be interpreted with some care.

Humans do have to break out of their animal nature. When animals act "like beasts," as nonmoral beings, nothing is amiss. To the contrary, spectacular values have been achieved over the evolutionary millennia. But if humans go no further, something is amiss; indeed, in theological terms, something is ungodly. They "fall" into evil, rather than rise to their destiny. This is not because their animal nature is selfish; the word "selfish" does not apply where there are no moral agents. Rather, trying to become human without emergence from the animal nature results in selfishness. That stagnates in animal nature. "The natural man [who] does not receive the gifts of the Spirit of God" (1 Corinthians 2.14) is not so much "fallen," as nonrisen, failing rather than falling, languishing in animal nature and falling away from his humane, godly ideal. That is the story parable of Genesis 1–3, a story that is both once upon a time, and once upon all times, aboriginal and perennial, the situation into which humans are now born, which also discloses the ancient past. That is the prologue, sketched mythically, and profoundly orienting the whole story of salvation to follow. What was and is in the animals a good thing becomes ("falls into") a bad thing when it is the only thing in human life. This arrests advancement to the next, the human, humane stage.

Is our genetic inheritance the source of the problem? Genetic processes conserve value, but they are, or ought to be, here surpassed. The impulses that give rise to sin (such as those for self-defense) are inherited, though they are not, biologically speaking, a defect in nonhuman lives, or even in human lives, unless and until the options for higher defenses of value arrive, behavioral possibilities, freedoms, from which humans do defect. Human cultural inheritance requires experiences super-to-the-genetic, super-to-the-natural, that is, beyond the previous attainment and power of biology. Those experiences come creatively, with struggle, with an arduous passage through a twilight zone of spirit in exodus from nature. This does not mean that nature is bad; nature is pronounced to be very good – not perfect, because culture is yet to come – but intrinsically good. Humans are made godward, to turn toward God, but shrink back and act like beasts. Genesis is the story not of the fall from perfection, but of the "fall" of the aboriginal couple from innocence into sin and of their awakening into this state. After the sin, "the eyes of both were opened, and they knew that they were naked" (Genesis 3.7).

The aboriginal couple, symbols of us all, rise out of innocence into the world of moral choice, which brings growth into responsibility, imaging God, but also, inevitably, falling into sin, as shown in Cain's killing of Abel and in the subsequent Genesis stories of the worsening human condition. Killing is not new in the world; primates have killed each other for millennia in the defense of their genetic lines. But murder is new in the world; the human has risen to an option to do otherwise and therefore ought to do otherwise. The murderer fails, falls back; his opportunity for humanity is now broken, and his society falls under a curse. The Earth cries out for justice. Society becomes a confused chaos, a babel.

Self-actualizing is a good thing for humans as well as animals. Self-interest is godly; the commandment, we remember, is to love others *as we do ourselves*. The garden is full of trees to eat; we pray for our daily bread. But concupiscence, the desire to possess and enjoy inordinately, is *not* a fitting form of life in the world. Natural selection does favor the "self-serving" individual, and there is no reason to deplore this process. The fauna and flora are checked in this possessive impulse by the limitations of their ecosystems – which provide a satisfactory place, a niche, for each specific form of life, but limit each species to its appropriate sector, where it has adapted fit. The human species is not so checked, but tempted by the fearful power of hand and mind to possess the whole. The human species has no natural niche, no limits by natural selection, that is relaxed progressively as the human species rises to culture as its niche, superposed on nature.

What religion warns, past the ethical aspirations (of Chapter 5), past the scientific aspirations (in Chapter 4), in critique of culture (Chapter 3), is that ethics and science, like all cultural activities, religion included, will be warped by human ambiguity, by the evil that besets their loftiest aspirations toward the good. Both morality and rationality, unredeemed from self-love, will prove dysfunctional and tragic. Both science and ethics need to be redeemed. Here the value crisis is taken to a new level. Symbolically put, those who wish themselves to be God fail tragically; those who wish to image God can become children of God, though made of the dust of Earth. The dusty beast reaches to be god; that is biology gone amok, the original sin.

In the Buddhist version of this story, our inordinate thirsts (*tanha*) make the world unsatisfactory (*dukkha*), and humans can be released only by enlightenment that transcends the self (*anatta*). In the Hindu version, human ignorance (*avidya*) mistakes the empirical self (*jiva*) for the true self (*atman*) and misses the universal (*Brahman*). There are important differences with the Hebrew-Christian vision, beyond our scope here. Meanwhile, each in its own way inhibits the genetically transmitted animal drives so that the cultural transmission checks and humanizes the genetic one. Genes may make one selfish, but it is not genes that make one Christian, Buddhist, Hindu, or religiously Jewish; rather, one converts to teachings that discipline and inhibit these genetically based drives.

Nature produces matter and energy, then objective life, then subjective life, then mind and culture. The fourth movement is mostly in a minor key – and beautiful for its conflict and resolution, for the struggling through to something higher. The evolutionary epic, when it comes to the human chapter at least, is the story of good and evil.

(4) Suffering and Creation

The story is of the evolution of suffering; this too is among the emergents. In chemistry, physics, astronomy, geomorphology, meteorology, nothing suffers; in botany life is stressed, but only in zoology does pain emerge. Genes do not suffer; organisms with genes need not suffer, but those with neurons do. One is not much troubled by seeds that fail, but it is difficult to avoid pity for nestling birds fallen to the ground. In every season, most of the sentient young starve; are eaten, abused, abandoned. Life is indisputably prolific; it is just as indisputably pathetic (Greek: *pathos*), almost as if its logic were pathos. The fertility is close-coupled with the struggle.

There is no moral agency in nature, no immoral selfishness; that was a category mistake. There are both intrinsic and shared values. Also, without doubt, there is suffering; there is no more certain fact than this disvalue. One is not going to get religion naturalized, or socialized, until one reckons with this. *Dukkha*, that the world is suffering, is the first noble truth of Buddhism. Genesis 1–2 begins with a good world, but by Genesis 3 it has fallen, and redemptive suffering is the critical theme of both faiths.

Suffering is a troubling fact, but the first fact to notice is that suffering is the shadow side of sentience, felt experience, consciousness, pleasure, intention, all the excitement of subjectivity waking up so inexplicably from mere objectivity. Rocks do not suffer, but the stuff of rocks has organized itself into animals who experience pains and pleasures, into humans whose *Existenz* includes anxiety and affliction. We may wonder why we suffer, but it is also quite a wonder that we are able to suffer. Something stirs in the cold, mathematical beauty of physics, in the heated energies supplied by matter, and there is first an assembling of living objects, and still later of suffering subjects. Energy turns into pain. The world begins with causes, mere causes; it rises to generate concern and care. Is this now ugliness emergent for the first time? Or a valuable good, sentient life, with its inevitable dark side? Suffering too involves the historical genesis of something in kind where nothing of that kind existed before.

Pain is objectively present in nature, and what is its connection with genesis? Struggle is the dark side of creativity, logically and empirically the shadow side of pleasure. One cannot enjoy a world in which one cannot suffer, any more than one can succeed in a world in which one cannot fail. The logic here is not so much formal or universal as it is dialectical and narrative. In natural history, the pathway to psychosomatic consciousness, the only kind of experience we know, is through flesh that can feel its way through that world. An organism can have needs, which is not possible in inert physical nature. If the environment can be a good to it, that brings also the possibility of deprivation as a harm. To be alive is to have problems. Things can go wrong just because they can also go right.

Sentience brings the capacity to move about deliberately in the world, and also to get hurt by it. There might have evolved sense organs without any capacity to be pained by them. But sentience is not invented to permit mere observation of the world, rather to awaken some concern for protection of the kinesthetic core of an experiential life that can suffer. A neural animal can love something in its world and is free to seek this, a capacity greatly advanced over anything known in immobile, insentient plants. The appearance of sentience is the appearance of caring, when the organism is united with or torn from its loves. The story is not merely of goings on, but of going concerns, that is, of values that matter.

Pain is eminently useful in survival, and it will be naturally selected, on average, as functional pain. Natural selection requires pain as much as pleasure in its construction of concern and caring; pain is an alarm system in a world where there are helps and hurts through which a sentient organism must move. On the other hand, any population whose members are constantly in counterproductive pain will be selected against and go extinct or develop some capacities to minimize it. In this sense, natural selection, so far from needlessly increasing pain, rather trims it back in the system, so far as the system can remain vital, conservationist, and developmental. Pain is self-eliminating except insofar as it is instrumental of a subsequent, functional good. Intrinsic pain has no logical or empirical place in the system; neither does maladaptive pain.

The capacity to suffer is generally accompanied by possibilities of avoiding suffering, some freedom and self-assertion. The capacity to suffer, for instance, drives the capacity for learned behavior; it brings animal life to a central focus in sentient consciousness, as cannot happen in plants. Thought appears in order to prevent pain and to affirm well-being, but the thinker that cannot feel pain cannot figure out how to escape it. In humans, this evolution of thought seeking comfort drives the transition from nature to culture.

We cannot show this in the detail of every case; perhaps we need not expect it to be true in every case, and there are troublesome anomalies. Nevertheless, the system statistically must select for beneficial pain. The system historically uses pain for creative advance. Such is the biology of life. Theologically speaking, this position is not inconsistent with a theistic belief about God's providence; rather, it is in many respects remarkably like it. There is grace sufficient to cope with thorns in the flesh (2 Corinthians 12.7–9). Life is a table prepared in the midst of enemies, green pastures in the valley of deep darkness (Psalm 23).

The vast number of creatures sprouted, hatched, or born are, of necessity, more or less well-endowed genetically and emplaced in a more or less congenial environment, despite or including the fact that in their environment they are spurred to earn they way. Even though most will not live to maturity, they are competently programmed for their tasks. Organisms survive in about that proportion in which they are viable, so that life is sustained in any individual in relative proportion to its fitness for it. The community of life is continually regenerated, as well as creatively advanced, and this requires value capture as nutrients, energy, and skills are shuttled round the trophic pyramids. From a systemic point of view, this is the conversion of a resource from one life stream to another – the anastomosing of life threads that characterizes an ecosystem. The "waste" (as it first appears) is really the systematic interconversion of life materials; nature recycles. Death *in vivo* is death ultimately; death *in communitatis* is death penultimately but life regenerated over the millennia of species lines and dynamic biotic communities, millennia continuing almost forever.

Individual organisms must die. Species do not have to die; most, of course, do die. Ninety-eight percent of all species that have ever existed did go extinct, so there are high probabilities, but there is no law of nature or inevitability about species extinction. But here a puzzling aspect of the matter strikes us. By virtue of the smart genes, the death of the organism feeds into the nondeath of the species. Only by replacements can the species track the changing environment; only by replacements can they evolve into something else. Genera and species sometimes do die, that is, go extinct without issue, but they are often transformed into something else, new genera and species, and, on average, there have been more arrivals than extinctions – the increase of both diversity and complexity over evolutionary history. The loss of species in natural systems has meant more birth than death; perhaps there too it is tragic, but it is not unredeemed tragedy. The "birthing" metaphor is at the root of the concept of "nature"; here creativity comes only with "labor" and "travail."

Genes do not suffer but they do code this story of coping with suffering. They make the story possible, necessary for it, but they are not sufficient to interpret it. The world is not a paradise of hedonistic ease, but a theater where life is learned and earned by labor; in this struggle there is something demanding appropriate respect, something inviting reverence, something divine about the power to suffer through. The cruciform creation is, in the end, deiform, godly, just because of this element of struggle, not in spite of it. Among available theories, there is no coherent alternative model by which, in a painless world, there might have come to pass anything like these dramas of nature and history that have happened, events that in their central thrusts we greatly value.

Environmental necessity is the mother of cultural invention. An environment that was entirely hostile would slay us; neither life nor culture could ever appear there. A nature that was entirely irenic would stagnate us; human life could never have appeared there either. All human culture, in which our classical humanity consists, originated in the face of oppositional nature. Nature insists that humans work, and this laboring and even suffering is its fundamental power for genesis. Creativity is through conflict and resolution. We suffer, and lest we suffer more, we organize ourselves creatively. In that sense, humans owe all culture to the hostility of nature, provided we can keep in tension with this the support of nature that is truer still, the one the warp, the other the woof, in the weaving of what we have become.

Early and provident fear moves half the world. Suffering, far more than theory, principle, or faith, moves us to action. One should not posit the half-truth for the whole; we are drawn by affections quite as much as pushed by fears. These work in tandem reinforcement; one passes over into the other and is often its obverse. In this sense, pain is a prolife force. In the evolution of caring, the organism is quickened to its needs.

Nor in humans is there only physical pain. Spirits know affliction. In humans the relationship between bodily wounding or deprivation and pain is quite complex, involving cognitive factors such as cultural conditioning and psychological evaluation of the situation. Sin appears, as do guilt, insult, humiliation, reproach, grief, angst, alienation, remorse. This is why such things as rape and slavery have meanings in culture that simply do not transfer to bluebirds or ants. One becomes reflectively self-conscious about the values of which one is deprived, sometimes by nature but now even more by the exploitations of culture. One knows one's social status, not just one's physical or biological state, and the former may determine behavior more than the latter. The concept one has of oneself, the gap between one's perceived real and ideal, and the placing of responsibility for closing of that gap, become critical.

All this drives the religious life. In human spirits, the distinctive characteristics of spirit make tragedy and redemption possible. Birth is superseded by rebirth; the question of generation by the question of regeneration. Any adequate interpretation of this story of spirits fallen into tragedy and redeemed from this fall is going to be irreducibly religious. That is the essential theme of Christianity and Judaism, for example, that suffering love is divine – and we doubt whether there is any competence in biology to evaluate whether this is true or false, although biology has competence enough to document the struggle for survival, the sequence of life, death, and life renewed. Zoology, perhaps joined with psychology, can raise the problem of suffering, but its redemption is a religious issue.

The way of history too, like that of nature, only more so, is a *via dolorosa*. Since the beginning, the myriad creatures have been giving up their lives as a ransom for many. In that sense, Jesus is not the exception to the natural order, but a chief exemplification of it. The secret of life is seen now to lie not so much in the heredity molecules, not so much in natural selection and the survival of the fittest, not so much in life's informational, cybernetic learning. The secret of life is that it is a passion play. This is the labor of divinity, misperceived if only seen as selfish genes.

In this evaluation, we have not painted the world as better than it is in the interests of a philosophical metaphysics, nor worse either; rather we have tried to see into the depths of what is taking place in natural history. The view here is not panglossian; it is a tragic view of life, but one in which tragedy is the shadow of prolific creativity. That is the case, and the biological sciences with their evolutionary history can be brought to support this view, although neither tragedy nor creativity is part of their ordinary vocabulary.

2. RELIGION AND FERTILITY

Religion, we have been claiming, is a response to the prolific Earth. There is an alternate account of the connection between religion and fertility. There is no religious behavior in nature, just as there is no moral behavior among the animals and plants. But humans behave religiously in culture, almost invariably so in classical cultures, and extensively still in modern cultures. Why so? For those wishing to explain, within the framework of biology, the genesis of religion in its connections with the genes, the evident way is to apply natural selection. Persons who are religious leave more offspring than those who are not. Persons who practice religion x leave more offspring than those who practice religion y. Religion a produced better adaptive fit in hunter–gatherer cultures; religion b produces better adaptive fit in agricultural cultures; religion c, in technological cultures. These begin to sound like claims that could be formulated in a statistical, mathematical model. Birth rates are measurable, though if one is to correlate them with religions one will need also to put numbers on the degrees and kinds of religious belief.

Such a religion-producing-offspring model might have been the way in which religion originated, or classically functioned, but no longer the way religions operate. In either case, the past numbers will now be hard to obtain. But this might also be elemental in all religion, and therefore the way religions operate today, in which case the theory might be more testable. Religious behavior in culture in any specific form is acquired, not innate, but there seems some genetic tendency to acquire some religion or other. The novelty is religious behavior, previously absent from all other fauna and flora, but, when religion emerges in humans, the fundamental biological rules still apply. The fittest – in this case, the religious – survive.

Wilson recognizes that religion is a critical test case:

> Religion constitutes the greatest challenge to human sociobiology and its most exciting opportunity to progress as a truly original theoretical discipline. . . . Religion is one of the major categories of behavior undeniably unique to the human species. The principles of behavioral evolution drawn from existing population biology and experimental studies on lower animals are unlikely to apply in any direct fashion to religion. (1978, p. 175)

There are reasons to believe that the deeper operations of religion will be concealed from its practitioners. Nevertheless:

When the gods are served, the Darwinian fitness of the tribe is the ultimate if unrecognized beneficiary. . . . The highest forms of religious practice, when examined more closely, can be seen to confer biological advantage. Above all they congeal identity. In the midst of the chaotic and potentially disorienting experiences each person undergoes daily, religion classifies him, provides him with unquestioned membership in a group claiming great powers, and by this means gives him a driving purpose in life compatible with his self-interest. (1978, pp. 184 and 188)

It is certainly true that fertility is a fundamental theme in primitive religions. Practitioners seek the fertility of fields and flocks. They worship the sun and its warmth, they pray for rain in drought, they dance to help the maize grow. They seek fertility in childbirth. They seek cures from diseases for themselves and their children. The etymological root of "salvation" (Latin: *salus*) is "health." "Elementary religions seek the supernatural for the purely mundane rewards of long life, abundant land and food, the avoidance of physical catastrophes, and defeat of enemies" (Wilson 1975a, p. 561).

Religious rituals and ethics get people to cooperate for their mutual good. Since people have to eat daily, reproduce each generation, and care for children throughout much of their adult lives, it is unsurprising that fertility – success in staying alive from one generation to the next – is pervasive in religions that have succeeded. Any religion persisting over the centuries will, necessarily, result in reproductive success. We know that before we look.

Such elemental fertility is there right at the origin of Hebrew monotheism. Yahweh's divine promise to Abraham, frustrated because he was childless, was "I will make of you a great nation" (Genesis 12.2). Abraham did not even have one child yet, and God promised to make of him a nation; that's real reproductive success! The Abrahamic covenant is sealed by circumcision, a genital sacrament if ever there was one! Moses at Sinai renewed this covenant and God commanded children, "Honor your father and your mother, that your days may be long in the land which the Lord your God gives you" (Exodus 20.12). God gave the Israelites a promised land, flowing with milk and honey.

Boldly stated, all religions are fertility religions. "The biology of religion," according to Vernon Reynolds and Ralph Tanner, "looks at religions in terms of their contributions to individual (and, though to a lesser extent, group) survival and reproductive fitness" (1983,

p. 267; 1995, pp. 38–40; Reynolds 1991). This is the analogue of the previous claims that science and ethics are to be understood, at deepest level, in terms of the fertility they produce. The general theory of religion and its practices is that "these rules and the actions resulting from them are adaptive in the sense that they are found in countries where the results they produce will tend to enhance the reproductive success of individuals following them. Religions thus act as culturally phrased biological messages" (1983, p. 294; 1995, p. 40).

This faith–fertility correlation can be cast into testable form and verified, so Reynolds and Tanner claim. They grade religions comparatively according to the impetus they give to reproductive activity. The result is a spectrum (Fig. 6.1) on which the maximally "pronatalist" religion is Islam and the least reproductive is Protestant Christianity (1983, p. 289; Reynolds 1991).

Some general positive correlation between religion and fertility sounds plausible. We cannot simply consider birth rates, of course, but must see how many children survive to reproductive age, and so general health and diet are critical. We need to know how religions contribute to sanitation, to parent–offspring and family caring and sharing, to solving of conflicts, to work initiative, what attitudes they have toward material necessities, which religions best stabilize societies and families sufficiently for the decades of child rearing, and so on. Reynolds and Tanner conclude that essentially "a religion is a

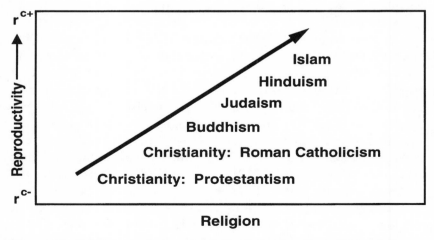

Figure 6.1. Religions and reproductivity (following Reynolds and Tanner 1995; 1983, p. 289).

primary set of 'reproductive rules', a kind of 'parental investment handbook' " (1995, p. 40; 1983, p. 294).

Different religions produce different sorts of comprehensive fitness appropriate to local circumstances. In some environments (typically the uncertain ones, with high mortality rates), it is advantageous to have many children, though each child has reduced survival likelihood, but in others (the more stable ones) it is advantageous to have few children with high survival probability. So it turns out that the pronatalist Muslims may not have the edge over the Protestant Christians after all, if the latter have lots of Yankee ingenuity or high medical, agricultural, and industrial technology and a Puritan work ethic or a stable democratic society.

There is a well-known model in ecology that describes differing reproductive strategies. On one end of a spectrum there are species with a reproductive strategy of numerous offspring in disturbed environments (r-selected species), and on the other are species producing fewer offspring in stable climax ecosystems (k-selected). The model needs to be adapted for humans; Reynolds and Tanner propose an analogous cultural variant r^c+ (high birth rate) model versus a r^c- (low birth rate) model (1983, pp. 11–17, and 269–270; 1995, p. 39; Reynolds 1991, p. 209). Protestants do not need to breed much; they are r^c- selected, not r^c+ selected.

Since it is difficult to compute quantitative judgments on all this, Reynolds and Tanner think that they can simplify the problem and get a fair estimate by looking at per capita energy consumption and gross national products, which are indicators of how prosperous a stable society has become (1983, pp. 290–295). Here the White Anglo-Saxon Protestants (WASPs) outconsume the poor Muslims and Hindus. That high consumption and production explains how the WASPs get their fewer children more often to reproductive age. At the same time the higher birth rates in the pronatalist religions, dominant in the lesser developed countries, explain the persistence of Islam and Hinduism. So now it turns out that all the religions on the spectrum, despite differentials in birth rates, have enough overall reproductive success to remain viable in their respective niches on the world scene.

Such an investigation may seem to have proved the religion/fertility thesis, but it may just as well be that it has assumed it. The only religions calibrated on the spectrum were religions that we knew before we started had supported substantial populations over generations. All that has been asked is whether differing religions, all successful, might succeed by varying their emphasis on number of children born relative to subsequent parental investment, depending on circumstances. Put that way, however, it seems entirely possible that one or more of these religions may, in different circumstances, allow differing reactions. Protestant Christians, for instance, in earlier centuries had higher birth rates. In 1800 in New England, when the birth rate was 7, compared to 2.1 in 1990, was the environment more or less stressful or stable than it is now? It is hard to say.

Christianity, over the centuries and around the globe, has persisted in remarkably diverse circumstances, as often nondemocratic as democratic, as often nontechnological as technological, as often in unstable as in stable environments, as also has Judaism. Buddhism has persisted in quite diverse environments, from ancient India to modern Japan. Reynolds claims that Christians, who are less pronatalist, do better in Europe because this is a less stressful environment than the Middle East, where the quite pronatalist Muslims flourish (1991, pp. 210–214). But Islam, Judaism, and Christianity, religions at both ends and the middle of the spectrum, all originated in the same place, the Middle East, where their originators presumably had about the same stressful or stable environments with which to cope. And all three spread widely. By this time we are beginning to lose any meaningful correlation between religion a and its r^c+ selected strategy or religion b and its r^c- selected strategy. All we are really left with is what we knew before inquiry, that the major world religions can encourage various behaviors enabling people to survive over generations in differing kinds of environments.

Wilson seems to think it embarrassing that seeking the "supernatural" brings "mundane" rewards, but this comes as no surprise to Jews or Christians. Moses urged, "You shall walk in all the ways which the Lord your God has commanded you, that you may live, and that it may go well with you, and that you may live long in the land which you shall possess. . . . And the Lord commanded us to do all these statutes, to fear the Lord our God, for our good always, that he might preserve us alive, as at this day" (Deuteronomy 5.33; 6.24). Jesus taught his disciples to pray for their "daily bread." "Therefore do not be anxious, saying, 'What shall we eat?' or 'What shall we drink?' or 'What shall we wear?' For the Gentiles seek all these things; and your heavenly Father knows that you need them all. But seek first his kingdom and his righteousness, and all these things shall be yours as well" (Matthew 6.31–32).

Meanwhile, we do not yet know whether fertility is the sole or chief determinant of religious beliefs and behaviors, nor what the relation may be between persisting religions, all of which must be functional in this regard, and their truth. The true ones, if there are any, might be equally fertile with the untrue ones, if there are any; or the true ones might be more, or even less fertile, than the untrue ones. Fertility and truth might be independent variables. To this relation among fertility, functionality, and truth we will later return (Section 4).

One religion, Judaism, might seem to provide a convincing example of the connection between genetics and religion. Judaism has been especially effective at keeping racial stock and religious convictions together; most Jews religiously are Jews genetically, and this pattern has persisted for thousands of years. But we no sooner note the Jewish example, positively corroborating the theory, than we run head on into a counterexample that seems decisively to falsify it. Jesus, a Jew, launched Christianity, which spread into the uncircumcised, Gentile world; gave up most of the distinctive ritual observances of Judaism; replaced previously existing religions all over the ancient Mediterranean, spread to Europe, and thence to many parts of the world. Today approximately one third of humans on Earth, well over a billion persons, are Christian, either in conviction or by heritage. Compared with these Christians, the Jews are minuscule in number.

Few Christians have any genetic relationship with the early Semitic Christians; Jesus had no offspring at all. The genetic survival value of Christianity, if there is any, is smeared out over thousands of different racial stocks in hundreds of countries. Christianity may be a parental investment handbook, but it seems that anybody can use it, around the globe and across the centuries, regardless of genetic origin. There is no identifiable relationship between this or that set of genes and Christian belief and behavior.

The Jewish genes-belief-behavior connection, if there is one, is only a fragment of the evidence, most of which dissociates specific genes, belief, and behavior. Even the Jews have long insisted on identifying what was specifically Jewish in their religion (the Abrahamic covenant, the Mosaic observances), separating this from what in Judaism also applied to the Gentiles and was to be a blessing to all nations (the Noachic covenant, ethical monotheism). Meanwhile, Christianity (as do all other persisting religions) has to result in Christians' regenerating themselves biologically over the generations. After all, one has to be born before one can be born again, even if being born again of the spirit then feeds back into the birth rate and results in more births in the flesh. If there ever were a divinely revealed religion by an ethical monotheist God, one would expect it to further the welfare of those to whom it was given. In that sense both the theological and the biological theories of the origins of religion predict the same results: prosperity over generations. If we observe such prosperity in actual history, either or both theories may be true.

Where religion brings such prosperity, reached as this must be by successful surviving through difficulty, by the creativity inseparable from suffering (Section 1[4]), this invites persons to return to reflection on the prolific Earth, the larger genesis into which human generativity is now incorporated. Perhaps, to some extent, religion results from and is generated by selection for fertility, a coping "myth" maximizing offspring or at least inspiring caring for offspring. In that respect, there is nothing ungodly about a religion that brings a fruitful life, including both prosperity and children.

But that truth will have to be put in a larger picture. Religion may arise as a coping myth ("the gods are for us and our children"), but what is one to say when the truth is found: that there has been successful coping over three and a half billion years, in which the local self now takes a part and plays a role. Perhaps it was once true, in the launching of religion, that the earliest humans mythologized these powers as sacred, personified them, and that this proved adaptive in child rearing. Perhaps religion is still useful in this way. But after that, these generative powers are in fact there surrounding us, past and present, to which any worldview must be suitably adapted. To see such creative process as sacred, to detect a Creator present, is as plausible an interpretive framework as any and is an explanation adequate to the results. These prevenient vital powers, sacred powers, numinous Presence in, with, and under the emergent phenomena, are, after we learn genetics, no less still there in and with the genetics underlying the genesis.

Religion results from and is generated by reflection over, as well as participation in, a prolific Earth. Genes generate a mind, which generates religion, which supports genetic survival. Such a mind, formulating its religion, protects its offspring, and also encounters the surrounding genesis in natural history, recorded in the genes of others, humans and nonhumans. Such a mind encounters also the human genius generating the myriad cultures, including that of its own heritage. The local self with its family line participates in wider communities of shared values, and an account of that too is significant in the religious challenge. We are back to the question of "others" and the sharing of values, now in religious form.

3. RELIGION AND ALTRUISM

Close analysis – Wilson, Reynolds, and Tanner were claiming – will show that all successful religion is really "selfish," in the sense of serving one's genes. In view of the fact that the most successful religions routinely urge altruism and censure selfishness, sociobiologists and behavioral psychologists will have to show that this altruism is only apparent and that these religions do in fact support biological selfishness. Or one will have to find some account(s) to give of this emphasis on altruism that can reinterpret it within the general biological theory. Or perhaps the altruism will be revealing counterevidence. Religion generates a social phenomenon that biology is incompetent to handle, either to explain or to evaluate. If so, such naturalistic accounts of the genesis of religion will be partial, at best. Religious accounts of the genesis of this altruism might be complementary or corrective to the biological accounts.

As with ethics before, we need to remember that religion is more than altruism. Religions too are concerned with justice, fairness, equitable sharing of resources, prudent care of oneself, a right relationship to God or the gods; with placating the spirits; or with reaching nirvana, or union with Brahman, and so on. In the Judeo-Christian tradition, the desired goal is often said to be a state of righteousness. In the vocabulary recommended here, this optimizes values, now religious values, both in the personal lives of believers and in the lives of those they benefit.

(1) Religion Generating Altruism

Many religions urge altruism; this is as frequent a theme as is increased fertility (Hefner 1993, chapters 11–12). Judaism summarized its ten commandments into two: love God and neighbor, which Jesus enthusiastically endorsed. New Testament writers prefer the Greek term *agape* over *eros*. *Eros* is an acquisitive love, responding to value in the other and being fulfilled by that other; *agape* is a giving love, offered regardless of value in the other and any reciprocating benefit. *Eros* may be a good thing in its place, but *agape* cares for the other more sacrificially. These writers did not think that *philea*, brotherly love, was profound enough to embody the Christian ideal. One ought to love the other as one does oneself. Augustine summed up the Christian ethic: love and do what you will. He contrasted the self-centered love, *concupiscence*, characteristic of Babylon, with an other-(altrus)-love, *caritas*, characteristic of Jerusalem.

Buddhism's first commandment is noninjury to others, *ahimsa*; the *bodhisattva* takes a vow of *karuna*, compassion on all beings:

> I have made the vow to save all beings. All beings I must set free. The whole world of living beings I must rescue from the terrors of birth-and-death. . . . My endeavours do not merely aim at my own deliverance. For with the help of the boat of the thought of all-knowledge, I must rescue all these beings from the stream of Samsarsa. (*Vajradhvaja Sutra*, 280–281)

The four noble truths locate the fundamental human disorder in thirst and clinging, in a grasping that feeds and satisfies the self; the route to salvation is by an enlightenment, *nirvana*, where one sees that the self is unreal, *anatta*.

> Contented, easily supported, with few duties, of simple liveli-hood, controlled in senses, discreet, not impudent, he [the Bo-dhisattva] should not be greedily attached to families. . . . Just as a mother would protect her only child even at the risk of her own life, even so let one cultivate a boundless heart towards all beings. Let one's thoughts of boundless love pervade the whole world – above, below and across – without any obstruction, without any hatred, without any enmity. (*Suttanipata*, I, 8)

That certainly doesn't sound like selfish genes. Neither *agape* nor *karuna* seems to be explicitly or implicitly promoting the self; this altruism runs counter to the fertility elsewhere vigorously sought in religion. What is one to make of religion exhorting altruism?

There is no problem when the altruism so promoted binds kin

group loyalty and facilitates tribal group reciprocity. This promotes inclusive fitness. Benefits redound to self and/or kin. But is this account of only-apparent altruism always plausible? Mother Teresa was certainly behaving with biological somatic altruism, since the food she fed to Indian children she herself could not eat (Muggeridge 1973). She was also behaving with genetic altruism, since Mother Teresa had no children herself, and many of the Indian children she fed have themselves grown up and reproduced. Nor is she related to these children.

This is hardly reciprocal altruism. Although she benefited nonrelatives, there was little resulting benefit to Mother Teresa's somatic self. There is no cause to think that Mother Teresa went to India because of the likely reciprocators there, or stayed many decades for this reason, or that her efforts in India were helping her relatives back in Yugoslavia and Albania to reproduce. Mother Teresa was not backscratching with unrelated others, nor did she expect these others to backscratch her nieces and nephews. There is no particular problem if Mother Teresa received an occasional bit of help in return for her charity, if for instance some young girl reciprocated and cooked food for Mother Teresa. But the net flow of benefits cannot be to Mother Teresa; the recipients of aid are, after all, the poor of India.

There is moral altruism, if we are able to give any credence to Mother Teresa's verbal reports. Whether or not we accept her reports, just observing her behavior alone, the biological selfishness interpretation is rather implausible. What is the evidence that she was not doing what she intended, helping nonrelated others, and doing it because of her religious convictions about divine love? Religion was also operating with the Good Samaritan. This joining gives us no cause to suspect either the morality or the religion.

Mother Teresa ate daily, conserving her biological somatic value, and there is no cause to censure her for selfishness in doing so. She also cared for the intrinsic values in those she fed. There is some reciprocity as those values are shared, backscratching. No religion protests when persons help each other out. In all this there results much conservation of biological genetic value. The Indian children live to reproduce. Mother Teresa knew, of course, that not everyone can or should be religiously celibate. There would be no next generation.

(2) Religion Generating Pseudoaltruism?

But, comes the protest, most persons are not like Mother Teresa; most look out for their own interests, and those of their next of kin. Most are sinners, few are saints, and, even among the saints, seldom do we find those as charitable and self-denying as Mother Teresa. What is protested is, one should notice, also what Mother Teresa herself, in her religion, taught: that by their first nature humans are selfish, and that such nature needs redeeming before humans can operate with this regenerated nature. That real sinners outnumber ideal saints has never been taken to discredit religion, although religion might be discredited if it could produce no working examples at all of the sorts of persons it recommends. Religion does in fact produce numerous such models. They are the myriad exemplars of the religious heritage, and the devout follow them with some measure of ideal mixed with real.

Alexander, Ruse, and Wilson claimed earlier, dealing with the Good Samaritan (Chapter 5, Section 3[1]), that the whole process works better if people are deceived about their deepest motives. We call this pseudo-pseudoaltruism, because not only is the *behavior* only apparently altruistic, really self-interest, but the *intention* too is apparently altruistic, as the selfish intent is screened off from the moral agent. Religion has a particular genius for inculcating this deception. The gods command this altruistic behavior. Loving God urges loving one's neighbor. If one can come to believe that, then there will be zeal indeed; one's real motives will be rationalized as obeying God. Religion is an especially powerful incentive reinforcing this (apparent) altruism. Perhaps altruism even originated in religion. Until modern times, most ethical behavior was entwined with religious behavior. The discussion in the previous chapter left something important out, assuming that ethics arose from mutual cooperative advantage, a social contract, without any serious look into its historical integration into religions. Ethics needs the sanction of religion to get established.

On this account, from here onward, one will not be able to ask religious people what they think. Humans are doubly mistaken, both about the altruism and about god(s). But one can watch what they do. These doubly mistaken humans are nevertheless productive; their mistake recouples religion and fertility in a surprising way. If the theory is true at a first level, one would expect ethics and religion to remain tribal, favoring kin selection, where the nearby genes are. The religions preach that "charity begins at home." Such charity, a misnomer, is actually genetic selfishness. Be that as it may, no classical religion teaches that charity stops at home, and yet this is the teaching one should expect, if religion were selected to promote a familial genome. Over recorded history, this has not been the trend at all. Religion and ethics are tribal at the start, but both together go universal, replacing tribal religions. Seemingly, that trend should have been selected against.

Charity in religion expands beyond the family, as was evident in Mother Teresa's actions, and now the genetic theory has to accommodate this by supposing that the charity is really reciprocal bargaining for benefits, gained by group association. The beyond-home charity ends up producing likely reciprocators. Recalling indirect, social altruism (Chapter 5, Section 2[4]), reciprocation does not have to be one on one; the religious operator is setting up a general cultural climate in which there is reciprocity. The individual does well in a Christian society, no matter whether the other Christians are genetically related, no matter whether the help received and given is in direct exchange, or statistically averaged out in a Christian community.

Religion is not adverse to one's being a good neighbor, or having good neighbors either. The organic model, one body with many members mutually supporting each other, is a favorite model of the church (1 Corinthians 12). But, in the end, when the question, Who is my neighbor?, is asked, the answer comes in terms of who is in need that I can help meet, not who is likely to reciprocate with net gain to myself. Universal morality has regularly been religiously based in the classical world religions – those religions that moved from tribal and national levels to become international and intergenerational faiths.

The pseudoaltruist will have to say that such moral persons were just setting up a world moral climate in which they themselves were most likely to prosper genetically. Charity is always a misnomer. Mother Teresa did not gain any personal benefit, but she did get the spread of the benefits of religion from India back to Yugoslavia, benefits that in her case started in Yugoslavia, her childhood home where she was reared religiously and later moved to India – benefits that started centuries before in Palestine and once moved to Yugoslavia, being shared by all who transmitted this religious altruism en route. That kind of religion is rather curiously selfish, setting up this pervasive cultural climate; indeed it has become almost indistinguishable from what the ethical monotheisms have taught. Nevertheless, Wilson is sure that, one way or another, Mother Teresa, even in her sainthood, remained "cheerfully subordinate" to her "biological imperatives" (1978, p. 166).

Mutatis mutandis, one can give the same account of the spread of Islam, with its ideas of universal brotherhood, and even of the spread of nonmonotheistic Buddhism from India to China and Japan, motivated by the *bodhisattva*'s ideal of universal love.

One can adamantly hang on to the selfishness paradigm, but this is a topsy-turvy kind of selfishness that has to act on universal altruism, and evangelize this faith to the world, that is, to share it with everybody else, before it works most efficiently to one's own benefit. It is odd that to serve their genetic interests people have to go to elaborate efforts to do just the opposite, to believe universal creeds, share them with others, act on universal altruism, build characters that are caring, fair, sympathetic, forgiving, magnanimous. One can say, if one insists, that all this is just reputation building, pretense that creates a climate in which the pretender and his kin prosper as a result of the reciprocity generated. But it is difficult to see how they prosper to the detriment of the others who are the beneficiaries of this allegedly pretended altruism. None of this is really very plausible anymore. Perhaps the charity isn't just apparent after all. Maybe it is time for a paradigm switch.

(3) Religion Generating Unsuccessful Altruism?

But first, perhaps one can save the general theory that religion is fertility-maximizing another way. There is also a negative, unsuccessful version of the theory: religion does indeed produce real altruism and this results in the genetic failure of such persons – contrary to all claimed in the previous section. Religious persons benefit the genes of others, who outreproduce them, and they themselves go extinct. This is a rather surprising conclusion, and one will have to find a convincing account of the anomalous persistence of religion although its practitioners are constantly failing genetically.

Struck by the degree and intensity of altruism exemplified in Mother Teresa, sociobiologists and behavioral psychologists may try a revised account. In addition to pseudoaltruism, there can be induced altruism (Chapter 5, Section 3[2]). This does indeed serve the interests of the helped at both somatic and genetic cost to the helper, but it is "an evolutionary mistake" (Alexander 1987, p. 191). This switches things around. Now believers are tricked into losing. Devoutly religious persons are conned into benefiting others and will have fewer offspring themselves. Priests and nuns fail to benefit either themselves or their blood lines genetically. They are dupes, but they do help others to succeed. They help some people directly, who have more offspring in result, and their ideal, though disastrous were everyone to practice it, produces enough spillover morality to help the lay reproducer. The Indian girls have children, Mother Teresa died childless, and her relatives in Yugoslavia and Albania do not have many children either.

If this is true, however, we will expect that the genes for becoming a priest or nun will disappear from the population, and society will be in worse shape, not having these clerical benefits spilling over to the lay reproducers – and similarly and proportionately for any lay believers duped into such assistance given to their fellow lay Christians. The masses of selfish people will exploit any altruism; everybody will cheat on altruists, and they are always losing. In just that proportion by which religious persons overdo their altruism, erring into induced rather than merely apparent altruism, their genes will be selected against. "In a world of egoists, the only one who suffers from exhortations that 'Everyone should try to be like Jesus,' is the one who succeeds" (Alexander 1987, p. 127): succeeds in being like Jesus, that is, but fails reproductively, as did Jesus, and so succeeds in becoming extinct.

Mother Teresas are one in a million, priests and nuns are one in ten thousand, but lay believers have children routinely and care for them with religious zeal. They may praise their saints, preachers, prophets, missionaries, but what they are really doing is exploiting them. Parishioners get direct help from them and such figures symbolize by exaggeration what everybody needs a little of for his or her own good. One will expect these genes for an overdose of altruism to be rare. But why should they be there at all? There is a ready explanation why most persons should be easily educable into a limited altruism; this in fact serves their genetic self-interests. But we have no explanation yet why these evolutionary mistakes should persist, rare though they are.

Perhaps they are just a repeated error, like Down's syndrome or (some say) homosexuality, neither of which facilitates reproduction. Such a tendency to repeated error might not be surprising, since there is a beneficial altruism (backscratching and winning) that is behaviorally almost indiscernible from a loser's overaltruism, distinguishable mostly by excess in degree, and everybody's intentions are screened off both from themselves and from others. The right kind of altruism, a fertility-maximizing pseudoaltruism, and a religion to exhort it, are good things genetically, though too much of them, universal morality and real altruism, is a mistake. Some unlucky mutants in every generation will edge over too far. Religion is like fertilizer; indeed religion is a kind of fertilizer. More is better up to a point, and more after that produces opposite results. A few people will regularly and counterproductively overfertilize.

Is this account plausible? Here one must remember that just this universal morality (an alleged overdose), religiously based, has been classically successful as a cognitive idea. Perhaps the one-in-a-million or one-in-ten-thousand mutant superbly exemplifies the idea, but leading intellectual traditions (Christianity and Judaism in the West; Buddhism in the East) have been conned into this idea, as an ideal though not as often real as it should be. The symbol catches on and convinces many. It is a quite a fertile idea, spreading globally, even if there is an overdose problem. The altruistic impulse does not just travel genetically from one generation to the next, passed down in modest amounts because of its reproductive success or in harmful amounts as a recurrent genetic error. The altruistic impulse is spread by conversion, by evangelism, by proselytizing; billions of persons come to hold it creedally if not behaviorally.

That is a strange mutant indeed, fertile though erroneous, one that arises rarely and harmfully, but that convinces intellectually though not behaviorally the many who become converts to the most successful religions globally. We need to explain why people around the world and across the centuries have been intellectually persuaded to accept a belief that they are not genetically disposed to adopt. Nor is this just a genetic problem of a few screwy mutants convincing millions of persons to believe what they are not disposed to believe; it is an intellectual one as well. Everybody has to have the wrong theory (universal, divinely willed altruism) to get the right result (fertility).

Holding onto the self-interest paradigm tenaciously, its defenders can reply that people everywhere need the overbelief (belief in too much altruism, commanded by God) in order behaviorally to act with a functionally minimum altruism (enough to produce reciprocation). The evolutionary mistake, manifested recurrently in these symbolic saints, provides the essential belief without which the masses cannot function well in their cultures. Maybe that is convincing. Maybe it is holding onto a theory when the evidence is beginning to mount against it. We are starting to wonder whether this altruism-really-selfishness thesis, modified into an altruism-really-mistake thesis, modified into an altruism-really-functionally-important-mistake thesis, modified into a wrong-theory-necessary-to-produce-right-results thesis is a paradigm proved true or a paradigm absorbing and eating up all the evidence. Perhaps it isn't a mistake at all; maybe culture needs different rules than genetic nature needs.

Perhaps we have been too generous to the saints. Already we have noticed subtle exploitation in religion, and this may be more widespread. Consider another kind of induced altruism, this time one by which the leaders gain and the multitudes lose. Religious leaders too can be self-aggrandizing; lay believers too can be the duped. The masses are conned into believing that God wills that they should faithfully obey the commandments, not to steal, or lie, or covet; to be honest, hardworking; to keep promises; to contribute sacrificially to the church. Such sanctified morality is really serving the interests of those in power ecclesiastically. Where the church supports the nation, as with an established religion, this can be also political exploitation.

"Religion is above all the process by which individuals are persuaded to subordinate their immediate self-interest to the interests of the group. Votaries are expected to make short-term physiological sacrifices for their own long-term genetic gains" (Wilson 1978, p. 176). Usually this does work to the longer-range benefit of the persons so subordinated. But it also means that such a subordinating tendency can easily be exploited by political and ecclesiastical leaders, who gather benefits from the subordinated. Now the morally faithful plebeian Christian citizens lose and have fewer children in result, while the leaders win. With the commoners kept in place contributing their support, the leaders outreproduce them. This deception too will work better if even the leaders are explicitly unaware of what is really going on. "Self-deception by shamans and priests perfects their own performance and enhances the deception practiced on their constituents" (Wilson 1978, p. 176). Both the exploiters and the exploited think that the divine command theory is true.

Wilson says:

> Religions, like other human institutions, evolve so as to further the welfare of their practitioners. Because this demographic benefit applies to the group as a whole, it can be gained in part by altruism and exploitation, with certain segments profiting at the expense of others. Alternatively, it can arise as the sum of generally increased individual fitnesses. The resulting distinction in social terms is between the more oppressive and the more beneficent religions. All religions are probably oppressive to some degree, especially when they are promoted by chiefdoms and states. The tendency is intensified when societies compete, since religion can be effectively harnessed to the purposes of warfare and economic exploitation. (1975a, p. 561)

No one would claim that religion has never been used for exploitation, least of all the seminal religious reformers, who are often intensely critical of ecclesiastical and political powers.

But this error of too much citizen-practitioner morality that lets commoners get suckered into serving bishop or king is exactly what should be selected against. The bishops were celibate and didn't have any children at all; kings, nobles, and chiefs were a minority, one in a hundred or one in a thousand. There is no evidence that the rulers outbred those they subjugated. It would be surprising if the machinations of small groups of elitist rulers could exploit whole populations into behavior that was to the commoners' breeding disadvantage and do this continually over the long millennia of human history. If so, the wrong genes (out there in the subjugated masses) have to be the most common ones. The theory doesn't predict that at all.

There is one thing the theory does predict, but one has to turn to the present and future to test this. If the theory is true, when believers find out about it, they will cease their religious behavior. They will start doing whatever it is that does increase their fertility or promote their own self-interest. A "rational" person will not want to be conned into producing benefits to others at cost to himself. Religion does not work unless it is well-disguised, and to find out the truth of the matter is to cease to be religious.

(4) Religion Generating Complementary Altruism

There is another possibility, developed by Donald T. Campbell (1991, 1975). Religion produces successful altruism, humanizing persons for the passage from nature to culture. Humans evolved with animal genes, selected to conserve values under the regimes of nature, where genetic transmission is virtually the sole process for the transmission of information. But humans form transmissible cultures, and the requirements of culture differ. Natural selection is relaxed, cooperation is intensified, educability is vital, acquired learning is essential. To elevate prehumans into humans, morality arises, almost always religion-based. Morality moves humans away from their merely genetic instincts toward more appropriate behavior in culture. "Social evolution has had to counter individual selfish tendencies which biological evolution has continued to select as a result of the genetic competition among the cooperators" (1975, p. 1115; 1991).

Genes are selected that are educable for culture, but the content of such education includes the moral heritage, supplied by cultural, not genetic transmission. This content urges altruism, and the urging has to overshoot to succeed. Those religions best succeed that most help humans to pull away from their genetic instincts toward the cooperative needs in culture. This best works if they preach not just tribal but universal altruism. When such altruism is preached, the result is behavioral change in the direction of more altruism, less selfishness.

To illustrate, Campbell imagines a sort of selfishness-altruism meter, with complete selfishness at one end of the scale (0 altruism) and complete altruism at the other (100 percent altruism) (Fig. 6.2). Complete selfishness is not successful even in the animal world, certainly not among social primates, who cooperate extensively. Kin altruism plus the limited amounts of reciprocal altruism of which primates are capable might put the biological optimum at 30 percent on the altruism scale. (The numbers are only illustrative, not empirically obtained.) For humans in their exodus to transmissible cultures this is not enough. The religious preachings (here scaled as 100 percent altruism in ideal, but see the caution later) are required to pull human behavior over toward the biosocial optimum for culture, which might be 60 percent altruism. Even the best religions are not so successful as would be operationally ideal; humans fall short of their fullest social possibilities, as a result of now counterproductive

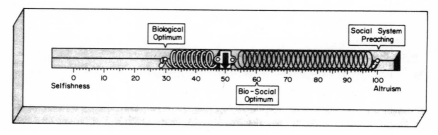

Figure 6.2. Meter illustrating tensions on a dimension of selfishness-altruism. From Donald T. Campbell, "On the Conflicts between Biological and Social Evolution and between Psychology and Moral Tradition," *American Psychologist* 30(1975): 1103–26, p. 1118. Copyright © 1975 by the American Psychological Association. Reprinted with permission.

tugging of their self-interests. The net result is that humans operate in culture with perhaps 50 percent selfishness, 50 percent altruism.

On this view, it is the religions, preaching altruism, that make culture possible; they humanize us. Without them, we are beasts. There is nothing pejorative about a beast acting like a beast, but a human ought to be something more. Beasts (primates) operate with a nonmoral, minimal (30 percent) altruism that is properly in their genetic interests, appropriate for the conservation of value at that level. Humans move toward a moral, more charitable altruism (60 percent), not only proper to but requisite for culture. At least in the behavior that religions produce, stretching humans away from our lingering, ancestral genetic dispositions, the religions are right. This is what ought to be in culture, following exodus from nature, appropriate for the conservation of emerging values at the cultural level.

The achievement of conscience, coupled with religious vision, is a surprising historical development making it possible to be human. That emergence is belied by the simplicity of the meter scale, suggesting only a quantitative where there is qualitative change. To think that the "selfishness" at the zero end is censurable is a category mistake. Altruism too changes its meaning as there is movement upscale; it enlarges its scope, universalizes, and becomes moral.

Religion now does produce a successful altruism, complementary to the biologically produced self-interest. Religions help humans to break away from what the genes, unaided, would otherwise produce. Religious ethics is superposed on the genes, facilitating the transposition to culture. Nevertheless, the religions preach a universal altruism contrary to our best interests in culture; they overshoot their mark. If their urging is heeded too enthusiastically, the result will not be optimum for culture. Here the saints may serve as symbols in the 80 percent and above range, beneficial because they move others up to 50 percent. More saints might be more beneficial, moving us nearer the 60 percent biosocial optimum.

Also, Judaism and Christianity couple what universal altruism they urge with a respect for the self, so it is not true that the religions urge sheer altruism and nothing else. As earlier noted, Christianity, Judaism, Islam, and others commend a righteousness that combines justice and mercy, in which altruism is only one component. Religions encourage self-actualizing, though they know that this is not the whole ideal. So the meter ideal is a doubtful 100 percent, since the Golden Rule recommends loving others as oneself, not instead of oneself. This would be 50 percent – 50 percent if one insists on scaling it, though the recommendation may be about the quality of this love as much as about the quantitative amount.

Such altruism, notice, is progressively less tightly coupled to the genes. Disciples need not have the genes of the prophets, seers, and saviors who launched these teachings. In a successful world religion, they seldom do. People do better with genes plastic enough to track the best religion, whether their blood kin launched it or not. When they convert to these better religions, people are moved to act not just by their genetic programming; nor are they moved to act only in the interests of self, family, and kin. They are moved to act by what makes culture possible, including their own satisfactory life in culture. They convert to, inherit, and reinherit over generations a motivating worldview, classically oriented by religion.

This makes possible the rearing of another generation of humans, because good religion brings cultural prosperity. But just this same good religion has to be universally shared; it generates concern for other humans near and far, relating to them with the moral values of justice, love, and respect. The commitment that one has to make transcends one's genetics, if one is to be stretched over, lifted up to sufficient altruism for high-quality social life. The fertility dimension, though it cannot and ought not to disappear, is subdued before the cultural enrichment theme. The biosocial optimum on the metered scale is not just to be measured by my progeny, not by escalating birth rates in my tribe, or population explosion in my nation, or even in the world, but by a harmonious society in which one generation successfully transmits its valued achievements to the next, leaving them open to new achievements.

This can only be done if the best religious and moral insights are among the skills transmitted. Religions will, if you like, be tested for their capacity to do this, and the best ones will survive. The claim is not so much that the genes are the secret of religion, as that religion is the secret that makes possible the human passage from genetic nature to transmissible culture. Religion is the key to our humanity. Religion is required for the genesis of culture; cultures are required to generate religion, else they go extinct.

Such an account makes the future problematic. If religion disappears, humans will revert to being beasts – unless they can find something else to do the job of religion. Perhaps a rational morality, autonomous from religion, can command such obedience. But there will not be much hope in looking to genetics to supply such ethics. What is needed is a culturally acquired motivational power that pulls away from genetics, that genetics itself cannot supply.

(5) Religion Converting Others

There can be too much focus on biological fertility. Religion has to be understood as reproduction cognitively, believers making more converts, as well as biologically, believers having babies. In analogy to science, a scientific idea outcompetes its competitors, and wins adherents, and they fare well in their world. Religions have fertile ideas, and people adopt them the better to cope. But the transmission process is neural, not genetic. One has to be indoctrinated into a religion.

"A form of group selection operates in the competition between sects. Those that can gain adherents survive, those that cannot fail" (Wilson 1975a, p. 561). We know before we ask that surviving religions must recruit adherents from one generation to the next. Those that can proselytize increase. That is tautological. If the claim is that those religions succeed that make the most converts (where success means making the most converts), who will doubt it? We do not need biology to be convinced of that.

Biologically speaking, the problem now is that, if this is effective proselytizing, the new adherents soon cease to have any genetic relationship to the proselytizer. Only a minuscule fraction of the billion or so persons who are Christians have the Jewish genes of Christianity's founders. Most do not have the Greek or Roman genes of the first generations of Christians either. What good are all these Christians around the world to the Semitic, Greek, or Roman launchers of Christianity, or their present-day descendants?

Wilson claims that the function of religion is to produce group loyalty for the local or tribal survival unit. Those who were indoctrinated with primitive myths and rituals had intense and unquestioning loyalty to their society, and they obeyed their leaders' decisions in situations in which it was more important to act in concert than to think critically and independently. Such concerted group action conveyed survival value on all, on average, so that it was to any individual's probable advantage to cooperate, even though he had some risk of losing (being killed in battle, for instance). Under the influence of such religions, persons acted altruistically, but this was really pseudoaltruism, because it was in their genetic self-interest to bond to others in this way.

Such an explanation has a certain plausibility dealing with tribal religions. Perhaps it explains certain contemporary phenomena, such as the kamikaze pilots of World War II, dying for the emperor. But it is powerless to explain the universalism in the major world faiths. The most successful world religions have spread widely, typically as a result of the missionary activity of their adherents. Christianity has spread from its origins in the Semitic Middle East throughout the Greek and Roman worlds, throughout Europe, North America, and even the world. Buddhism spread from India to China and Japan, to California.

The Muslim armies advanced outside the Semitic world, across North Africa, into Spain, into India. That makes sense if one is gaining plunder in one's group self-interest or inducing others to serve one, but it makes no sense if one is spreading a religion that benefits nonrelatives. Even this "religion of the sword" was as much spread, southward in Africa, by the Sufis with their mystical visions. It was entirely so spread in India, Indonesia, and the Philippines, going where Muslim armies never went. *Dar-al-Islam*, the household of Islam, joins in daily prayer millions facing Mecca in solidarity and equality under God. Every one of the five principal pillars – conversion by profession of faith, daily prayer, fasting, almsgiving, and the pilgrimage to Mecca, joining other Muslims from around the earth in common submission to God – violates genetics in the name of universal fraternity.

If the function of a religion is to provide fervent loyalty for a tribal group, urging one's religion on aliens is exactly the wrong behavior. Missionary activity is helping to ensure the replication of genes unlike one's own. If one has a religion that serves his genes, holds his society together well, and produces numerous offspring, then the last thing he wants to do is share this religion with others. He would be giving the secret away. That would be altruism of the most self-defeating kind! Proselytizing those with foreign genes is the worst religious mistake you can make from a genetic viewpoint, and yet it has been the secret of success of all the world's great religions: evangelism in Christianity, or the *bodhisattva*'s vow in Buddhism.

Even Judaism, the one classical faith that might first seem best to fit the religion-genes theory, belies it. "I will make of you a great nation and . . . by you all the families of the earth shall bless themselves" (Genesis 12.1–3). On the genetic explanation, the second half of the promise undoes the first half. To be a people chosen by Yahweh to prosper in a promised land and to have descendants as numerous as the stars – that is to have a religion that leaves one with many offspring. But to be chosen by God to launch a religion in which everybody else also gets blessed? That is no genetic gain at all. That is a self-defeating religion, foolishly altruistic, and it will be selected against. On the genetic view, the first half of the promise has caused Judaism to survive for three thousand years; the second half of the promise is paradoxically antithetical to the first half. But surely, outside the genetic view, it is quite plausible to argue that ethical monotheism has had benefits that many others could share, and that this happened when (via Christianity), the Romans, the Germans, the British, the Americans adopted it, 99 44/100 percent of them without any Jewish genes at all.

The tribal group does need to reach a functionally efficient size, and one might want to indoctrinate enough others, preferably kindred or at least those of the same race, to reach this critical size. Beyond that, why should one send missionaries abroad to convert the Gentiles? They live in other nations and are not part of one's own political or economic survival unit. This preaching to the unconverted is not predicted by the theory, nor explained retrodictively. The Great Commission is, "Go therefore and make disciples of all nations" (Matthew 28.19). But the "catholicism" is counterproductive to any leaving of more Semitic genes in the next generation.

These outsiders from afar coming into the faith will convey incoming benefits only if they are needed reciprocators or can be exploited. They have to be made allies or colonies politically or economically. But evangelism is not always covert politics or economics. In the classical religions, the question ceases to be what tribe or clan a person is from, whether he or she is ally or enemy. The question is, Can he or she be saved? The secret of success is preaching a universal concern. Should we then argue that a group, covertly or tacitly to defend itself, discovers a faith that it shares with as many others as it can persuade around the world? That seems odd, that group self-defense requires proselytizing the world. A world mission is not covert intergroup altruism. If intergroup altruism must become intragroup altruism in order to retain its intergroup altruism, we can begin to doubt whether all these other groups evangelized are really nothing but needed reciprocators or subtly exploited aliens. No doubt one nation benefits when another is converted to a just and charitable form of life. Every nation benefits from harmonious international relations, to which religiously based ethical convictions about "the brotherhood of man" or "universal human rights" or "loving your neighbor as yourself" may contribute. But there is no reason to think that this expanding of altruism to the ecumenical limit is maximizing the group interest of those who launched Christianity.

Behavioral psychologists generally hold that humans are genetically inclined to xenophobia. The gene-fertility theory easily predicts this inclination in animals and has found it confirmed. "This xenophobic principle has been documented in virtually every group of animals displaying higher forms of social organization" (Wilson 1975a, p. 249, cf. pp. 286–287). This may have carried over to our early human ancestors. Possibly for millions of years natural selection favored those genes that caused the protohumans to be altruistic toward members of their own group but intolerant of outsiders. Possibly, humans today still have that innate tendency. Possibly, primitive religions are of this xenophobic kind; some sectarian religions today remain partisan. Possibly the principle works in politics. "Xenophobia becomes a political virtue" (Wilson, 1975a, p. 565). Love your neighbors and hate your enemies.

The one thing impossible is a xenophobic universal altruism. "The essential characteristic of a tribe is that it should follow a double standard of morality – one kind of behavior for in-group relations, another for out-group" (Wilson 1975a, p. 565).[1] But the major world faiths have escaped this, not only in ideal but also in the real proportionately to their success. And it seems impossible to explain this "xenophilia" on the basis of genetics. Somehow, somewhere, they reached insight into a better standard of what is right.

"You have heard that it was said, 'You shall love your neighbor and hate your enemy.' But I say to you, Love your enemies" (Matthew 5.43). Putting it another way, we even have to "hate" our families to be disciples of this universal love. "If anyone comes to me and does not hate his own father and mother and wife and children and brothers and sisters, yes, and even his own life, he cannot be my disciple" (Luke 14.26). That certainly doesn't sound like promoting one's genetic interests. "Whatever living beings there may be – feeble or strong, long or tall, stout, medium, short, small, or large, seen or unseen, those dwelling near or far, those who are born and those who are yet to be born – may all beings, without exception, be happy-minded" (*Suttanipata*, I, 8). If Christian and Buddhist say this universalism came by divine insight, prophecy, revelation, or mystic vision, there is nothing in genetic theory to gainsay such claims.

One converts to a religion culturally. "The idea of God" has "high survival value" in the pool of memes, as Dawkins puts it. "The idea of God is copied . . . readily by successive generations of individual brains. God exists, if only in the form of a meme with high survival value, or infective power, in the environment provided by human culture" (1989, p. 193). Large numbers of peoples have adopted religions that did not come with their ancestral genetic sets. One does not need Semitic genes to be a Christian, any more than Plato's genes to be a Platonist, nor Einstein's genes to adopt the theory of relativity. Religious beliefs overleap genes. But that does not confirm the religion/fertility hypothesis; it falsifies it.

Perhaps there is a competition between religions – some win, some lose – and in result people often convert to a faith originated by somebody other than their progenitors. Nevertheless people settle in on some religious belief that promotes their fertility. But this faith reached by conversion in the parents has to be transmitted to the children, who will be, as were their parents before, subject to proselytizing. A religion, to stay around, has to have a reproductive capac-

---

[1] Quoting Garrett Hardin.

ity cognitively. These beliefs must be transmitted nongenetically, though, when adopted, they promote fertility. But here the cognitive content of the successful religions is universalistic, and if so, the children will soon be spreading this fertility-producing faith to nonfamily and nongroup aliens. Perhaps they will be going off on missionary journeys again or contributing money to support such missions. Once we allow that a faith spreads by its persuasive powers, and that a vital element is universal altruism, it becomes impossible to keep the benefits local and in-group.

The function of religion is not simply to produce unthinking group loyalty, but to deal with many aspects of human nature that need to be curbed if optimal social cooperation is to be achieved, for example, selfishness, pride, greed, dishonesty, covetousness, anger, jealousy, sloth – aspects of human nature that are not genetically specific but are ubiquitous problems of *Homo sapiens*. These problems indisputably have some roots in our genetic past, but these shortcomings are common to all flesh. The religions that have stood the test of time have unanimously taught that humans must discipline and inhibit many tendencies in human nature. Here biology "frames" religion only in the sense that generic and genetically based traits have to be addressed by religion, but the solutions are supragenetic. The identity question has shifted from genetic identity (Chapter 2, Section 1) to religious identity, a nongenetic level. Christians around the world, confessing a common creed, share a cognitive identity. Values – now valuable answers – are getting shared again, rather than being something selfishly defended.

One is no longer dealing with just the logic of the genes. These religions criss-cross races, nations, and centuries; they operate in diverse times and cultures and involve some logic of the mind that is tracking what is transgenetically right or of value, no matter whether one has this or that set of genes. Genetic success is necessary but not sufficient to explain this universalism. It makes more sense to say that such religions were discovering what is transtribally, transculturally valuable. Something has emerged for which biology is not giving us a convincing account.

The rules change. Values are no longer defended at the level of natural selection, primarily. The value activity is now at a level that is culturally enjoyed and transmitted. The dominant monkey who feeds first and thereby protects his genes in his kindred may indeed be leaving superior monkeys in subsequent generations. But a human who grabs food from a neighbor is not improving the human genome at all (contributing superior grabbing genes to the gene pool), because the human genome functions in culture, where there are sharing and general educability. The human who shares food with a neighbor, where this contributes to their mutual survival – should this behavior be genetically based and selected for – is improving the human genome. This capacity for sharing behavior is essential in culture where things can be acquired during lifetimes, where values are transmitted nongenetically – knowledge, skills, resources, language, traditions, ideas, scientific discoveries, ethical convictions, and religious beliefs.

What happens in the monkey case and what happens in the human case are radically different conceptually. If humans have some elements of wild nature left in them genetically, these dispositions will frequently remain functional. When one gets hungry, one goes in search of food. But there may be other wild dispositions that humans have to rise above, if they are to rise into culture. Humans may fail to rise to their moral possibilities, fail to share, and lapse into sheer selfishness. Although self-defense is proper and valuable in animals, and proper and valuable also for persons, when self-defense passes over into selfishness, this is improper and disvaluable in culture. Then there will appear what the theologians call sin, and this historic and perennial lapsing has sometimes been called original sin. Religions deal with this tendency; they regenerate humans for successful life in culture.

Such sharing capacities do indeed produce human prosperity (fertility), but tight connections to the genes – this behavior linked to this genetic coding – have been left behind in the exodus from nature to culture. Natural selection is relaxed. By the time one encounters the universal altruism taught in the world faiths, there is no genetic leash at all. Rather, religious values are, to recall words used earlier, being "distributed," "dispersed," "allocated," "proliferated," "divided," "multiplied," "recycled," "shared," deliberately and out of conviction that this is good and right. And if some of these persons say that "God commands this altruism," that this kind of suffering love is divine, there seems no reason yet forthcoming from the biologists to think otherwise. To the contrary, this genesis of religion with its capacity to generate the generous altruism requisite for culture still needs adequate explanation.

4. FUNCTIONAL AND TRUE RELIGION

Religion, then, must function to generate innovative ethical behavior, unknown otherwise in natural history, which makes possible the human genius (*Geist*), which cannot exist outside the social covenant. When this happens, the human genius is still more fertile. Reflective religion comes further to serve the function of explaining the creative genesis in both natural and cultural history, both describing and evaluating it. Such explanations have to work; they are the backing for the ethic. But we must press the further question of whether they also need to be true. Religion can generate ideologies that help persons cope. Religion can generate altruism sufficient for cultural survival. Can religion generate truth? Is that too part of the human genius?

For a pragmatist there is no further question, since whatever works is ipso facto true. Our question is more realist, about facts as well as functions. Since we believe that the genesis needing explanation in both nature and culture is the actual fact of the matter, events that have taken place in history, we also seek an accurate explanation and evaluation. Is it also the fact of the matter, the way things are, to say that God is in, with, and under such genesis? We first look at some doubts, en route to a more positive conclusion.

(1) Survival Value with and without Truth

"Traditional religious beliefs have been eroded, not so much by humiliating disproofs of their mythologies as by the growing awareness that beliefs are really enabling mechanisms for survival" (Wilson 1978, p. 3). That is a rather strange disproof itself.

1. If S (survival-enabling), then not T (true).
2. S.
3. Therefore not T.

The logic is valid, but are the premises sound? Religious beliefs enable survival in the general sense of helping people manage over the generations, as no one wishes to deny, although we just also concluded that religious universalism extensively overreaches the genetic survival of particular practitioners. So the second premise is half-true. Meanwhile the first premise is curious in its connecting of coping with untruth.

If eyes enable humans to cope, then what they see is not true.
If ears enable humans to cope, then what they hear is not true.
If mathematics enables humans to cope, then what this calculates is not true.
If medicine enables humans to cope, then medical theories are not true.
If science enables humans to cope, then science is not true.
If ethics enables humans to cope, then ethics is not true.
If religion enables humans to cope, then religion is not true.

The presumption, usually, is that survival-enabling mechanisms track something there in the world – eyes seeing a predator, ears hearing a friend, mathematics balancing income and expenditures, science making a medicine that kills germs, ethics distributing re-sources fairly. This presumption has to be overridden in religion. Anomalously, religion is a coping myth.

Neither perception nor conception is infallible; both are often true at a native range from which one cannot extrapolate too far. The eyes see what is there (trees, tigers, houses), but also what is not there (the flat earth, setting sun, green trees, blue sky); sometimes a super-ficial correctness is at depth illusory. Appearance is not reality. Even science, as philosophers constantly warn nowadays, is pragmatic and not ultimately descriptive, not descriptive of anything ultimate. Hu-mans frequently know how to manipulate things with little knowl-edge of what is really going on, as when people bake bread with no idea of the chemistry involved.

Still it is hard to see how science makes medicines that kill germs unless the germs are there and the drug in fact is toxic to them. Some scientists do know the bread chemistries. Humans know about many events rather far removed from our native range – astronomical ones, such as a round Earth orbiting a sun, or supernovae, and microscopic ones, such as DNA coding and covalent bonding. Such knowledge partly is and partly is not survival-enabling; it partly retains and partly modifies native-range impressions that are survival-enabling. So, although it does not demean a science to realize that it has sur-vival value, the relation between truth and survival value is not straightforward even there.

It is reasonable to begin with the assumption that acting on true beliefs will bring success. If I believe rightly that there are deer in the valley and go there to hunt them, I may well succeed. If I believe wrongly, I will fail. Even on the pragmatic theory, survival selects for native-range truth. Scientists need descriptive truth when they go after the world, just as much as hunters who go after deer. There is no cause to expect that whatever meets the needs of practice is going to be theoretically wrong.

Ruse thinks that even mathematics just conveys survival advantage. "The human who believes that '2 + 2' *really* equals '4' is going to act upon it without question, as are his/her fellows. And this will give them a selective advantage over those who question the basic premises of logic and mathematics, sometimes disobeying them." This is the way it is, but it might have been otherwise. "Selection cares only about keeping us alive and our passing on of our genes. . . . Thus, if we benefit biologically by being deluded about the true nature of formal thought, so be it" (1986, p. 172). But can we imagine that humans who randomize for the outcome of 2 + 2 might have been selected for? Or that those prosper who, in the interest of equity, rotate the outcome through 1 to 10 on successive days? Hardly, because these procedures are logically wrong, and therefore they will fail in empirical application. It is implausible that life should have evolved a bad computational logic that is a good adaptive fit. A theoretically mistaken mathematics might meet the needs of practice, if it nevertheless provided good approximations, but not if the formal mistake really misinformed the practice.

Those who cope well need a worldview that represents rather reliably what the world is like, at least those sectors of it through which they have to move. There cannot be too much gap between appearance and reality. That works with sense perception, with science, even in ethics at everyday empirical ranges. People use religion too to operate at everyday empirical ranges. Nevertheless in religion a problem arises because a modern person, whether monotheist believer or secular scientist, will soon enough encounter beliefs that seem to enable persons to function reasonably well that bear no resemblance to world facts. Even believers in the monotheist God of Judaism and Christianity have a lot to explain away in the enormous variety of "pagan" beliefs and practices. Shinto believers held that the emperor was descended from the sun goddess; an Australian Aranda may think he is descended from the kangaroos. Divinely descended leaders and totemism may inform practice successfully, but both are misinformed theoretically. How can it be that in religion, unlike mathematics, misinformation forms successful practice?

Darwin found himself beset by "the horrid doubt" whether the convictions of a man's mind are any more trustworthy than those of a monkey's mind (Chapter 4, Sec. 7). But there is every reason to trust a monkey's mind about whether raptors, snakes, or tigers are present. There is every reason to trust any creature's mind in the niche in which it is adapted to survive. The human niche is culture as well as nature. When one finds that human minds are disposed to beliefs of both conscience and religion, what then? We are first inclined to trust such minds and then startled to find them full of ideas that seem incredible. How can this happen?

C. S. Peirce claims:

> Logicality in regard to practical matters . . . is the most useful quality an animal can possess, and might, therefore, result from the action of natural selection; but outside of these it is probably of more advantage to the animal to have his mind filled with pleasing and encouraging visions, independently of their truth; and thus, upon unpractical subjects, natural selection might occasion a fallacious tendency of thought. (1960, vol. 5, sec. 366)

Lionel Tiger adds: "Optimism has been central to the process of human evolution. . . . Making optimistic symbols and anticipating optimistic outcomes is as much a part of human nature, of the human biology, as are the shape of the body, the growth of children, and the zest of sexual pleasure" (1979, p. 15; cf. Taylor 1989). Arnold Ludwig agrees: "Fantasy, then, often represents a convenient way for man to temporarily lie to himself in order to make life more palatable" (1965, pp. 179).

Now the logic is different:

> If a practical matter, then thought must be accurate (true).
> If an impractical matter, then thought must be pleasant (not true).

The animal knows the truth when it is vital; the animal feels good when it isn't important; this combination is better for survival than truth all around. It is as though the animal operates in the real world by day and has pleasant dreams at night, which relax it for work again the next day.

If a scientific matter, then thought must be accurate (true).
If a religious matter, then thought must be pleasant (not true).

A problem with this account is that any religion so explained has to be impractical, and there is no evidence that religion is unimportant or irrelevant in the lives of these myriad believers. To the contrary, their life practices are oriented by religion. Almost by definition, religion is what one is "bound to" (Greek: *re-ligio*, redoubled binding). The biologists have been insisting that religion has survival value just because it congeals group loyalty, demands short-term sacrifices in the interests of long-term benefits, offsets biological self-interest to develop the more altruistic virtues necessary for culture, results in more offspring in the next generation, and so on.

If Cinderella thinks about mice in the pantry, she must be accurate.
If Cinderella thinks about Prince Charming, she must be pleased.

Religion is a Cinderella story that helps girls function when scrubbing the floors, it fulfills Cinderella's psychological "needs," but it is really out of touch with reality. Here she needs a pleasant appearance that veils her harsh reality.

So, to project the Cinderella parable into metaphysics, the world is really harsh and meaningless, but we humans invent religions to save us *from* the truth, rather than to help us discover the real truth. Here people need to get it wrong. Religiously, we need an illusion in order to keep our spirits up, though practically, we must have correspondence between appearance and reality in order to operate successfully (Rue 1994). Practically, Cinderella needs a science to tell it like it is; mythically, she needs a religion to tell it like it isn't.

If that is true, science is bad news, eroding these traditional mythologies, as Wilson thinks it does. Science is disabling these enabling mechanisms, and how can humans then survive, so disabled? So science, which we have earlier found to be quite enabling for survival, is now discovered to be only penultimately so, and ultimately disabling. Fortunately, or unfortunately (?), scientists, who get it right, are likely to be disabled most, and those who continue the traditional mythologies, and get it wrong, will outreproduce them!

Meanwhile, Peirce's account fails to reckon with how religion has to be functional in society corporately, not just provide a relaxing dream for individuals at night or on Sundays when there is no work. Cinderella has to relate to her stepmother and sisters religiously and thereby to manage in her real world, not just the imaginary one of her dreams. That is where her real, social need fulfillment must take place. Prince Charmings come rarely or never. Religion cannot be just episodic retreats from reality. Religion too has to keep her in contact with reality; she must return to the operating world and work there practically. Humans do have to get some values right, if they are to succeed, and young girls who wait around for handsome princes to rescue them have few actual needs fulfilled and even fewer offspring. Fantasy is seldom functional. People who are wrong about causal connections (bad science) will fail, but people who are wrong about what is valuable in the world (bad religion) will just as surely fail.

Religions do need to distinguish between ideal and real (which is not the same as appearance and reality). Often persons need an ideal toward which they reshape the real. It can certainly be adaptive to have ideals to which one aspires, even if one falls short. All of us are like that, in our better moments. That is what makes us better. Religion and ethics both prescribe what ought to be even when it isn't, as well as describe what is. The question turns around what these ideals are. If the sociobiologists and evolutionary psychologists are right, the ideals, sooner or later, tacitly if not explicitly, are the most offspring in the next generation. But we have been recognizing various other ideals ("myths") – universal love, evangelism, redemption, self-actualizing, justice, fairness, honesty – that seem maladaptive for maximizing offspring. Perhaps Cinderella thought her rights were being violated. If she could assert her rights, she would probably have more offspring, and religion might support human rights.

Humans absolutely must get their social functioning act together. A religion has to have enough realism about the human condition, real and ideal; about the values that motivate persons to behave in interpersonal relationships; about parenting responsibilities; and so forth, to get a whole society from one generation to the next, and the next, across generations. There is nothing impractical about that. Fairy tales, like fables, embody occasional bits of wisdom, but can you operate a whole society on fairy tales and other pleasant fantasies? What one really needs to examine is the relationship between this social functioning, necessary for any persisting religion, and its truth.

(2) Testing Religions Socially

This is not a problem first conjured up by the biologists; the sociologists have been troubled by it for years – although not in the genetic form. Religions have regularly claimed that they were good for the well-being of society. Finding such a connection is, ipso facto, no cause to cease to behave religiously. To the contrary, it is an excellent reason to continue observance. One's reasons for observance might be weakened, though, if it were shown that observance is "nothing but" behavior for the good of society, that is, that there is no further reason to think that any of these beliefs is true.

Those reasons will have to come from some other dimension of human experience and will force asking whether functioning is a test for truth in religion. Can there be functional religions that are not true? Can there be true religions that are not functional? What is one to make of the error in religion? To be wrong in religion seems to mean more than just not functional. For it is certainly true that religion has been (and often continues to be) filled with beliefs that no scientifically minded person can seriously entertain to be true, and that these beliefs are sometimes more or less functional.

What separates out religion as anomalous, is the feeling, by modern persons, that the religions just cannot all be true – they are too diverse, conflicting, and fantastic – so one must look for some other account of them, which preserves their functionality without requiring their truth. Humans with their eyes and ears, humans with their mathematics and sciences, reach much consensus on what is true, but in philosophy, in ethics, and especially in religion, they do not. Wilson cites with approval Anthony F. C. Wallace's estimate that humans have produced 100,000 religions (Wallace 1966, p. 3; Wilson 1978, p. 169). If there are that many they all must really be the same thing, some essence beneath the fluff; the explicit cognitive claims cannot be right; there must be a tacit, functional explanation. Like a kaleidoscope that produces 100,000 patterns with a simple mirror arrangement reflecting the contingent falls of bits of glass, there is really only one mechanism driving religion, and the particulars are frills on the universal.

Animals that misperceive their environments do not survive; an adapted fit cannot be based on false information about the world. But with animals there is no problem with conflicting worldviews: the deer are unanimous about whether that is an approaching panther, whether it is dangerous, and which direction it is coming from. Even humans are unanimous about such empirical facts. But humans differ widely in their worldviews, and they cannot all be right about these. "The enduring paradox of religion is that so much of its substance is demonstrably false, yet it remains a driving force in all societies. Men would rather believe than know." It is hard to believe that "such force could really be extracted from 'a tissue of illusions' " (Wilson, 1975a, p. 561). The solution is that religion, though a tissue of illusions cognitively, is a strange kind of error that evolution promotes, since the outcome of religion that is visible to natural selection is its survival value, and natural selection selects for that, regardless of truth.

Can evolution promote error? Usually no, but sometimes yes. There can be myths that insulate, pacify, sedate, or unify; stimulate, inspire, or engender other useful behaviors. The anomaly is that such a fantastic point of view results in an adequate response to the environment. It is as though humans live in a bad world – one where they must all compete with each other genetically and for resources, where nature is red in tooth and claw, where they must band together selfishly yet also cooperatively in culture, where they will all eventually lose. Paradoxically, this world is so bad that only the optimistic will survive. Usually, you can function well only if you know the truth about your world, but here you can only function well if you do not. So one must be deceived to succeed. This becomes a self-fulfilling prophecy. Those who believe the gods are on their side make it through. The realists, if there are any, go extinct. If this is true, then one might ask whether we want to know otherwise. These biologists will not be bringing us a truth that sets humans free, but a truth that triggers our extinction.

This is such an anomaly that one will want to examine the paradigm that is framing it. Perhaps this anomaly is only an artifact of a particular theory. Such a picture comes from selfish genes, random variation, blind selection, evolutionary history as a random walk, survival of the fittest, and so forth. But there is the alternative account: of widely distributed and conserved (shared) values both in nature and in culture. There are dimensions of struggle, suffering, and tragedy in this picture too, but there are genesis and creativity, generation and regeneration. Accompanying a human exodus from nature into culture, some religious persons discover, over time, the virtues of cooperation and altruism that make culture possible (more sharing). Some of them detect the divine in their midst endorsing such virtue. From this perspective we have not really been given any reason to think that those who make these sorts of claims in religion, though survival-enabling, are not right because they have too rosy an account of the bleak world. Whatever account one may have to give of the many myths in religion, in this part of it, about the divine will in, with, and under the socially functional and universal altruism, we are getting a clash of philosophical views. These views are being superimposed on the facts; indeed people are seeing differing facts as a result of differing interpretive gestalts.

We need a better account of whether there is sometimes cognitive truth in religion, and, if there is, whether such truth might not be quite compatible with survival value. Then one might be in a position better to understand why and how such believers not only make it through but evangelize others.

(3) Testing Religions Cognitively

Are all the 100,000 religions just turns on a socially functional kaleidoscope, with the differences between them insignificant? That might be true of many religions, but there is something more to be said. The religions that have spread worldwide, that persist and develop over the centuries, are quite few: about ten religions form the chapters in a typical world religions textbook. So all are not equally socially functional on a world scale; less than a dozen were exported outside their originating tribes to become global faiths. What account is one to give of the few that were?

We are confronting universalism again, now in religion as well as in ethics, and wondering whether the universalism in the ten survivors is just more covert social functioning promoting covert genetic survival, from which one must dismiss any cognitive content as irrelevant fluff. Perhaps this long-continuing, ever-widening social functioning is linked to some insightful cognitive content, especially where the classical religions are so anomalously transtribal and transgenetic.

An alternative account, recalling the "generate and test" model, holds that the 100,000:10 selection effect has been a trial-and-error learning process. The creeds (theories) that remain have survived because they have a good deal of corroboration and have not yet been falsified. They have a staying power in the face of arguments and evidence. Analogously, humans have produced 100,000 theories in science, of which, again, perhaps 10 are leading survivors (relativity theory, quantum theory, atomic theory, the chemical periodic table, the geological epochs and cycles, plate tectonics, evolutionary theory). There is a consensus on such scientific theories not yet reached by the remaining, often competing religions. Nevertheless, the winnowing of religions is a testing by which some survive and others do not. Mostly, there are chaff, dross, noise, but sometimes there are grain, gold, information.

Along the diverse routes of religious development, humans will often have constructed beliefs that are functional in some local and limited context, though they are not true. Sometimes erroneous scientific theories stayed around quite a while (fixity of species, spontaneous generation of life, phlogiston), but eventually better ones supplanted them. The better adapted survived – that is, the better ones with which to adapt survived – and we also believe that the prevailing theories are more approximately true. The history of astronomy is beset with astrology, the history of mathematics is beset with curious numerology, and it will not discredit all religion to find such things as demonology, angelology, and superstition in some of it. One can employ a developmental, trial-and-error, generate-and-test account of religion too. Often these trials will work briefly, or in elementary circumstances. But they will not survive the cross-critical sifting that the world religions do manage to survive.

The process is generate and test again, but this time it is also regenerate and test. What is tested now is not genes, not just hypotheses, but persons who embody creeds, who may be "saved" as those creeds inform appropriate behavior for managing in the world. One does not want to dismiss the survival-enabling component, but to search for an appropriate lived experience that lives successfully in the world because it detects the bigger forces operating there, as a compass detects invisible global forces. The fantasies will be selected out in critical insight into what really orients in the world, and these tested (and true) insights will cumulate over the millennia of the religious heritages. Once again, natural selection is relaxed, this time in favor of religious selection. Believers, like ethicists and scientists, must have offspring, but believers have to have disciples, whether their children or proselytes, whom they can persuade to adopt the style of life their religion commends.

Some religious claims that are functional will perhaps remain long in place because these conserve what some local society values, un-contested by any rival claims. As long as such a claim stays isolated, it can be retaught over generations; perhaps no indigenous believer has the imagination to challenge it. Natural selection in wild nature often leaves locally endemic species in odd niches, surviving more by isolation than by competitive success.

But when such once-isolated claims do face challenges, in the conflict with missionary arrivals, or in the dialogue of world faiths, or in the effort to proselytize others with differing faiths, or in the encounter with science, the functional test is no longer good enough. The naive religious claim must meet skeptics, resist invasion, invade the status quo, displace vested interests, win debates, regenerate sinners, as well as make better parents. The claim must have rational defensibility proportionate to relevant evidence; it must deploy to cover an expanding set of evidence. That has been the core problem with the myriad nonexportable indigenous faiths. None of their theses could survive the onslaughts of ecumenical criticism. Thus Shinto never left Japan, nor could it; and the Australian Aranda who thinks himself descended from the kangaroos has convinced no one outside tribal Australia. Those beliefs vanish in the modern world because they cannot make converts.

Only the universalist, synoptic creeds have proved exportable, globally functional, because they speak to the common condition of humankind, a necessary condition of success. They do not simply offer fertility, nor even doing well in life, if this means survival. They offer persons the promise that they can understand the fundamental structure of reality (what *is*) and tap powers here for the redemption of life in its brokenness (what *ought* to be). They invite critical self-assessment and reformation; they promise enlightenment and freedom. And they are tested against each other in the fulfillment of these promises. Of the functional faiths, only those with the theses that are the most defensible rationally, as well as the most operational experientially, and those that give life the most meaning, are competent to survive.

Kitcher concludes:

> Just as a detailed history of arithmetical concepts and counting practices might show us a succession of myths and errors, yet would not lead us to question the objectivity of the arithmetical statements we now accept, so too reconstructions of the historical development of ethical ideas and practices do not preclude the

> possibility that we have now achieved a justified system of moral precepts. Wilson is far too hasty in assuming that the evolutionary scenario he gives for the emergence of religious ideas – a scenario that stresses the adaptive advantages of religious beliefs and practices – undercuts the doctrine that religious statements are true. Even if Wilson's scenario were correct, the devout could reasonably reply that, like our arithmetical ideas and practices, our religious claims have become more accurate as we have learned more about the world. (1985, p. 419)
>
> Wilson mentions religious systems of morality only to dismiss them; his reason is spurious. "If religion . . . can be systematically analyzed and explained as a product of the brain's evolution, its power as an external source of morality will be gone forever." The argument turns on a critical ambiguity. If religious concepts are nothing but products of our brains, then, of course, religion is just a story. If, however, the history of religious belief shows human beings gaining knowledge of entities that actually exist, then there are no grounds for Wilson's conclusion. . . . There is no quick argument for debunking religion (or mathematics) on the grounds that it has a checkered history. (1985, p. 424)

Every set of human ideas – science, ethics, religion, or whatever – has a history that connects with the brain that has evolved, that has been used for building a culture and getting along in the world, but what one wants to know is whether some of these ideas, gained with this evolved brain, are true and correct and others false and incorrect.

The basic theoretical model is variation, selection, and retention. Biology uses one version, naturalized in genetics. Religion, as does science, uses a socialized version that goes beyond genetics. Not only is all religion culturally transmitted, but some is transmitted by universal proselytizing, urging universal altruism. The first round of selection is pragmatic, socially functional survival value, but the second round of selection is critical and cognitive. Both rounds are evaluative, testing a religion for what it is worth. The only ones that are able to survive and flourish over the millennia are the universal ones, able to win by proselytizing and universal altruism. That does not sound like *selfish* genes. It sounds like truths that have got loose transgenetically and are being *shared* around the world, that is, in the religious imagery, truth that is "blessing" all nations.

We first think that function is underdetermining truth, but the truth may rather be that progressively powerful functioning is corroborating truth. There is nothing particularly biological about a culling process that generates and tests variations on religion, any more than there is about a similar systematic methodology in science or ethics. Indeed it seems countergenetic when a principal criterion for a religion's survival is a universal scope that discredits the tribal religions that have been unable to generate exportable, globally true theses.

5. GENESIS AND GOD

There is a metaphysical version of the if-functional-then-not-true argument. Wilson argues that if something has evolved in natural history, then it cannot be the work of transcendent deity.

1. If E (evolved), then not T (transcendent).
2. E.
3. Therefore not T.

"No species, ours included, possesses a purpose beyond the imperatives created by its genetic history. . . . We have no particular place to go. The species lacks any goal external to its own biological nature" (1978, pp. 2–3). "There is no transcendental guide or extrasomic set of universal principles to follow" (Wilson 1980a, p. 70). As before, the logic is valid, but are the premises sound? "If x emerged in historical time, then x is not divine"? If from genes, then not from God?

That fails to consider whether one purpose of God might be this Earth history: the creation and its redemption. This amounts to claiming, in the traditional vocabulary of theologians, that immanence cannot combine with transcendence, that the beyond cannot be in our midst. Theologians almost unanimously think otherwise, on the evidence of religious experience, critically evaluated. So one will need to know what it is about biology, about genetics, that authorizes this conclusion that the historical cannot be the immanent location of a transcendent divine presence. We humans do not particularly want some goal "external to our biological nature"; we wish one consistent with it, but we might want to maintain that, metaphysically, neither our biological nor our sociological natures are self-explanatory.

Wilson insists, "The evolutionary epic is probably the best myth we will ever have" (1978, p. 201). We agree, but the question is whether the dramatic events on this Earth contain no hint of larger, more universal powers in which they are embedded. Perhaps, rather, culture and biology are finding out in their historical domains what Kurt Gödel found for the much simpler domains of mathematics and logic, that systems to be completely understood require reference to other systems at a higher level of organization. Against the reductionists, religious persons have to be compositionists, to move up, not down, to get the interpretive level needed to frame and complete lower level truths. Nature and history have been creative, making more out of less. The essential characteristic of narrative is that events have to be understood in the light of the complexities to which they lead, not just in the light of the origins from which they flow. The event structures toward which things climb, their endings, are as significant as the matter–energy out of which they arise, their beginnings.

We have no cause to think that the startling genesis on Earth, recorded in the genes, recorded in the cultural heritages, including the religions, is not sacred; nor that humans, funded by their evolved perceptual and cognitive equipment, can never detect that sacred presence. The idea of God has been among the most fertile in shaping history. That is the fertility that ultimately needs to be explained.

That returns us to the global claims of religion, claims that are transcendent at least in the sense of detecting a divine power in, with, and under the genesis on Earth. Contra Wilson, does biology leave space for such claims or even invite such claims as complementary explanations? Genes record only a portion of the history that has taken place: they do not, for instance, record the prelife cosmological story; nor do they record the postgenetic cultural story. Still, vital to the Earth epic is this fertility intimately linked with the genes, the means by which all the more complex structures on Earth, living things, are formed. There are no such genes on the moon, nor Jupiter, nor Mars. Genes remember, research, and recompound discoveries, and the storied achievements, the values achieved, rise, over several billion years, to spectacular levels of attainment and power. The cosmic universals give way to the particulars of Earthen natural history.

(1) Actual and Possible Natural History

What can we say about how the possible becomes actual over evolutionary time? Here, one must increasingly pass from bioscience to

348

metaphysics. We return, at the end, to questions faced earlier, about the increase of complexity and diversity, about contingency and inevitability in such increase, about progress, now with questions looming about the possibility of divine presence. This is the fertility question in its metaphysical form, the generation of the actual out of the possible, and the generation of those possibilities, and even a Generator of such possibilities. The possibility route to be found is not so much logical, or empirical, or even physical; it is historical. What possibility spaces are needed to get from beginnings to where we have now arrived, in Earth history?

At the other extreme from those emphasizing the contingency, there are eminent biologists – though they tend to be molecular biologists rather than paleontologists – who find this storied natural history to be inevitable, at least in outline, and therefore predictable. Christian de Duve concludes: "Life was bound to arise under the prevailing conditions, and it will arise similarly wherever and whenever the same conditions obtain. There is hardly any room for 'lucky accidents' in the gradual, multistep process whereby life originated." After life arises there is contingency as to its directions and species, but this is "constrained contingency" so that the general trends in the development of life – cellular organisms, multicellular organisms, solar energized organisms, increasingly diverse and complex organisms, and intelligent organisms – are likewise inevitable. "Life and mind emerge not as the results of freakish accidents, but as natural manifestations of matter, written into the fabric of the universe. I view this universe [as] . . . made in such a way as to generate life and mind, bound to give birth to thinking beings" (1995, pp. xv–xvi and xviii).

"This universe breeds life inevitably," concludes George Wald (1974, p. 9). Life is an accident waiting to happen, because it is blueprinted into the chemicals, rather as sodium and chlorine are preset to form salt, only much more startlingly so because of the rich implications for life and because of the openness and information transfer also present in the historical life process. Whatever place dice throwing has in its appearance and maturation, life is something arranged for in the nature of things. The dice are loaded.

When the predecessors of DNA and RNA appear, enormously complex molecules appear; bearing the possibility of genetic coding and information, they are conserved, writes Melvin Calvin, "not by accident but because of the peculiar chemistries of the various bases and amino acids. . . . There is a kind of selectivity intrinsic in the structures." The evolution of life, so far from being random, is "a logical consequence" of natural chemistries (1975, pp. 176 and 169). Manfred Eigen concludes that "the evolution of life . . . must be considered an *inevitable* process despite its indeterminate course" (1971, p. 519; 1992). Life is destined to come as part of the narrative story, although the exact routes it will take are open and subject to historical vicissitudes. Kauffman agrees: "I believe that the origin of life was not an enormously improbable event, but law-like and governed by new principles of self-organization in complex webs of catalysts" (1993, p. xvi; 1995).

Such accounts suggest that the possibilities are always there, latent in the physics and chemistry, although the resulting Earth history is not so "fine-tuned" as astrophysics and nuclear physics have found in their cosmologies. But even in Earthen biology, the possibilities must, or almost must, become actual. Alternately put, there are few possibilities beyond those that do actualize. But of course all such possibilities are seen only retrospectively. What does happen, can happen. But we are wondering how it comes about that these events can happen. If, *per impossibile*, some scientist had under observation the elementary particles forming after the first three minutes, nothing much in them suggests anything specific about the coding for life that would take place, fifteen billion years later, on Earth. After Earth forms, the lifeless planet is irradiated by solar energy, as are other planets as well. The events in physics and chemistry there are to a considerable extent lawlike and predictable, at least statistically, although in geology and meteorology the system is quite complex as a result of shifting initial conditions, possibly even at times chaotic. Still, in orogeny and erosion, or the shifting of the tectonic plates, the possibilities always seem there.

At the microscopic levels, quantum physics depicts an open system and nested sets of possibilities, but, at first, all the atoms and molecules take nonliving tracks. Only later do some atoms and molecules begin to take living tracks, called forth as interaction phenomena when cybernetic organisms appear. If there is some "inside order" to matter that makes it prolife, it is in the whole system and not just in the particles. Despite the anthropic principle, such order is not generally evident in the systemic astronomy, since by far the vastest parts of the universe are lifeless. Life is an Earth-bound probability. Nor, on Earth, are the meteorological or geomorphological systems all that suggestive of inevitable life. They mostly seem kaleidoscopic variations on geophysical and geochemical processes.

Only in biology do there open up entirely unprecedented levels of achievement and power. Such possibilities are not inside the atoms and molecules apart from their systemic location, since atoms and molecules would not even be collected into a "thin hot soup" except for the Earth world in which this is possible, nor can this or that sequence of DNA code for anything unless there is an environment in which to behave this way or that, with a niche to fill. Even if there is some "selectivity intrinsic in the structures," this does not rule out a universe of myriad options, only some of which are realized.

Physics and chemistry, unaided, do not get us very near to life and mind. There really isn't much in the physics and chemistry of atoms and molecules, prior to their biological assembling, that suggests that they have any tendencies to order themselves up to life. Even after things have developed as far as the building blocks of life, there is nothing in a "thin hot soup" of disconnected amino acids to predict that they will connect themselves or be selected along upward, negentropic though metastable courses into proteins, nor that they will arrange for DNA molecules in which to record the various discoveries of structures and metabolisms specific to the diverse forms of life.

All these events may occur naturally, but they are still quite a surprise. Recent microbiology has been revealing their enormous complexity. We do not know that life, if it occurs on some other planet, there built too of the same atoms, must select these same biochemistries, although the amino acids found on meteorites and the prebiotic molecules guessed to be present in interstellar dust clouds can suggest that the potential for life is omnipresent in matter. Laws are important in natural systems, whether extraterrestrial or terrestrial. But natural law is not the complete explanatory category for nature, any more than are randomness and chance. In nature, especially on this historical Earth, there is creativity by which more comes out of less.

Science does not handle historical explanations very competently, especially where there are emergent novelties; science prefers lawlike explanations in which there are no surprises. One predicts, and the prediction comes true. If such precision is impossible, science prefers statistical predictions, probabilities. One predicts, and, probably, the prediction comes true. Biology, meanwhile, though prediction is often possible, is also full of unpredictable surprises – like calcium endoskeletons in vertebrates after millennia of diatomaceous silica and chitinous arthropod exoskeletons. A main turning point in the history of life fused once-independent organisms into the cell and its mitochondria, which became the powerhouses for life. Another critical symbiosis introduced free-living chloroplasts into the plant cell, again producing the energy vital for all life.

There is no induction (expecting the future to be like the past) by which one can expect, even probably, trilobites later from prokaryotes earlier, or dinosaurs still later by extrapolating along a regression line (a progression line!) drawn from prokaryotes to trilobites. There are no humans invisibly present (as an acorn secretly contains an oak) in the primitive eukaryotes, to unfold in a lawlike or programmatic way. The ancient ancestral forms are not protovertebrates or preterrestrials, nor are gymnosperms about-to-be angiosperms, as though the descendant forms were latent among the functions of the predecessors. Originating events often become what they become only retrospectively: "Vertebrates began (possibly) with the notochords of primitive chordates." "Eyes began with. . . ." Nevertheless, there is the epic story – eukaryotes, trilobites, dinosaurs, primates – swarms of wild creatures in seas and on land, followed by humans who arrive late in the story.

Making this survey, can one insist that the probabilities, or at least the possibilities, must always have been there? Can one claim that what did actually manage to happen must always have been either probably probable, or, minimally, improbably possible all along the way? Push this to extremes, as one must do, if one claims that all the possibilities are always there, latent in the dust, latent in the quarks. Such a claim becomes pretty much an act of speculative faith, not in present actualities, since one knows that these events took place, but in past probabilities always being omnipresent. Is the claim some kind of induction or deduction, or most-plausible-case conclusion from present actualities? Speculation about such possibilities that are always there is easy, provided one does not have to specify any of the details. But this perennial and vast library of possibilities is mostly imaginary.

For in fact, on Earth, there really isn't anything in rocks that suggests the possibility of *Homo sapiens*, much less the American Civil War, or the World Wide Web, and to say that all these possibilities are lurking there, even though nothing we know about rocks, or carbon atoms, or electrons and protons suggests this is simply to let possibilities float in from nowhere.[2] Unbounded possibilities that one posits ad hoc to whatever one finds has in fact taken place – possibilities of any kind and amount desired in one's metaphysical enthusiasm – can hardly be said to be a scientific hypothesis. This is hardly even a faith claim with sufficient warrant. It is certainly equally credible, and more plausible, and no less scientific to hold that new possibility spaces open up en route.

Karl Popper concludes that science discovers "a world of propensities," open to historical innovation, the possibility space ever enlarging.

> In our real changing world, the situation and, with it, the possibilities, and thus the propensities, change all the time. . . . This view of propensities allows us to see in a new light the processes that constitute our world: the world process. The world is no longer a causal machine – it can now be seen as a world of propensities, as an unfolding process of realizing possibilities and of unfolding new possibilities. . . . New possibilities are created, possibilities that previously simply did not exist. . . . Especially in the evolution of biochemistry, it is widely appreciated that every new compound creates new possibilities for further new compounds to synthesize: possibilities which previously did not exist. The possibility space . . . is growing. . . . Our world of propensities is inherently creative. (1990, pp. 17–20)

The result is the evolutionary drama. "The variety of those [organisms] that have realized themselves is staggering." "In the end, we ourselves become possible" (1990, p. 26, p. 19).

But – the reply comes – since all those things did come in subsequent evolutionary and cultural history, their possibilities must have been there all along. You were not listening when we discovered that matter is self-organizing, autopoietic. That posits enormous possibilities, there from the start, and nothing in the historical drama ought to take by all that much surprise one who believes in self-organizing nature. Thomas R. Cech, a molecular biologist, reviews the origin of life:

> If intrinsic to these small organic molecules is their propensity to self-assemble, leading to a series of events that cause life forms

---

[2] Against the caution of Alfred North Whitehead (1929, p. 46).

to originate, that is perhaps the highest form of creation that one could imagine. . . . At least from the perspective of a biologist, I have given an account of how possibilities did, in times past, become actual. When this happened, life originated with impressive creativity, and it does not seem to me that possibilities floated in from nowhere; they were already present, intrinsic to the chemical materials. (1995, p. 33)

True, matter – energized as it is on Earth – is now self-organizing. But that leaves open the question whether, on the adaptive landscapes on which organisms struggle to increase their fitness for survival, landscapes which themselves shift as the organisms make their discoveries, there are changing possibility spaces coming in through evolutionary history. In creating themselves, the creatures need possibility space, opportunity space, transformational space. Evolving into *Homo sapiens* is, we can suppose, in the possibility space of *Homo habilis* (or whatever the hypothetical ancestor). But it takes considerable imagination to find *Homo* in the possibility space of trilobites (or whatever the remote ancestor in that epoch). The creatures do have, over time, the possibility of speciating and respeciating. But it is not so clear that the creatures, in their self-actualizing, do have, or generate all by themselves, all these other kinds of selves into which they are transformed. There is enormously more out of less, and enormous space for the introduction of novelties that do not seem "up to" the faculties of the organism. One can say, if one likes, that a dinosaur is lurking in the possibility space of a microbe, or that microbes self-transform into dinosaurs, which self-transform into primates. But that really is not a claim based on anything we know about the biology or ontology of microbes.

The self-creating is more a holistic, systemic affair; it is what happens to microbes when they are challenged in their habitats and after a very long time. This requires the creation of new possibility spaces. From a God's-eye view, perhaps the possibilities are always there,[3] but we humans have no such viewpoint. We do view results and know that the possibilities both got there and got actualized, but it is quite as much an act of faith to see dinosaurs in the possibility spaces of quarks as to see dinosaurs in the possibility space of God.

---

[3] "My frame was not hidden from thee, when I was being made in secret, intricately wrought in the depths of the earth. Thy eyes beheld my unformed substance" (Psalm 139.15–16).

Looking at a pool of amino acids and seeing dinosaurs or *Homo sapiens* in them is something like looking at a pile of alphabetical letters and seeing *Hamlet*. In fact *Hamlet* is not lurking around a pile of *A – Z*'s; such a play is not within their possibility space – not until Shakespeare comes around, and in Shakespeare plus a pile of letters, *Hamlet* does lurk. By shaking a tray of printer's type, one can get a few short words, which are destroyed as soon as they are composed. If sentences begin to appear (an analogue of the long, symbolically coded DNA molecules and the polypeptide chains) and form into a poem or a short story (an analogue of the organism), one can be quite sure there are some formative, even irreversible, constraints on the sorting and shaking that are catching the upthrusts and directionally organizing them.

It hardly seems coherent to hold that nonbiological materials are randomly the more and more derandomized across long structural sequences and thus ordered up to life. That is quite as miraculous as walking on water. Something is introducing the order, and, further, something seems to be introducing layer by layer new possibilities of order, new information achieved, not just unfolding the latent order already there from the start in the setup.

Some will reply that all actual events materialize in a global possibility space, and while the former become over time, the latter does not. The possibility space is always there. There is no such thing as the creation of possibilities that were not there. New doors may open but only into rooms that previously existed, albeit unoccupied and with no furniture. One does not need to get possibilities from nowhere because there are infinite possibilities everlastingly, or at least since the Big Bang. The proof of this lies in what has subsequently happened.

But surely the possibility space of serious alternatives does enlarge and shrink. There are times of opportunity, in which taking one direction opens up new possibilities and taking another shuts them out. Along the way, new possibility space for genetic engineering is brought into the picture, and this is linked with the appearance of new information, to which we next turn.

(2) The Genesis of Information

The story becomes memorable – able to employ a memory – only with genes (or comparable predecessor molecules). The story becomes cumulative and transmissible. The fertility possibilities are a hundred times recompounded. If the DNA in the human body were uncoiled and stretched out end to end, that slender thread would reach to the sun and back over half a dozen times.[4] That conveys some idea of the astronomical amount of information soaked through the body. In nature, in the Newtonian view there were two metaphysical fundamentals: matter and energy. Einstein reduced these two to one: matter-energy. In matter in motion, there is conservation of matter, also of energy; neither can be created or destroyed, although each can take diverse forms, and one can be transformed into the other. In the biological sciences, as we have emphasized, the novelty is that matter-energy is found in living things in diverse information states. The biologists still claim two metaphysical fundamentals: matter-energy and information. Norbert Wiener insists, "Information is information, not matter or energy" (1948, p. 155).

In living things, concludes Manfred Eigen, this is "the key-word that represents the phenomenon of complexity: information. Our task is to find an algorithm, a natural law that leads to the origin of information. . . . Life is a dynamic state of matter organized by information" (1992, p. 12, p. 15). Bernd-Olaf Küppers agrees: "The problem of the origin of life is clearly basically equivalent to the problem of the origin of biological information" (1990, p. 170). George C. Williams is explicit:

> Evolutionary biologists have failed to realize that they work with two more or less incommensurable domains: that of information and that of matter. . . . Matter and information [are] two separate domains of existence, which have to be discussed separately in their own terms. The gene is a package of information, not an object. . . . Maintaining this distinction between the medium and the message is absolutely indispensable to clarity of thought about evolution. (in Brockman 1995, p. 43)

John Maynard Smith says: "Heredity is about the transmission, not of matter or energy, but of information. . . . The concept of information is central both to genetics and evolution theory" (1995, p. 28). The most spectacular thing about planet Earth, says Dawkins, is this

---

[4] Estimated from data in Orten and Neuhaus (1982, pp. 8 and 154).

"information explosion," even more remarkable than a supernova among the stars (1995, p. 145). And, adds, Klaus Dose,

> More than 30 years of experimentation on the origin of life in the fields of chemical and molecular evolution have led to a better perception of the immensity of the problem of the origin of life on Earth rather than its solution. . . . We do not actually know where the genetic information of all living cells originates. (1988, p. 348)

When sodium and chlorine are brought together under suitable circumstances, anywhere in the universe, the result will be salt. This capacity is inlaid into the atomic properties; the reaction occurs spontaneously. Energy inputs may be required for some of these results, but no information input is needed. When nitrogen, carbon, and hydrogen are brought together under suitable circumstances anywhere in the universe, with energy input, the spontaneous result may be amino acids, but it is not hemoglobin molecules or lemurs – not spontaneously. The essential characteristic of a biological molecule, contrasted with a merely physicochemical molecule, is that it contains vital information. Its conformation is functional. With the typical protein, enzyme, lipid, or carbohydrate this is structural, keyed by the coding in DNA. The coding here is information about coping in the macroscopic world that the organism inhabits. The information (in DNA) is interlocked with an information producer-processor (the organism) that can transcribe, incarnate, metabolize, and reproduce it. All such information once upon a time did not exist but came into place; this is the locus of creativity.

Nevertheless, on Earth, there is this result during evolutionary history. The result involves significant achievements in cybernetic creativity, essentially incremental gains in information that have been conserved and elaborated over evolutionary history. The know-how, so to speak, to make salt is already in the sodium and chlorine, but the know-how to make hemoglobin molecules and lemurs is not secretly coded in the carbon, hydrogen, and nitrogen. Life is a local countercurrent to entropy, an energetic fight uphill in a world that typically moves thermodynamically downhill (despite some negentropic eddies, and despite irreversible thermodynamics). Thermodynamics need be nowhere violated, because there is a steady "downhill" flow of energy, as energy is irradiated onto Earth from the sun, and, eventually, reradiated into space.

But some of this energy comes to pump a long route uphill. This is something like an old-fashioned hydraulic ram, where the main downstream flow is used to pump a domestic water supply a hundred yards uphill through a pipe to a farmhouse – except of course that the ram pump is deliberately engineered and the "life pump" spontaneously assembled itself as an open cybernetic system several thousand times more complex and several billion years long. Life is a river that runs uphill, and even if it nowhere runs uphill very steeply (if we look at its incremental assembly bit by bit), the river as a whole runs far uphill, and each living creature in the stream is quite highly ordered. Some forces are present, some force, some Force! that sucks order in superseding steps out of disorder. Organisms must be constructed along a long negentropic pathway. This requires the continual introduction of information not previously present.

The central dogma of molecular and evolutionary biology is that random variations are introduced into the replication of this information, that rarely such variations prove beneficial in the sense that they improve performance with the result that more offspring are produced, and that such variations in result increase proportionately in the gene pool. The classical view emphasizes that such variations occur at random and without regard to the needs of the organisms. Contemporary genetics is increasingly inclined to interpret this process as a kind of information search using random variations in problem solving and to see the search space as more constrained by the prior achievements of the organism; nevertheless the random element remains prominent. Here is where possibilities lie and where actual novelties are generated out of such possibilities.

John Maynard Smith and Eörs Szathmáry analyze "the major transitions in evolution" with the resulting complexity, asking "how and why this complexity has increased in the course of evolution." "Our thesis is that the increase has depended on a small number of major transitions in the way in which genetic information is transmitted between generations." Critical innovations have included the origin of the genetic code itself, the origin of eukaryotes from prokaryotes, meiotic sex, multicellular life, animal societies, and language, especially human language. But, contrary to de Duve, Eigen, Calvin, Kauffman, or Cech, they find "no reason to regard the unique transitions as the inevitable result of some general law"; to the contrary, these events might not have happened at all (1995, p. 3). So what makes the critical difference in evolutionary history is increase in the information possibility space, which is not something inherent in the precursor materials, nor in the evolutionary system, nor something for which biology has an evident explanation, although these events, when they happen, are retrospectively interpretable in biological categories. The biological explanation is modestly incomplete, recognizing the importance of the genesis of new information channels.

The philosophical, metaphysical, and theological challenge, left over after the current scientific accounts, is the query what is the most adequate account of the origin of these information channels and the genetic information thereby discovered. In the course of evolutionary history, one would be disturbed to find matter or energy spontaneously created, but here is information floating in from nowhere. For the lack of better explanations, the usual turn here is simply to conclude that nature is self-organizing (autopoiesis), though, since no "self" is present, this is better termed spontaneously organizing. An autopoietic process can be just a name, like "soporific" tendencies, used to label the mysterious genesis of more out of less, a seemingly scientific name that is really a sort of mystic chant over a miraculously fertile universe.

What is inadequately recognized in the "self-organizing" accounts is that, though no new matter or energy is needed for such spontaneous organization, new information is needed in enormous amounts and that one cannot just let this information float in from nowhere. Over evolutionary history, something is going on "over the heads" of any and all of the local, individual organisms. More comes from less, again and again. A more plausible explanation is that, complementing the self-organizing, there is a Ground of Information, or an Ambience of Information, otherwise known as God.

(3) The Genesis of Value

Another way of interpreting this genesis of information arises from looking at its result: the generation, transmission, and deepening of values. Scientists and philosophers have been much exercised about the generation of values, about how an *ought* comes out of an *is*, but it seems pretty much fact of the matter that, over evolutionary history, values have been generated, startling though this may also be. "Survival value" figures large in evolutionary theory. Something is always dying, and something is always living on. For all the struggle, violence, and transition, there is abiding value. The question is not whether Earth is a well-designed paradise for all its inhabitants, nor whether it was a former paradise from which humans were anciently expelled. The question is whether it is a place of significant value achievement.

Scientists have sometimes tried to portray nature as a valueless place, and that can seem so in the emptiness of outer space, or the frozen wastes of Antarctica, or the sands of the Sahara. But where there is life, value is always at stake. Once humans might have thought that even biological nature is valueless, with value lighting up as, and only as, humans take an interest in what is going on. But such anthropocentrism has become increasingly incredible in Darwin's century. The same evolutionary science that discovered nature red in tooth and claw discovered the value in teeth and claws, the vitality flowing in the blood, the world as a sphere of the contest of values, generated in this perpetual contest. These biological scientists and their evolutionary and ecological sciences are a witness to the genesis of values, in the biodiversity they describe and wish to protect, in the insights into human origins and possibilities they seek to gain, in the morality they urge, at the same time that their theory is incompetent to warrant, support, or appraise such values.

Evaluating Earth, the appropriate category is not *moral* goodness, for there are no moral agents in nature; the appropriate category is some one or more kinds of *nonmoral* goodness, better called its *value*, its worth. One must evaluate phenomena such as the achievement of diversity and complexity out of simplicity; the discovery of sentience, cognition, experience; the mixture of order and contingency, of autonomy and interdependence. This epic of vital ascent is the rare expression point, on Earth, of a peculiar power in cosmic nature. Something divine is embodied (incarnate) in the story. Any struggle and suffering can only be interpreted in the context of such creativity.

According to a long dominant paradigm, there is no value without an experiencing valuer, just as there are no thoughts without a thinker, no percepts without a perceiver, no deeds without a doer, no targets without an aimer. Valuing is felt preferring by human choosers. Possibly, extending this paradigm, sentient animals may also value, using their teeth and claws, or maybe even plants can value as they, nonconsciously, defend their lives with thorns and propagate their kind with seeds. But, in an evolutionary account, the value story becomes systemic, more holistic, ecological, global. Earth is a value-generating system, value-genic, valuable, value-able, that is, able to generate values that are widely "distributed," "dispersed," "allocated," "proliferated," "divided," "multiplied," "transmitted," "recycled," and "shared" over the face of the Earth.

It is true that humans are the only evaluators who can reflect about what is going on at this global scale, who can evaluate what has happened in natural and cultural history, who can deliberate about what they ought to do conserving these events. When humans do this, they must set up the scales, and humans are the measurers of things. Animals, organisms, species, ecosystems, Earth cannot teach us how to do this evaluating. But they can display what it is that is to be valued and evaluated. The axiological scales we construct do not constitute the value, any more than the scientific scales we erect create what we thereby measure.

Humans are not so much lighting up value in a merely potentially valuable world as they are psychologically joining ongoing planetary natural history in which there is value wherever there is positive creativity. Although such creativity can be present in subjects with their interests and preferences, it can also be present objectively in living organisms with their lives defended, and in species that defend an identity over time, and in systems that are self-organizing and that project storied achievements. The valuing subject in an otherwise valueless world is an insufficient premise for the experienced conclusions of those who value natural history. Conversion to an evolutionary and ecological view seems truer to world experience, more logically compelling, better informed.

From this more objective viewpoint, there is something subjective, something philosophically naive, and even something hazardous in a time of ecological crisis, for humans to continue to live (as in an age of science they have often done) as though nature were valueless and everything previously generated in natural history were only to be evaluated relative to its potential to produce benefit for humans. When Earth's most complex product, *Homo sapiens*, becomes intelligent enough to reflect over this earthy wonderland, everyone is left stuttering about the mixtures of accident and necessity out of which we have evolved. But nobody has much doubt that this is, recalling the way that the astronauts phrased it, "a small pearl in a thick sea of black mystery" (Mitchell), "to be treasured and nurtured, something precious that *must* endure" (Collins; Section 1[1]). Almost as if to dispute Wilson's claim that nothing Earth-bound can be transcendent, Mitchell adds, "My view of our planet was a glimpse of divinity" (Kelley 1988, at photograph 52).

Those are astronauts, not biologists, but what they see is the home planet, the living planet in all its startling possibilities, of which evolutionary history is the most indisputable evidence. We have earlier heard Edward Wilson celebrating that biodiversity, finding it in its own way "miraculous,"[5] and urging its conservation, even when he could find no such divinity. Here again is the fertility, which generates religion. Earth is dirt, all dirt, but we find revealed what dirt can do when it is self-organizing under suitable conditions with water and solar illumination. We will not be valuing Earth objectively until we appreciate this marvelous natural history.

Life persists because it is provided for in the ecological Earth system. Earth is a kind of providing ground, where the life epic is lived on in the midst of its perpetual perishing, life arriving and struggling through to something higher. One may think, as we near a conclusion, that biology produces many doubts; here are two more: I doubt whether one can take biology seriously, the long epic of life on Earth, the prolific fecundity that surrounds us on this planet, without a respect for life, and the line between respect for life and reverence for life is one that I doubt that you can always recognize. If anything at all on Earth is sacred, it must be this enthralling generativity that characterizes our home planet. "The world is sacred." That is the conclusion of even so resolute a naturalist as Daniel Dennett, which not even Darwin's "universal acid" can dissolve, dissolve God though this acid can (1995, pp. 520–521). So the secular – this present, empirical epoch, this phenomenal world, studied by science – does not eliminate the sacred after all; to the contrary, the secular evolves into the sacred. If there is any holy ground, any land of promise, this promising Earth is it.

But then why not say that here, if anywhere, is the brooding Spirit of God? One needs an adequate explanation for generating the sacred out of the secular. Indeed, why not even go on to say that this genesis of value is the genesis of grace, since the root idea in "grace" (Latin: *gratia*) is pleasing, favorable, praiseworthy; essentially, again, the idea of something valuable, now also a given. In this genesis, nature is a sequence of gifts; we are given what has "sprung forth"

---

[5] "The flower in the crannied wall – it *is* a miracle" (Wilson 1992, p. 345).

and find that, in this springing forth, values are created. Whatever else has happened, there has been the genesis of values; each of us is a remarkable instance of that.

"The essence of religion," said Harald Höffding, "consists in the conviction that value will be preserved" (1906, p. 14). That helps us to understand Mayr's remark that most biologists are religious. If one finds a world in which value is given and persists over time, one has a religious assignment. A central function of religion is the conservation of value, and value generated and conserved is the first fact of natural history, as well as the principal task of culture. Frederick Ferré defines religion, "One's religion . . . is one's way of valuing most intensively and most comprehensively" (1970, p. 11). At the metaphysical level, science neither describes nor evaluates the genesis of value adequately, although the descriptions of biological science – those of evolutionary history eventuating in cultural history – present an account that demands evaluating, intensively and comprehensively. Religion is about the finding, creating, saving, redeeming of such persisting sacred value in the world. In this sense, whatever the quarrels between religion and biology, there is nothing ungodly about a world in which values persist in the midst of their perpetual perishing. That is as near as Earthlings can come to an ultimate concern; such benefit, such "blessing," is where, on Earth, the Ultimate might be incarnate.

(4) Detecting the Transcendent

The universe existed for ten or fifteen billion years without any biological information present, so far as we know. The divine presence in that epoch will need to be found in the setup, in the fine-tuned universe, or, along the way, in, with, and under the physics, astrophysics, and chemistry. Such presence continues during the biological epoch on Earth. But now the creativity is more notably that generating the information vital to life. Again, one can appeal to the set-up. In our corner of the universe, the interplay of matter and energy accumulated into a solar system with one lucky planet. Perhaps there are other such planets; we do not know whether they are common or rare. But at least there is this one.

Located at a felicitous distance from the sun, Earth has liquid water; atmosphere; a suitable mix of elements, compounds, minerals; and an ample supply of energy. Radioactivity deep within the Earth produces enough heat to keep its crust constantly mobile in counter-action with erosional forces, and the interplay of such forces generates and regenerates landscapes and seas – mountains, canyons, rivers, plains, islands, volcanoes, estuaries, continental shelves. Geo-chemistry is as relevant as chemistry. The properties of the elements – hydrogen, carbon, and so on – are necessary but not sufficient. The properties of the Earth system, a kind of cooking pot, are also necessary, and, together with the physicochemical properties, perhaps these are sufficient to make life probable, even inevitable.

Detecting the transcendent asks whether God underlies that setup. God lies in, with, and under the forces that created Earth as the home (the ecosystem) that could produce all those myriads of kinds. God, the Ground of the Universe, is also the Good Fortune of the Planet. "Let the earth bring forth living creatures according to their kinds" (Genesis 1.24). The Earth-system does prove to be prolife; the story goes from zero to five million species in five billion years, passing through perhaps five billion species that have come and gone en route. The setup, first on cosmological levels and later on planetary levels, mixes chance and order in creative ways. If, once, there was a primitive planetary environment in which the formation of living things had a high probability, for such living things to become actual would require not so much interference by a supernatural agency as the recognition of a marvelous endowment of matter with a propensity toward life. So the molecular biologists were earlier arguing. Such a natural performance could be congenially seen, at a deeper level, as the divine creativity.

But one still has to give an account of the information appearing ex nihilo, that is, where no such information was present before. One may indeed need a fortunate endowment of matter with a life propensity (helped perhaps by the anthropic principle in astrophysics) and at the same time still need something to superintend the possibilities during evolutionary history. That there are complementary explanations does not always mean that one is superfluous. Here one can posit God as a countercurrent to entropy, a sort of biogravity that lures life upward. God would not do anything in particular but be the background, autopoietic force energizing all the particulars. The particulars would be the discoveries of the autonomous individuals. God would be the lift-up (more than the setup) that elevates the creatures along their paths of cybernetic and storied achievement. God introduces new possibility spaces all along the way. What theologians once termed an established order of creation is rather an order that dynamically creates, an order for creating.

One should posit, says Daniel Dennett, "cranes," not "skyhooks," for the building up of evolutionary history (1995, pp. 73–80). That contrast of metaphor seems initially persuasive, appealing to causes more natural than supernatural, more immanent than transcendent. When we pinpoint the issue, however – what account to give of this remarkable negentropic, cybernetic self-organizing that characterizes the life story on Earth – the metaphor becomes more pejoratively rhetorical than analytically penetrating. There is the repeated discovery of information how to redirect the downhill flow of energy upward for the construction of ever more advanced, higher forms of life, built on and supported by the lower forms. Up and down are rather local conditions (down, up a few miles); it does not matter much which direction we imagine this help as coming from – east or west, from the right or left, from below or above, high or deep, immanence or transcendence, skyhooks or cranes. The Hebrew metaphor was that one needs "wind" as well as "dirt." The current metaphor is that one needs "information" as well as "matter" and "energy."

Stripped of the rhetoric, what the "skyhook" metaphor means, Dennett says, is explanations that are more "mindlike," and the "cranes" metaphor posits "mindless, motiveless mechanicity." Dennett holds that Darwinian science, extrapolated philosophically, has discovered cranes upon cranes "all the way down" and building up and up with "creative genius." "There is simply no denying the breathtaking brilliance of the designs to be found in nature" (1995, pp. 76, 155, and 74). But if the secret of such creativity is information possibilities opening up and information searched and gained, then the kind of explanation needed can as plausibly be said to be mindlike as mindless mechanicity.

One might look to the potential deep in matter, "cranes all the way down." There is a kind of bottomless bootstrapping, as if lifting oneself up and up by one's own bootstraps were not remarkable, matter lofting itself up into mind. Such cranes, piling up higher and higher, are still pretty "super," quite imposing with their endless superimposing of one achievement on another. One can just as well look to some destiny toward which such matter is animated and inspired (skyhooks). Even after an infinite regress of cranes, or a regress ending in nothing at all, or in informationless matter-energy, or in a big bang, one might not find that explanations are over. The issue is where the information comes from by which matter and energy become so superimposingly informed across evolutionary history that this brilliant, "sacred" (Dennett) output arises from a beginning in mindless chaos; how "out of next to nothing the world we know and love created itself" (p. 185).

In this "world of propensities," concludes Karl Popper, the "inherently creative" process with its "staggering" biodiversity is neither mechanistic nor deterministic. "This was a process in which both *accidents* and *preferences*, preferences of the organisms for certain possibilities, were mixed: the organisms were in search of a better world. Here the preferred possibilities were, indeed, allurements" (1990, pp. 26 and 20). Cranes or skyhooks, evolutionary development is "attracted to" (in the current "chaos" metaphor) cumulating achievements in both diversity and complexity, and this attraction needs explanation. Attractors, or, at a more metaphysical level, even an Attractor, seem quite rational explanations.[6]

Returning to the metaphor of the alphabet and Shakespeare, the question is whether, in the introduction of these possibilities, one needs an author as well as an alphabet. What is required to get *Hamlet* is a great deal of information input into the letters. Perhaps the alphabet–author analogy is flawed. That analogy places all the creativity in the author working with an inert alphabet. One needs rather to posit a self-organizing alphabet, and a maker to start up and sustain such a self-organizing alphabet. Still, the elemental materials are not evidently an alphabet from the beginning; they have to be taken over for alphabetic functions. Some story has to be generated with these materials-become-alphabet. That requires information input

---

[6] "To me the most fascinating property of the process of evolution is its uncanny capacity to mirror *some* properties of the human mind (the intelligent Artificer) while being bereft of others" (Dennett 1987, p. 299). It seems important to Dennett that the design is a mirage. Or, more accurately, the design isn't a mirage, for there is a designing system, but that there is a Designer of the designing system is a mirage. One needs no supernature, and the evidence for this is that we can plunge into subnature, and subsubnature, and subsubsubnature, simplifying all the way down until there is nothing at all. Although creativity is forbidden from above, it is welcomed from below. But set aside the above-below imagery, still the "attraction" to something out of chaos, the "genesis" of something out of nothing, of more out of less – such brute fact remains as evident as ever, and as demanding of explanation.

into such alphabetic materials, or, if not "input," information generation in some way or other. The skeptic will protest that there is no need for an author at all. One can have law without lawgivers, history without historians, creativity without creators, information without an informer, and stories without storytellers.

Change the analogy: the elements are more like "seeds" than "letters." The root meaning of "nature" is "generating," and nature has all these possibilities "seeded" into it. The problem with such a model is that we now know what is in seeds as the secret of their possibilities – information – and there is no such information inside amino acids, much less hydrogen and carbon atoms, much less electrons and protons. The creation of matter, energy, law, history, stories, of all the information that generates nature, to say nothing of culture, does need an adequate explanation: some sources, source, or Source competent for such creativity. Seeds need a source. In the materializing of the quantum states, bubbling up from below; in the compositions of prebiotic molecules; in the genetic mutations, there are selective principles at work, as well as stabilities and regularities, forming and in-forming these materials, which principles order and order up the story.

This portrays a loose teleology, a soft concept of creation, one that permits genuine, though not ultimate, integrity and autonomy in the creatures. We have in the life adventure an interaction phenomenon, where a prolife principle is overseeing the affairs of matter. The divine spirit is the giver of life, pervasively present over the millennia. God is the atmosphere of possibilities, the metaphysical environment in, with, and under first the natural and later also the cultural environment, luring the Earthen histories upslope. God orchestrates such self-organizing, steadily elevating the possibilities, making for storied achievements, enriching the values generated.

God could sometimes also be in the details. The general picture is not one of divine micromanagement; rather of secular integrity and creaturely self-organizing. The extent to which divine inspiration enters into particulars might be difficult to know, especially if God operated with the resolve to maximize the creaturely autonomy, to prompt rather than to command. Dennett concedes, for example, that no Martian biologists, examining "a laying hen, a Pekingese dog, a barn swallow, and a cheetah," could prove, simply from an examination of the organisms, that the former were the product of deliberate, engineered artificial selection, as well as of natural selection, and the latter were the product of natural selection only. "If the engineers chose to conceal their interventions as best they could . . . there may be no foolproof marks of natural (as opposed to artificial) selection" (1987, pp. 284–285).

If there has been divine selection, this will not be detectable as any gap in or perforation of the natural order; it might be detectable in the resulting genesis, or creativity. If the roulette wheels at Las Vegas spin at random most of the time, but once a year God loaded the dice, that would be difficult to detect. Chance is an effective mask for the divine action. Still, God could be slipping information into the world. One might suspect such divine presence if the resulting story, in the lotteries of natural history, produced the epic adventures that have in fact actually managed to happen. An "information explosion" on our Earth, rare in the universe, might be a clue that "inspiration" is taking place.

Perhaps it is a mistake to look for God in the particulars of information discovery. God does not intervene as a causal force in the world, not at least of such kind as science can detect. "God" is not among the entries to be found in the index of a biology text. God perennially underlies the causal forces in the world, and God gives meaning to the world, which science is incompetent to evaluate. That does require the introduction of channels for information, and information in those channels, which arrives in the particulars of genetic trial and error. Such information is not a mere cause, not in any physicochemical sense, but a novel "cause" that puts meanings into events, that generates all the richness of evolutionary history.

God is an explanatory dimension[7] for which contemporary biology leaves ample space, as we have seen as biologists stutter over the origins of the information that generates complexity and diversity, over any selection for progress, over what to make of randomness, over the introduction of possibilities. If one adds the desire of a Creator not so much to conceal such complementing selective activity as to optimize the integrity, autonomy, and self-creativity of the creatures – letting them do their thing, generating and testing, discarding what does not work and keeping what does – with divine coaching on occasion, then a conclusion that there is a divine presence underneath natural history becomes as plausible as that there is

---

[7] A cause in the Aristotelian, though not the scientific sense.

not. The question becomes not so much a matter of conclusive proof as of warranted faith.

There once was a causal chain that led to vertebrae in animals, where there were none before, an incremental chain no doubt, but still a chain by which the novelty of the vertebral column was introduced on Earth. Such a chain is constructed with the emergence of more and more information; this information, coded in DNA, informs the matter and energy so as to build the vertebral cord. The cord is constructed because it has a value (a significance, here a precursor of meaning) to the organism. It makes possible the diverse species of life that the vertebrate animals defend. Continuing the development of the endoskeleton, it makes possible larger animals with mobility, flexibility, integrated neural control. When such construction of valuable biodiversity has gone on for millennia, the epic suggests mysterious powers that signal the divine presence.

The question, the biologists will say, is of the selective forces. Yes, but the answer comes, partly at least, from seeing the results, with ever more emerging from what is earlier less and less. One seeking to detect the divine inspiration will notice how there are occasions – seasons, contexts, events, episodes, whatever they are called – during which critical information emerges in the world, breakthroughs, as it were, incremental and cumulative though these can also be. This will be true in culture, perhaps the inspiration that underlies the Ten Commandments or the Sermon on the Mount. It can as well be true in nature, in some inspiration that first animates matter and energy into life; or launches replication and genetic coding, or eukaryotes, or multicellular life, or sexuality; or energizes life with mitochondria and chloroplasts, or glycolysis and the citric acid cycle; or moves life onto land; or invents animal societies or acquired learning; or endows life with mind; and inspires culture, ethics, religion, science.

The skeptic's reply is always to emphasize that evolution is not elegant. It is wasteful, blundering, struggling. Evolution works with what is at hand and makes something new out of it. The creatures stumble around, and if there is a God who "intervenes," God ought to do better than that. There is only a "blind watchmaker" (Dawkins 1986). Still, consider again the remarkable results, and the providence appropriate to a God who celebrates an Earth history, who inspires self-creativity. The word "design" nowhere occurs in Genesis,[8]

---

[8] The word "design" also seldom occurs in this book, by design.

though the concept of creativity pervades the opening chapters. There is divine fiat, divine doing, but the mode is an empowering permission that places productive autonomy in the creation. It is not that there is no "watchmaker"; there is no "watch." Looking for one frames the problem the wrong way. There are species well adapted for problem solving, ever more informed in their self-actualizing. The watchmaker metaphor seems blind to the problem that here needs to be solved: that informationless matter–energy is a splendid information maker. Biologists cannot deny this creativity; indeed, better than anyone else biologists know that Earth has brought forth the natural kinds, prolifically, exuberantly over the millennia, and that enormous amounts of information are required to do this.

The achievements of evolution do not have to be optimal to be valuable, and if a reason that they are not optimal is that they had to be reached historically along story lines, then we rejoice in this richer creativity. History plus value as storied achievement in creatures with their own integrity is better than optimum value without history, autonomy, or adventure in superbly designed marionettes. That is beauty and elegance of a more sophisticated form, as in the fauna and flora of an ancient forest. The elegance of the thirty-two crystal classes is not to be confused with the grace of life renewed in the midst of its perpetual perishing, generating diversity and complexity, repeatedly struggling through to something higher, a response to the brooding winds of the Spirit moving over the face of these Earthen waters.

References

Abele, Lawrence G., and Sandra Gilchrist, 1977. "Homosexual Rape and Sexual Selection in Acanthocephalan Worms," *Science* 197:81–83.

Alexander, Richard D., 1975. "The Search for a General Theory of Behavior," *Behavioral Sciences* 20:77–100.

Alexander, Richard D., 1979. *Darwinism and Human Affairs*. Seattle: University of Washington Press.

Alexander, Richard D., 1987. *The Biology of Moral Systems*. New York: Aldine de Gruyter.

Alexander, Richard D., 1993. "Biological Considerations in the Analysis of Morality." Pages 163–196 in Nitecki and Nitecki, eds., *Evolutionary Ethics*. Albany: University of New York Press.

Alumets, J., R. Hakanson, F. Sundler, and J. Thorell, 1979. "Neuronal Localisation of Immunoreactive Enkephalin and Beta-Endorphin in the Earthworm," *Nature* 279:805–806.

Axelrod, Robert, 1984. *The Evolution of Cooperation*. New York: Basic Books.

Axelrod, Robert, and Douglas Dion, 1988. "The Further Evolution of Cooperation," *Science* 242:1385–1390.

Axelrod, Robert, and W. D. Hamilton, 1981. "The Evolution of Cooperation," *Science* 211:1390–1396.

Ayala, Francisco, J., 1974. "The Concept of Biological Progress." Pages 339–355 in Francisco Jose Ayala and Theodosius Dobzhansky, eds., *Studies in the Philosophy of Biology*. New York: Macmillan.

Ayala, Francisco J., 1978. "The Mechanisms of Evolution," *Scientific American* 239, no. 3, September: 56–69.

Ayala, Francisco, J., 1982. *Population and Evolutionary Genetics: A Primer*. Menlo Park, CA: Benjamin-Cummings.

Ayala, Francisco, J., 1988. "Can 'Progress' Be Defined as a Biological Concept?" Pages 75–96 in Matthew H. Nitecki, ed., *Evolutionary Progress*. Chicago: University of Chicago Press.

Ayala, Francisco, J., 1995. "The Difference of Being Human: Ethical Behavior as an Evolutionary Byproduct." Pages 117–135 in Holmes Rolston, III, ed., *Biology, Ethics, and the Origins of Life*. Boston: Jones & Bartlett.

Bainton, Dorothy F., 1981. "The Discovery of Lysosomes," *Journal of Cell Biology* 91, no. 3, pt. 2:66s–76s.

References

Barash, David, 1977. "Sociobiology of Rape in Mallards (*Anas platyrhynchos*): Responses of the Mated Male," *Science* 197:788–789.

Barash, David, 1979. *The Whisperings Within*. New York: Harper & Row.

Barkow, Jerome H., 1989. "Overview: Evolved Constraints on Cultural Evolution," *Ethology and Sociobiology* 10:1–10.

Barnett, S. A., 1988. *Biology and Freedom: An Essay on the Implications of Human Ethology*. Cambridge: Cambridge University Press.

Barrow, John D., and Frank J. Tipler, 1986. *The Anthropic Cosmological Principle*. New York: Oxford University Press.

Batson, C. Daniel, and Laura L. Shaw, 1991. "Evidence for Altruism: Toward a Pluralism of Prosocial Motives," *Psychological Inquiry* 2:107–122.

Baumann, Donald, J., Robert B. Cialdini, and Douglas T. Kenrick, 1981. "Altruism as Hedonism: Helping and Self-Gratification as Equivalent Responses," *Journal of Personality and Social Psychology* 40:1039–1046.

Bell, Graham, 1982. *The Masterpiece of Nature: The Evolution and Genetics of Sexuality*. Berkeley: University of California Press.

Bell, Graham, and A. Burt, 1990. "B-Chromosomes: Germ-Line Parasites Which Induce Changes in Host Recombination," *Parasitology* 100: S19–S26.

Benacerraf, Baruj, 1991. "When All Is Said and Done . . ." *Annual Review of Immunology* 9:1–26.

Benton, M. J. 1995. "Diversification and Extinction in the History of Life," *Science* 268:52–58.

Beveridge, W. I. B., 1957. *The Art of Scientific Investigation*, revised edition. New York: W. W. Norton.

Bock, Kenneth, 1980. *Human Nature and History: A Response to Sociobiology*. New York: Columbia University Press.

Boggess, Jane, 1984. "Infant Killing and Male Reproductive Strategies" in Langurs (*Presbytis entellus*)." Pages 283–310 in Glenn Hausfater and Sara Blaffer Hrdy, eds., *Infanticide: Comparative and Evolutionary Perspectives*. New York: Aldine.

Bonner, John Tyler, 1988. *The Evolution of Complexity by Means of Natural Selection*. Princeton, NJ: Princeton University Press.

Boyd, Robert, and Peter J. Richerson, 1985. *Culture and the Evolutionary Process*. Chicago: University of Chicago Press.

Bradie, Michael, 1986. "Assessing Evolutionary Epistemology." *Biology and Philosophy* 1:401–459.

Bradie, Michael, 1994. *The Secret Chain: Evolution and Ethics*. Albany: State University of New York Press.

Breuer, Georg, 1982. *Sociobiology and the Human Dimension*. Cambridge: Cambridge University Press.

Brisson, A., and P. N. T. Unwin, 1985. "Quaternary Structure of the Acetylcholine Receptor," *Nature* 315 (6 June):474–477.

Brockman, John, 1995. *The Third Culture*. New York: Simon & Schuster.

Brown, Donald E., 1991. *Human Universals*. Philadelphia: Temple University Press.

Bueler, Lois E., 1973. *Wild Dogs of the World*. New York: Stein & Day.

Burnham, C. R. and J. T. Stout, 1983. "Linkage and Spore Abortion in Chromosomal Interchanges in *Datura stramonium* L. Megaspore Competition?" *American Naturalist* 121:385–394.

Buss, David, 1989. "Sex Differences in Human Mate Preferences: Evolutionary Hypotheses Tested in 37 Cultures," *Behavioral and Brain Sciences* 12:1–49.

Buss, David, et al., 1990. "International Preferences in Selecting Mates: A Study of 37 Cultures," *Journal of Cross-Cultural Psychology* 21: 5–47.

Byrne, Richard, 1995. *The Thinking Ape: Evolutionary Origins of Intelligence*. Oxford: Oxford University Press.

Cairns, John, Julie Overbaugh, and Stephan Miller, 1988. "The Origin of Mutants," *Nature* 335:142–145.

Calvin, Melvin, 1975. "Chemical Evolution," *American Scientist* 63:169–177.

Campbell, Donald T., 1974. "Evolutionary Epistemology." Pages 413–463 in P. A. Schilpp, ed., *The Philosophy of Karl Popper*, Book 1. LaSalle, IL: Open Court.

Campbell, Donald T., 1975. "On the Conflicts Between Biological and Social Evolution and Between Psychology and Moral Tradition," *American Psychologist* 30:1103–1126.

Campbell, Donald T., 1991. "A Naturalistic Theory of Archaic Moral Orders," *Zygon* 26:91–114.

Campbell, Jeremy, 1982. *Grammatical Man: Information, Entropy, Language, and Life*. New York: Simon & Schuster.

Campbell, John H., 1983. "Evolving Concepts of Multigene Families." Pages 401–417 in *Isozymes: Current Topics in Biological and Medical Research*. Volume 10. *Genetics and Evolution*. New York: Alan R. Liss.

Campbell, Mary K., 1991. *Biochemistry*. Philadelphia: Saunders.

Cavalli-Sforza, L. L., and M. W. Feldman, 1981. *Cultural Transmission and Evolution: A Quantitative Approach*. Princeton, NJ: Princeton University Press.

Cech, Thomas R., 1995. "The Origin of Life and the Value of Life." Pages 15–37 in Holmes Rolston, III, ed., *Biology, Ethics, and the Origins of Life*. Boston: Jones & Bartlett.

Cheney, Dorothy L., and Robert M. Seyfarth, 1990. *How Monkeys See the World*. Chicago: University of Chicago Press.

Chomsky, Noam, 1986. *Knowledge of Language: Its Nature, Origin, and Use*. New York: Praeger Scientific.

Collins, Michael, 1980. "Foreword," in Roy A. Gallant, *Our Universe*. Washington, DC: National Geographic Society.

Cosmides, Leda, John Tooby, and Jerome H. Barkow, 1992. "Introduction: Evolutionary Psychology and Conceptual Integration." Pages 3–15 in Jerome H. Barkow, Leda Cosmides, and John Tooby, eds., *The Adapted Mind: Evolutionary Psychology and the Generation of Culture*. New York: Oxford University Press.

Cram, Donald J., 1988. "The Design of Molecular Hosts, Guests, and Their Complexes," *Science* 240:760–767.

Crowley, Philip H., and R. Craig Sargent, 1996. "Whence Tit-for-Tat?" *Evolutionary Ecology* 10:499–516.

Darwin, Charles, 1858. "Letter to J. D. Hooker." Pages 114–115 in Francis Darwin and A. C. Seward, eds., *More Letters of Charles Darwin*, Volume 1. London: John Murray, 1903.

Darwin, Charles, [1859], 1964. *On the Origin of Species*. Cambridge, MA: Harvard University Press (First Edition).

Darwin, Charles, [1872a] 1962. *The Origin of Species*. New York: Collier Macmillan (Sixth Edition).

Darwin, Charles, [1872b] 1903. "Letter to Alpheus Hyatt." Pages 341–344 in Francis Darwin and A. C. Seward, eds., *More Letters of Charles Darwin*, Volume 1. New York: D. Appleton.

Darwin, Charles, [1874] 1895. *The Descent of Man*. New York: D. Appleton (Second Edition).

Darwin, Charles, [1881], 1897. "Letter to W. Graham." Pages 284–286 in Francis Darwin, ed., *The Life and Letters of Charles Darwin*, Volume 1. New York: D. Appleton.

Davis, Lawrence, ed., 1987. *Genetic Algorithms and Simulated Annealing*. Los Altos, CA: Morgan Kaufmann Publishers.

Dawkins, Richard, 1976. *The Selfish Gene*. New York: Oxford University Press.

Dawkins, Richard, 1982. "Replicators and Vehicles." Pages 45–64 in King's College Sociobiology Group, Cambridge, *Current Problems in Sociobiology*. Cambridge: Cambridge University Press.

Dawkins, Richard, 1983. *The Extended Phenotype*. New York: Oxford University Press.

Dawkins, Richard, 1986. *The Blind Watchmaker*. New York: W. W. Norton.

Dawkins, Richard, 1989. *The Selfish Gene*, new edition. New York: Oxford University Press.

Dawkins, Richard, 1995. *River Out of Eden: A Darwinian View of Life*. New York: Basic Books, HarperCollins.

de Beer, Gavin, 1962. *Reflections of a Darwinian*. London: Thomas Nelson and Sons.

de Duve, Christian, 1995. *Vital Dust: Life as a Cosmic Imperative*. New York: Basic Books.

Delbrück, Max, 1966. "A Physicist Looks at Biology." Pages 9–22 in John Cairns, Gunther S. Stent, and James D. Watson, eds., *Phage and the Origins of Molecular Biology*. Cold Spring Harbor, NY: Cold Spring Harbor Laboratory of Quantitative Biology.

Delbrück, Max, 1978. "Mind from Matter?" *American Scholar* 47:339–353.

Delbrück, Max, 1986. *Mind from Matter? An Essay on Evolutionary Epistemology*. Palo Alto, CA: Blackwell Scientific.

Dennett, Daniel C., 1983. "Intentional Systems in Cognitive Ethology: The 'Panglossian Paradigm' Defended," *The Behavioral and Brain Sciences* 6: 343–390.

Dennett, Daniel C., 1987. *The Intentional Stance*. Cambridge, MA: MIT Press.

Dennett, Daniel C., 1995. *Darwin's Dangerous Idea*. New York: Simon & Schuster.

DeVore, Irven, and Scott Morris, 1977. "The New Science of Genetic Self-Interest," *Psychology Today* 10, no. 9 (February):42–51, 84–88.

de Waal, Frans, 1989. "Food Sharing and Reciprocal Obligations Among Chimpanzees," *Journal of Human Evolution* 18:433–459.

de Waal, Frans, 1996. *Good Natured: The Origins of Right and Wrong in Humans and Other Animals.* Cambridge, MA: Harvard University Press.

Diamond, Jared M., 1992. *The Third Chimpanzee: The Evolution and Future of the Human Animal.* New York: HarperCollins.

Dickemann, Mildred, 1979. "The Ecology of Mating Systems in Hypergynous Dowry Societies," *Social Science Information* 18:163–195.

Dickerson, R. E., 1971. "The Structure of Cytochrome *c* and the Rates of Molecular Evolution," *Journal of Molecular Evolution* 1:26–45.

Dobzhansky, Theodosius, 1956. *The Biological Basis of Human Freedom.* New York: Columbia University Press.

Dobzhansky, Theodosius, 1963. "Anthropology and the Natural Sciences: The Problem of Human Evolution," *Current Anthropology* 4:138, 146–148.

Dobzhansky, Theodosius, 1974. "Chance and Creativity in Evolution." Pages 307–338 in Francisco Jose Ayala and Theodosius Dobzhansky, eds., *Studies in the Philosophy of Biology.* New York: Macmillan.

Donald, Merlin, 1991. *Origins of the Modern Mind.* Cambridge, MA: Harvard University Press.

Dose, Klaus, 1988. "The Origin of Life: More Questions Than Answers," *Interdisciplinary Science Reviews* 13:348–356.

Dovidio, John F., 1984. "Helping Behavior and Altruism: An Empirical and Conceptual Overview." Pages 361–427 in L. Berkowitz, ed., *Advances in Experimental Social Psychology*, vol 17. New York: Academic Press.

Drake, John W., 1991. "Spontaneous Mutation," *Annual Review of Genetics* 25: 125–146.

Dugatkin, Lee Alan, 1997a. *Cooperation Among Animals: An Evolutionary Perspective.* New York: Oxford University Press.

Dugatkin, Lee Alan, 1997b. "The Evolution of Cooperation," *BioScience* 47:355–362.

Durham, William H., 1991. *Coevolution: Genes, Culture, and Human Diversity.* Stanford, CA: Stanford University Press.

Eibl-Eibesfeldt, Irenäus, 1989. *Human Ethology.* New York: Aldine de Gruyter.

Eigen, Manfred, 1971. "Selforganization of Matter and the Evolution of Biological Macromolecules," *Die Naturwissenschaften* 58:465–523.

Eigen, Manfred, with Ruthild Winkler-Oswatitsch, 1992. *Steps Towards Life: A Perspective on Evolution.* New York: Oxford University Press.

Einstein, Albert, 1954. *Ideas and Opinions.* New York: Crown.

Eiseley, Loren, 1960. *The Firmament of Time.* New York: Atheneum.

Eldredge, Niles, 1995a. "Mass Extinction and Human Responsibility." Pages 64–87 in Holmes Rolston, ed., *Biology, Ethics, and the Origins of Life.* Boston: Jones & Bartlett.

Eldredge, Niles, 1995b. *Reinventing Darwin.* New York: John Wiley.

Farber, Paul, 1994. *The Temptations of Evolutionary Ethics.* Berkeley: University of California Press.

Farquhar, Marilyn G., 1983. "Multiple Pathways of Exocytosis, Endocytosis, and Membrane Recycling: Validation of a Golgi Route," *Federation of American Societies for Experimental Biology Proceedings* 42:2407–2413.

Ferré, Frederick, 1970. "The Definition of Religion," *Journal of the American Academy of Religion* 38:3–16.

Fitch, Walter M., and Emanuel Margoliash, 1967. "Construction of Phylogenetic Trees," *Science* 155:279–284.

Forrest, Stephanie, 1993. "Genetic Algorithms: Principles of Natural Selection Applied to Computation," *Science* 261:872–878.

Fox, Charles W., Monica S. Thakar, and Timothy A. Mousseau, 1997. "Egg Size Plasticity in a Seed Beetle: An Adaptive Maternal Effect," *American Naturalist* 149:149–163.

Frank, Robert H., 1988. *Passions Within Reason: The Strategic Role of the Emotions*. New York: W. W. Norton.

Friedberg, Errol C., 1985. *DNA Repair*. New York: W. H. Freeman.

Friedberg, Errol C., Graham C. Walker, and Wolfram Siede, 1995. *DNA Repair and Mutagenesis*. Washington, DC: American Society for Microbiology Press.

Gardner, Eldon J., and D. Peter Snustad, 1981. *Principles of Genetics*, 6th edition. New York: John Wiley & Sons.

Geertz, Clifford, 1973. *The Interpretation of Cultures*. New York: Basic Books.

Ghiselin, Michael T., 1974. *The Economy of Nature and the Evolution of Sex*. Berkeley, CA: University of California Press.

Gibbard, Allan, 1990. *Wise Choices, Apt Feelings: A Theory of Normative Judgment*. Cambridge, MA: Harvard University Press.

Glass, Bentley, 1974. "The Long Neglect of Genetic Discoveries and the Criterion of Prematurity," *Journal of the History of Biology* 7:101–110.

Goffman, William, and Vaun A. Newill, 1964. "Generalization of Epidemic Theory: An Application to the Transmission of Ideas," *Nature* 204:225–228.

Goffman, William, and Vaun A. Newill, 1968. "Communication and Epidemic Processes." *Proceedings of the Royal Society* A 298:316–334.

Goldberg, David E., 1989. *Genetic Algorithms in Search, Optimization, and Machine Learning*. Reading, MA: Addison-Wesley.

Gorczynski, R. M., and E. J. Steele, 1980. "Inheritance of Acquired Immunological Tolerance to Foreign Histocompatibility Antigens in Mice," *Proceedings of the National Academy of Sciences* 77:2871–2875.

Gould, Stephen Jay, 1977a. *Ever Since Darwin*. New York: W. W. Norton.

Gould, Stephen Jay, 1977b. "Eternal Metaphors of Palaeontology." Pages 1–26 in A. Hallam, ed., *Patterns of Evolution as Illustrated by the Fossil Record*. New York: Elsevier Scientific.

Gould, Stephen Jay, 1979. "Darwin's Middle Road," *Natural History* 88, no. 10, December:27–31.

Gould, Stephen Jay, 1980. *The Panda's Thumb*. New York: W. W. Norton.

Gould, Stephen Jay, 1983. "Extemporaneous Comments on Evolutionary Hope and Realities." Pages 95–103 in Charles L. Hamrum ed., *Darwin's Legacy, Nobel Conference XVIII*, San Francisco: Harper & Row.

References

Gould, Stephen Jay, 1989. *Wonderful Life: The Burgess Shale and the Nature of History*. New York: W. W. Norton.

Gould, Stephen Jay, 1996. *Full House: The Spread of Excellence from Plato to Darwin*. New York: Harmony Books.

Greene, Erick, 1989. "A Diet-Induced Developmental Polymorphism in a Caterpillar," *Science* 243(1989):643–646.

Hadamard, Jacques, 1949. *The Psychology of Invention in the Mathematical Field*. Princeton, NJ: Princeton University Press.

Haldane, J. B. S., 1932. *The Causes of Evolution*. London: Longmans.

Hamilton, William D., 1964. "The Genetical Evolution of Social Behavior. Parts I and II." *Journal of Theoretical Biology* 7:1–52.

Harcourt, Alexander H., and Frans B. M. de Waal, eds., 1992. *Coalitions and Alliances in Humans and Other Animals*. Oxford: Oxford University Press.

Hausfater, Glenn, 1984. "Infanticide in Langurs: Strategies, Counterstrategies, and Parameter Values." Pages 257–281 in Glenn Hausfater and Sarah Blaffer Hrdy, eds., *Infanticide: Comparative and Evolutionary Perspectives*. New York: Aldine.

Hefner, Philip, 1987. "Sociobiology and Ethics." Pages 115–137 in Frank T. Birtel, ed., *Science, Religion, and Public Policy*. New York: Crossroad.

Hefner, Philip, 1993. *The Human Factor: Evolution, Culture, and Religion*. Minneapolis: Fortress Press.

Hegel, G. W. F., 1888. *Lectures on the Philosophy of History*, trans. J. Sibree. London: George Bell and Sons.

Hempel, Carl G., 1965. *Aspects of Scientific Explanation*. New York: Free Press.

Hempel, Carl G. 1966. *Philosophy of Natural Science*. Englewood Cliffs, NJ: Prentice-Hall.

Höffding, Harald, 1906. *The Philosophy of Religion*. London: Macmillan and Co.

Holland, John H., 1975. *Adaptation in Natural and Artificial Systems*. Ann Arbor: University of Michigan Press.

Holland, John H., 1980. "Adaptive Algorithms for Discovering and Using General Patterns in Growing Knowledge Bases," *International Journal of Policy Analysis and Information Systems* 4:245–268.

Holland, John H., 1992. "Genetic Algorithms," *Scientific American* 267, no. 1, July:66–72.

Hrdy, Sarah Blaffer, 1977a. "Infanticide as a Primate Reproductive Strategy," *American Scientist* 65:40–49.

Hrdy, Sarah Blaffer, 1977b. *The Langurs of Abu*. Cambridge, MA: Harvard University Press.

Hull, David, 1980. "Individuality and Selection," *Annual Review of Ecology and Systematics* 11:311–332.

Hull, David, 1988. *Science as a Process: An Evolutionary Account of the Social and Conceptual Development of Science*. Chicago: University of Chicago Press.

Huxley, Thomas H., 1880, 1897. "The Coming of Age of *The Origin of Species*." Pages 227–243 in Thomas H. Huxley, *Darwiniana: Essays*. New York: D. Appleton. 1897.

Jacob, François, 1977. "Evolution and Tinkering," *Science* 196:1161–1166.

References

Kauffman, Stuart A., 1991. "Antichaos and Adaptation," *Scientific American* 265, no. 2, August: 78–84.

Kauffman, Stuart A., 1993. *The Origins of Order: Self-Organization and Selection in Evolution*. New York: Oxford University Press.

Kauffman, Stuart A., 1995. *At Home in the Universe: The Search for Laws of Self-Organization and Complexity*. New York: Oxford University Press.

Kaye, Howard L., 1986. *The Social Meaning of Modern Biology*. New Haven: Yale University Press.

Keddy, Paul A., 1989. *Competition*. London: Chapman and Hall.

Keddy, Paul A., 1990. "Is Mutualism Really Irrelevant to Ecology?" *Bulletin of the Ecological Society of America* 71:101–102.

Kelley, Kevin W., ed. 1988. *The Home Planet*. Reading, MA: Addison-Wesley.

Kimura, Motoo, 1961. "Natural Selection as the Process of Accumulating Genetic Information in Adaptive Evolution," *Genetical Research* 2:127–140.

King, Mary-Claire, and A. C. Wilson, 1975. "Evolution at Two Levels in Humans and Chimpanzees," *Science* 188:107–116.

Kitcher, Philip, 1985. *Vaulting Ambition: Sociobiology and the Quest for Human Nature*. Cambridge, MA: The MIT Press.

Kitcher, Philip, 1996. *The Lives to Come: The Genetic Revolution and Human Possibilities*. New York: Simon & Schuster.

Knoll, Andrew H., 1986. "Patterns of Change in Plant Communities Through Geological Time." Pages 126–141 in Jared Diamond and Ted J. Case, eds., *Community Ecology*. New York: Harper & Row.

Kohlberg, Lawrence, 1981. *Essays on Moral Development*. Volume 1. *The Philosophy of Moral Development*. San Francisco: Harper & Row.

Koza, John R., 1992. *Genetic Programming: On the Programming of Computers by Means of Natural Selection*. Cambridge, MA: MIT Press.

Kroeber, A. L., and Clyde Kluckhohn, 1963. *Culture: A Critical Review of Concepts and Definitions*. New York: Vintage Books, Random House.

Kuhn, Thomas S., 1970. *The Structure of Scientific Revolutions*, 2nd edition. Chicago: University of Chicago Press.

Kummer, Helmut, 1980. "Analogs of Morality Among Nonhuman Primates." Pages 31–47 in Gunther Stent, ed., *Morality as a Biological Phenomenon*. Berkeley: University of California Press.

Küppers, Bernd-Olaf, 1990. *Information and the Origin of Life*. Cambridge, MA: The MIT Press.

Lake, James A., 1981. "The Ribosome," *Scientific American* 245, no. 2, August: 84–97.

Landman, Otto E., 1991. "The Inheritance of Acquired Characteristics," *Annual Review of Genetics* 25:1–20.

Latour, Bruno, 1987. *Science in Action*. Cambridge, MA: Harvard University Press.

Leigh, Egbert G., Jr. 1971. *Adaptation and Diversity: Natural History and the Mathematics of Evolution*. San Francisco: Freeman, Cooper.

Leslie, John, 1989. *Universes*. London: Routledge.

Lewontin, R. C., 1972. "The Apportionment of Human Diversity," *Evolutionary Biology* 6:381–398.

References

Lewontin, R. C., 1982. *Human Diversity*. San Francisco: W. H. Freeman.

Lewontin, R. C. 1987. "The Shape of Optimality." Pages 151–159 in John Dupré, ed., *The Latest on the Best: Essays on Evolution and Optimality*. Cambridge, MA: The MIT Press.

Lewontin, R. C., 1991. *Biology as Ideology: The Doctrine of DNA*. New York: HarperCollins.

Ludwig, Arnold M., 1965. *The Importance of Lying*. Springfield, IL: Charles C. Thomas.

Lumsden, Charles J., and Edward O. Wilson, 1981. *Genes, Mind, and Culture*. Cambridge, MA: Harvard University Press.

Lumsden, Charles J., and Edward O. Wilson, 1983. *Promethean Fire: Reflections on the Origins of Mind*. Cambridge, MA: Harvard University Press.

Margulis, Lynn, 1993. *Symbiosis in Cell Evolution*, 2nd edition. San Francisco: W. H. Freeman.

Marianoff, Dimitri, 1944. *Einstein: An Intimate Study of a Great Man*. Garden City, NY: Doubleday, Doran.

Maturana, Humberto R., and Francisco J. Varela, 1980. *Autopoiesis and Cognition: The Realization of the Living*. Dordrecht, Boston: D. Reidel.

Maynard Smith, John, 1972. *On Evolution*. Edinburgh: University of Edinburgh Press.

Maynard Smith, John, 1976. "A Short-Term Advantage for Sex and Recombination Through Sib-Competition," *Journal of Theoretical Biology* 63:245–258.

Maynard Smith, John, 1978. *The Evolution of Sex*. Cambridge: Cambridge University Press.

Maynard Smith, John, 1982a. *Evolution and the Theory of Games*. Cambridge: Cambridge University Press.

Maynard Smith, John, 1982b. "Introduction." Pages 1–4 in King's College Sociobiology Group, Cambridge, *Current Problems in Sociobiology*. Cambridge: Cambridge University Press.

Maynard Smith, John, 1984. "Science and Myth," *Natural History* 93, no. 11, November:11–24.

Maynard Smith, John, 1986. "Natural Selection of Culture?" *New York Review of Books* 33(17):11–12, November 6.

Maynard Smith, John, 1991. "The Way Forward." Pages 238–239 in Connie Barlow, ed., *From Gaia to Selfish Genes*. Cambridge, MA: The MIT Press.

Maynard Smith, John, 1995. "Life at the Edge of Chaos?" *New York Review of Books* 52, no. 4, March 2, 1995:28–30.

Maynard Smith, John, and N. Warren, 1982. "Models of Cultural and Genetic Change," *Evolution* 36:620–627.

Maynard Smith, John, and Eörs Szathmáry, 1995. *The Major Transitions in Evolution*. New York: W. H. Freeman.

Mayr, Ernst, 1963. *Animal Species and Evolution*. Cambridge, MA: Harvard University Press.

Mayr, Ernst, 1976. *Evolution and the Diversity of Life*. Cambridge, MA: Harvard University Press.

Mayr, Ernst, 1982. *The Growth of Biological Thought*. Cambridge, MA: Harvard University Press.

Mayr, Ernst, 1985a. "The Probability of Extraterrestrial Intelligent Life." Pages 23–30 in Edward Regis Jr., ed., *Extraterrestrials: Science and Alien Intelligence*. New York: Cambridge University Press.

Mayr, Ernst, 1985b. "How Biology Differs from the Physical Sciences." Pages 43–63 in David J. Depew and Bruce H. Weber, eds., *Evolution at a Crossroads*. Cambridge, MA: The MIT Press.

Mayr, Ernst, 1988. *Toward a New Philosophy of Biology*. Cambridge, MA: Harvard University Press.

Mayr, Ernst, 1991. *One Long Argument: Charles Darwin and the Genesis of Modern Evolutionary Thought*. Cambridge, MA: Harvard University Press.

Mayr, Ernst, 1994. "Does It Pay To Acquire High Intelligence?" *Perspectives in Biology and Medicine* 37:337–338.

McKusick, Victor A., and Frank H. Ruddle, 1977. "The Status of the Gene Map of the Human Chromosomes," *Science* 196:390–405.

McShea, Daniel W., 1991. "Complexity and Evolution: What Everybody Knows," *Biology and Philosophy* 6:303–324.

Mead, Margaret, 1959, 1989. "Preface." Pages xi–xiv in Ruth Benedict, *Patterns of Culture*. Boston: Houghton Mifflin.

Mitchell, Melanie, 1996. *An Introduction to Genetic Algorithms*. Cambridge, MA: The MIT Press.

Mitchell, Melanie, and Stephanie Forrest, 1994. "Genetic Algorithms and Artificial Life," *Artificial Life* 1:267–289.

Mithen, Steven, 1996. *The Prehistory of the Mind*. London: Thames and Hudson.

Monod, Jacques, 1972. *Chance and Necessity*. New York: Random House.

Monroe, Kristen R., 1996. *The Heart of Altruism*. Princeton, NJ: Princeton University Press.

Moynihan, Martin, 1976. *The New World Primates*. Princeton, NJ: Princeton University Press.

Muggeridge, Malcolm, 1973. *Something Beautiful for God: Mother Teresa of Calcutta*. New York: Ballantine Books.

Mühlenbien, Heinz, M. Gorges-Schleuter, and O. Krämer, 1988. "Evolution Algorithms in Combinatorial Optimization," *Parallel Computing* 7:65–85.

Munro, Robin. E., 1974. "Interpreting Molecular Biology." Pages 103–120 in John Lewis, ed., *Beyond Chance and Necessity*. Atlantic Highlands, NJ: Humanities Press.

Myers, Norman, 1997. "Mass Extinction and Evolution," *Science* 278:597–598.

Nee, Sean, and Robert M. May, 1997. "Extinction and the Loss of Evolutionary History," *Science* 278:692–694.

Newell, Norman D., 1963. "Crises in the History of Life," *Scientific American* 208, no. 2, February: 76–92.

Niklas, Karl J., 1986. "Large-Scale Changes in Animal and Plant Terrestrial Communities." Pages 383–405 in D. M. Raup, D. Jablonski, eds., *Patterns and Processes in the History of Life*. New York: Springer-Verlag.

Niklas, Karl J., 1997. *The Evolutionary Biology of Plants*. Chicago: University of Chicago Press.

References

Nitecki, Matthew H., ed., 1988. *Evolutionary Progress*. Chicago: University of Chicago Press.

Nitecki, Matthew H., and Doris V. Nitecki, eds., 1993. *Evolutionary Ethics*. Albany: State University of New York Press.

Nomura, Masayasu, 1984. "The Control of Ribosome Synthesis," *Scientific American* 250, no. 1, January:102–114.

Nowak, Martin A., Robert M. May, and Karl Sigmund, 1995. "The Arithmetics of Mutual Help," *Scientific American* 272, no. 6, June:76–81.

Nowak, Martin A., and Karl Sigmund, 1992. "Tit for Tat in Heterogeneous Populations," *Nature* 355:250–253.

Nowak, Rachel, 1994. "Mining Treasures from 'Junk DNA,' " *Science* 263:608–610.

Oreskes, Naomi, Kristin Shrader-Frechette, and Kenneth Belitz, 1994. "Verification, Validation, and Confirmation of Numerical Models in the Earth Sciences," *Science* 263:641–646.

Orgel, L. E., and F. H. C. Crick, 1980. "Selfish DNA: The Ultimate Parasite," *Nature* 284 (17 April):604–607.

Ortega y Gasset, José, 1961. *History as a System, and Other Essays Toward a Philosophy of History*. New York: Norton.

Orten, J. M., and O. W. Neuhaus, 1982. *Human Biochemistry*, 10th edition. St. Louis: C. V. Mosby.

Peirce, Charles Sanders, 1960. *Collected Papers of Charles Sanders Peirce*. Cambridge, MA: Harvard University Press.

Perry, D. A., M. P. Amaranthus, J. G. Borchers, S. L. Borchers, and R. E. Brainerd, 1989. "Bootstrapping in Ecosystems," *BioScience* 39:230–237.

Peters, Ted, 1997. *Playing God: Genetic Determinism and Human Freedom*. New York: Routledge.

Pielou, E. C., 1975. *Ecological Diversity*. New York: John Wiley & Sons.

Pilbeam, D., 1972. *The Ascent of Man: An Introduction to Human Evolution*. New York: Macmillan.

Plomin, Robert, 1990. "The Role of Inheritance in Behavior," *Science* 248:183–188.

Popper, Karl R., 1968. *Conjectures and Refutations: The Growth of Scientific Knowledge*. New York: Harper & Row.

Popper, Karl R., 1972. *Objective Knowledge: An Evolutionary Approach*. Oxford: Clarendon Press.

Popper, Karl R., 1990. *A World of Propensities*. Bristol, England: Thoemmes.

Queller, D. C., 1983. "Kin Selection and Conflict in Seed Maturation," *Journal of Theoretical Biology* 100:153–172.

Ralls, Katherine, Jonathan D. Ballou, and Alan Templeton, 1988. "Estimates of Lethal Equivalents and the Cost of Inbreeding in Mammals," *Conservation Biology* 2:185–193.

Raup, David M., 1991. *Extinction: Bad Genes or Bad Luck?* New York: W. W. Norton.

Raup, David M., and J. John Sepkoski, Jr., 1982. "Mass Extinctions in the Marine Fossil Record," *Science* 215:1501–1503.

Rawls, John, 1971. *A Theory of Justice*. Cambridge, MA: Harvard University Press.

Reynolds, Vernon, 1991. "Socioecology of Religion." Pages 205–222 in Mary Maxwell, ed., *The Sociobiological Imagination*. Albany: State University of New York Press.

Reynolds, Vernon, and Ralph Tanner, 1983. *The Biology of Religion*. New York: Longman.

Reynolds, Vernon, and Ralph Tanner, 1995. *The Social Ecology of Religion*. New York: Oxford University Press.

Richerson, P. J., and R. Boyd, 1989. "The Role of Evolved Predispositions in Cultural Evolution," *Ethology and Sociobiology* 10:195–219.

Ridley, Matt, 1997. *The Origins of Virtue*. London: Penguin Books.

Robertson, John A., 1994. *Children of Choice: Freedom and the New Reproductive Technologies*. Princeton, NJ: Princeton University Press.

Rolston, Holmes, III, ed., 1995. *Biology, Ethics, and the Origins of Life*. Boston: Jones & Bartlett.

Rosen, Robert, 1991. *Life Itself: A Comprehensive Inquiry into the Nature, Origin, and Fabrication of Life*. New York: Columbia University Press.

Rosenberg, Alexander, 1980. *Sociobiology and the Preemption of Social Science*. Baltimore: Johns Hopkins University Press.

Rothman, James E., 1985. "The Compartmental Organization of the Golgi Apparatus," *Scientific American* 253, no. 3, September: 74–89.

Rue, Loyal, 1994. *By the Grace of Guile: The Role of Deception in Natural History and Human Affairs*. New York: Oxford University Press.

Runciman, W. G., John Maynard Smith, and R. I. M. Dunbar, 1996. *Evolution of Social Behaviour Patterns in Primates and Man*. Oxford: Oxford University Press.

Ruppert, Edward E., and Robert D. Barnes, 1994. *Invertebrate Zoology*, 6th edition. Fort Worth, TX: Saunders College Publishing, Harcourt Brace College Publishers.

Ruse, Michael, 1984. "Review of Peter Singer, *The Expanding Circle*," *Environmental Ethics* 6:91–94.

Ruse, Michael, 1986. *Taking Darwin Seriously*. Oxford: Basil Blackwell.

Ruse, Michael, 1989. "Is Darwinism Forever?" *Institute on Religion in an Age of Science Newsletter* 37, no. 2 (15 January):6–8.

Ruse, Michael, 1994. "Evolutionary Theory and Christian Ethics: Are They in Harmony?" *Zygon* 29(1994):5–24.

Ruse, Michael, 1995. "Evolutionary Ethics: A Defense." Pages 89–112 in Rolston, ed., *Biology, Ethics, and the Origins of Life*. Boston: Jones & Bartlett.

Ruse, Michael, 1996. *Monad to Man: The Concept of Progress in Evolutionary Biology*. Cambridge, MA: Harvard University Press.

Ruse, Michael, and Edward O. Wilson, 1985. "The Evolution of Ethics," *New Scientist* 108, no. 1478 (17 October):50–52.

Ruse, Michael, and Edward O. Wilson, 1986. "Moral Philosophy as Applied Science," *Philosophy: Journal of the Royal Institute of Philosophy* 61:173–192.

Sahlins, Marshall, 1976. *The Use and Abuse of Biology: An Anthropological Critique of Sociobiology*. Ann Arbor: University of Michigan Press.

Salthe, Stanley N., 1993. *Development and Evolution: Complexity and Change in Biology*. Cambridge. MA: The MIT Press.

Salvini-Plawen, L. V., and Ernst Mayr, 1977. "On the Evolution of Photoreceptors and Eyes." Pages 207–263 in Max K. Hecht, William C. Steere, and Bruce Wallace, eds. *Evolutionary Biology*, Volume 10. New York: Plenum Press.

Sapp, Jan, 1994. *Evolution by Association: A History of Symbiosis*. New York: Oxford University Press.

Saunders, P. T. and M. W. Ho, 1976. "On the Increase in Complexity in Evolution," *Journal of Theoretical Biology* 63:375–384.

Schull, Jonathan, 1990. "Are Species Intelligent?" *Behavioral and Brain Sciences* 13:63–75.

Searls, David B., 1992. "The Linguistics of DNA," *American Scientist* 80:579–591.

Serres, Michael, 1995. *The Natural Contract*. Ann Arbor: University of Michigan Press.

Service, Robert F., 1997. "Microbiologists Explore Life's Rich, Hidden Kingdoms," *Science* 275:1740–1742.

Sesardic, Neven, 1995. "Recent Work on Human Altruism and Evolution," *Ethics* 106:128–157.

Seyfarth, R. M., D. L. Cheney, and P. Marler, 1980. "Monkey Responses to Three Different Alarm Calls: Evidence of Predator Classification and Semantic Communication," *Science* 210:801–803.

Shrader, Douglas, 1980. "The Evolutionary Development of Science," *Review of Metaphysics* 34:273–296.

Sibley, Charles G., and Jon E. Ahlquist, 1984. "The Phylogeny of the Hominoid Primates, as Indicated by DNA-DNA Hybridization," *Journal of Molecular Evolution* 20:2–15.

Sides, Stanley D., and Harold Meloy, 1971. "The Pursuit of Health in the Mammoth Cave," *Bulletin of the History of Medicine* 45, no. 4, July-August: 367–379.

Simon, Herbert A., 1969. *The Sciences of the Artificial*. Cambridge, MA: The MIT Press.

Simon, Herbert A., 1990. "A Mechanism for Social Selection and Successful Altruism." *Science* 250:1665–1668.

Simonton, Dean Keith, 1988. *Scientific Genius: A Psychology of Science*. New York: Cambridge University Press.

Simpson, George Gaylord, 1966. "The Biological Nature of Man," *Science* 152: 472–478.

Simpson, George Gaylord, 1967. *The Meaning of Evolution*, revised edition. New Haven, CT: Yale University Press.

Singer, Peter, 1981. *The Expanding Circle: Ethics and Sociobiology*. New York: Farrar, Straus & Giroux.

Singer, Peter, 1994. *Ethics*. New York: Oxford University Press.

Skagestad, Peter, 1978. "Taking Evolution Seriously: Critical Comments on D. T. Campbell's Evolutionary Epistemology," *Monist* 61:611–621.

Sober, Elliott, 1993a. *Philosophy of Biology*. Boulder, CO: Westview Press.

Sober, Elliott, 1993b. "Evolutionary Altruism, Psychological Egoism, and Morality: Disentangling the Phenotypes." Pages 199–216 in Nitecki and Nitecki, *Evolutionary Ethics*. Albany: State University of New York Press.

Stanley, Steven M., 1979. *Macroevolution: Pattern and Process*. San Francisco: W. H. Freeman.

Stent, Gunther, ed., 1980. *Morality as a Biological Phenomenon*. Berkeley: University of California Press.

Sterelny, Kim, 1995. "The Adapted Mind," *Biology and Philosophy* 10:365–380.

Stevens, Charles F., 1985. "AChR Structure: A New Twist in the Story," *Trends in NeuroSciences* 8, no. 1, January:1–2.

Stevens, Charles F., 1987. "Channel Families in the Brain," *Nature* 328:198–199.

Strutt, Robert J., [1924] 1968. *Life of John William Strutt, Third Baron Rayleigh*. Madison: University of Wisconsin Press.

Stryer, Lubert, 1975. *Biochemistry*. San Francisco: W. H. Freeman.

Stryer, Lubert, 1995. *Biochemistry*, 4th edition. New York: W. H. Freeman.

Symons, Donald, 1992. "On the Use and Misuse of Darwinism in the Study of Human Behavior." Pages 137–159 in Jerome H. Barkow, Leda Cosmides, and John Tooby, eds., *The Adapted Mind: Evolutionary Psychology and the Generation of Culture*. New York: Oxford University Press.

Tamarin, Robert H., 1996. *Principles of Genetics*, 5th edition. Dubuque, IA: William C. Brown.

Taton, R., 1957. *Reason and Chance in Scientific Discovery*. New York: Philosophical Library.

Taubes, Gary, 1997. "Computer Design Meets Darwin," *Science* 277:1931–1932.

Taylor, Shelley A., 1989. *Positive Illusions: Creative Self-Deception and the Healthy Mind*. New York: Basic Books.

Thaler, David S., 1994. "The Evolution of Genetic Intelligence," *Science* 264:224–225.

Thompson, Paul, ed., 1995. *Issues in Evolutionary Ethics*. Albany: State University of New York Press.

Thornhill, Nancy W., ed., 1993. *The Natural History of Inbreeding and Outbreeding*. Chicago: University of Chicago Press.

Tiger, Lionel, 1979. *Optimism: The Biology of Hope*. New York: Simon & Schuster.

Tooby, John, and Leda Cosmides, 1992. "The Psychological Foundations of Culture." Pages 19–136 in Jerome H. Barkow, Leda Cosmides, and John Tooby, eds., *The Adapted Mind: Evolutionary Psychology and the Generation of Culture*. New York: Oxford University Press.

Toulmin, Stephen E., 1967. "The Evolutionary Development of Natural Science," *American Scientist* 55:456–471.

Toulmin, Stephen E., 1972. *Human Understanding*. Princeton, NJ: Princeton University Press.

Trivers, Robert L., 1971. "The Evolution of Reciprocal Altruism," *Quarterly Review of Biology* 46:35–57.

Trivers, Robert L., 1985. *Social Evolution*. Menlo Park, CA: Benjamin-Cummings.

Tylor, E. B., 1903. *Primitive Cultures*, 4th edition, 2 volumes. London: John Murray.

Valentine, James W., 1969. "Patterns of Taxonomic and Ecological Structure of the Shelf Benthos During Phanerozoic Time," *Palaeontology* 12:684–709.

Valentine, James. W. 1973. *Evolutionary Paleoecology of the Marine Biosphere*. Englewood Cliffs, NJ: Prentice-Hall.

van Lawick-Goodall, Jane, 1971. *In the Shadow of Man*. Boston: Houghton Mifflin.

Van Valen, Leigh M. 1991. "How Far Does Contingency Rule?" *Evolutionary Theory* 10:47–52.

Vogel, Christian, and Hartmut Loch, 1984. "Reproductive Parameters, Adult-Male Replacements, and Infanticide Among Free-Ranging Langurs (*Presbytis entellus*) at Jodhpur (Rajasthan), India." Pages 237–255 in Glenn Hausfater and Sarah Blaffer Hrdy, eds., *Infanticide: Comparative and Evolutionary Perspectives*. New York: Aldine.

Wald, George, 1959. "Light and Life," *Scientific American* 201, no. 4, October: 92–108.

Wald, George, 1974. "Fitness in the Universe: Choices and Necessities." Pages 7–27 in J. Oró et al., eds., *Cosmochemical Evolution and the Origins of Life*. Dordrecht, Netherlands: D. Reidel.

Wallace, Anthony F. C., 1966. *Religion: An Anthropological View*. New York: Random House.

Washburn, S. L., 1978. "Animal Behavior and Social Anthropology." Pages 53–74 in Michael S. Gregory, Anita Silvers, and Diane Sutch, eds., *Sociobiology and Human Nature*. San Francisco: Jossey-Bass.

White, Michael J. D., 1978. *Modes of Speciation*. San Francisco: W. H. Freeman.

Whitehead, Alfred North, [1929] 1978. *Process and Reality*, corrected ed. New York: Free Press.

Whitley, D., T. Starkweather, and C. Bogart, 1990. "Genetic Algorithms and Neural Networks: Optimizing Connections and Connectivity," *Parallel Computing* 14:347–361.

Whittaker, R. H., 1972. "Evolution and Measurement of Species Diversity," *Taxon* 21:213–251.

Wicken, Jeffrey S., 1987. *Evolution, Thermodynamics, and Information*. New York: Oxford University Press.

Wiener, Norbert, 1948. *Cybernetics*. New York: John Wiley.

Williams, George C., 1975. *Sex and Evolution*. Princeton, NJ: Princeton University Press.

Williams, George C., 1985. "Comments by George C. Williams on Sober's *The Nature of Selection*," *Biology and Philosophy* 1:114–122.

Williams, George C., 1988. "Huxley's Evolution and Ethics in Sociobiological Perspective," *Zygon* 23:383–407, and reply to critics, 437–438.

Williams, George C., 1993. "Mother Nature Is a Wicked Old Witch." Pages 217–231 in Nitecki and Nitecki, *Evolutionary Ethics*. Albany: State University of New York Press.

Wills, Christopher, 1989. *The Wisdom of the Genes: New Pathways in Evolution*. New York: Basic Books.

Willson, Mary F., and Nancey Burley, 1983. *Mate Choice in Plants*. Princeton, NJ: Princeton University Press.

Wilson, Edward O., 1975a. *Sociobiology: The New Synthesis*. Cambridge, MA: Harvard University Press.

Wilson, Edward O., 1975b. "Human Decency Is Animal," *New York Times Magazine*, October 12, pp. 38–50.

Wilson, Edward O., 1978. *On Human Nature*. Cambridge, MA: Harvard University Press.

Wilson, Edward O., 1980a. "Comparative Social Theory." Pages 51–73 in Sterling M. McMurrin, ed., *The Tanner Lectures on Human Values, 1980*, vol. I. Salt Lake City: University of Utah Press.

Wilson, Edward O., 1980b. "The Relation of Science to Theology," *Zygon* 15: 425–434.

Wilson, Edward O., 1984. *Biophilia*. Cambridge, MA: Harvard University Press.

Wilson, Edward O., 1992. *The Diversity of Life*. Cambridge, MA: Harvard University Press.

Wilson, Edward O., 1994. *Naturalist*. Washington, DC: Island Press.

Wilson, Edward O., 1998. "The Biological Basis of Morality," *The Atlantic Monthly* 281, no. 4 (April): 53–70.

Wilson, Edward O., et al., 1977. *Life: Cells, Organisms, Populations*. Sunderland, MA: Sinauer Associates.

Wimsatt, William C., 1982. "Reductionist Research Strategies and their Biases in the Units of Selection Controversy." Pages 155–201 in Esa Saarinen, ed., *Conceptual Issues in Ecology*. Dordrecht, Holland: D. Reidel.

Wolf, Larry, 1975. " 'Prostitution' Behavior in a Tropical Hummingbird," *Condor* 77:140–144.

Wrangham, Richard W., W. C. McGrew, Frans B. M. De Waal, and Paul G. Heltne, eds., 1994. *Chimpanzee Cultures*. Cambridge, MA: Harvard University Press.

Wuketits, Franz M., 1984. "Evolutionary Epistemology: A Challenge to Science and Philosophy." Pages 1–33 in Franz M. Wuketits, ed., *Concepts and Approaches in Evolutionary Epistemology*. Dordrecht: D. Reidel.

Index

Index

Index

crustaceans, 12, 38, 42, 82, 190
crystals, 14
culpability, 213–214; *see also* ethics
culture, x–xi, 14, 45, 50; 108–159, 302, 304;
    contrasted with nature, 108–112, 157,
    227, 279–280, 281, 333–334, 342; cul-
    tural selection, 139; culturgens, 124–
    130, 143; cumulative transmissible
    culture, x, 109–112, 135, 137, 141,
    170, 215, 244, 260, 292, 299, 325, 333;
    ethics in, 212, 260, 265; differing
    roles in, 148; genetic determinants in,
    120–135, 264; on genetic leash, 120–
    122, 124, 130, 131, 141, 147, 151, 161,
    199, 202, 244, 264, 334; oral and liter-
    ate, 129–130, 294; panculturalism,
    217, 220, 287–291, 333, 343; and re-
    built environments, 144, 191
culturgens, 124–130
cybernetics, 23–24, 27, 37, 52, 63–64, 68,
    80, 114, 119, 169, 180, 181, 208, 238,
    307, 350, 357, 364, 365; *see also* so in-
    formation
cytochrome *c,* 46–47, 48, 149

Darwin, Darwinism, x, xv, 12, 15–16, 27,
    64, 65, 120–121, 132, 136, 139, 147,
    160, 161, 162, 165, 167–168, 179, 186,
    191, 203, 206, 213, 226, 265, 338, 362,
    365; ethics Darwinized, 214, 217, 244,
    249–280, 289; science Darwinized,
    168–170, 189–200; social Darwinism,
    194; religion Darwinized, 308–347;
    ultra-Darwinism, 97, 187
Davis, Lawrence, 34
Dawkins, Richard, 26, 47, 70, 71, 74, 79,
    83, 86, 95, 145, 146, 169, 185, 186,
    229, 233, 250, 251, 265, 267, 271, 275,
    332, 356
death, 56–57, 299, 305, 359–360
deBeer, Gavin, 27
deception, 197, 256–260, 269, 308–309; self-
    deception, 252–253, 255, 256, 269, 318–
    320, 322, 323, 335–347; *see also* illu-
    sion
de Duve, Christian, 106, 112–113, 349, 358
defection, 230, 240–241; *see also* cheating
Delbrück, Max, 51, 163–164
deliberation: *see* intentionality
delusion: *compare* deception; illusion
democracy, 221, 239, 311
Dennett, Daniel, 362, 365–367
deoxyribonucleic acid (DNA), 28–29, 31,
    43–44, 47, 64, 65, 66, 68–70, 75, 111,
    127, 149, 170, 177, 181, 185, 207, 263,
    286, 297, 336, 349–350, 351, 355, 356,
    357, 369; in mitochondria, 77, 100;
    junk DNA, 33, 47, 74, 79; purpose of,
    79; selfish DNA, 71
Descartes, René, 140, 158
design, 31, 365, 369–370

detection, of God, 344, 348, 363–370
determinism, biological, 141, 264, 268,
    270, 279, 320, 349–350; *see also* tran-
    scendence
development, sustainable, 161
DeVore, Irvin, 121, 151–152, 240
de Waal, Frans, 186–187, 212, 230
Diamond, Jared, 8, 47
Dickemann, Mildred, 192
Dickerson, R. E., 46
dilemma, is/ought, 232, 246, 254, 259, 263–
    272, 281, 345, 359
dinosaurs, 202, 208, 275, 289, 299, 352,
    354, 355
Dion, Douglas, 229
disorder, 17, 358; *see also* chance; chaos;
    contingency; randomness
diversity, biological, 1–53, 106, 112, 161,
    162, 180, 181, 201, 208, 223, 224, 266,
    268, 289, 297, 305, 349, 360, 369, 370;
    see also species, speciation
Dobzhansky, Theodosius, 13–14, 111, 117–
    118
docility, and ethics, 260–263
dogs, wild, 103
dolphins, 8
Donald, Merlin, 115
Dose, Klaus, 357
Dovidio, John, 276
Down's syndrome, 101, 321
Drake, John, 28
drift, 16, 23, 27, 227, 299; climatic 16, 68,
    129; continental, 3, 5, 129; genetic,
    194
Druids, 119, 136
Dugatkin, Lee Alan, 103, 186, 231, 236
Dunbar, R. I. M., 132
Durham, William, 141
Durkheim, Émile, 226
duty, 214, 232, 255; *see also* ethics

Earth (planet), 1, 51–52, 71, 112–114, 119,
    140, 149, 151, 155, 160, 188, 201, 203,
    208, 213, 215, 225, 268, 278, 280, 283,
    287, 288, 293, 301, 336, 347, 348, 350,
    357; information explosion on, 356–
    357, 368; prolific Earth, 295, 296–298,
    308, 314, 350, 363–364; seen from
    space, 297–298, 362–362
ecology, behavioral, 96, 135, 320
ecology: *see* ecosystems
economics, xi, 109, 134, 138, 142, 153, 155,
    230, 238–239, 240, 324, 330
ecosystems, 30, 42, 43, 53, 54, 97, 153, 158,
    174–175, 183, 186, 209, 306, 311, 361,
    364; identity in, 66–68, 225, 285; or-
    ganisms in, 93–96, 285, 301–302, 306;
    trophic pyramids, 12, 94, 150, 201,
    285, 305
Edison, Thomas, 147

Index

Index